Recent Advancements in Connected Autonomous Vehicle Technologies

Volume 1

Series Editors

This Springer book series shows control theory applications and the latest achievements in the field of the autonomous vehicle. It emphasizes the practical application of various control methods, with real-world experimental validations. New and developing control algorithms, including data-driven control, robust control, cooperative control, and predictive control, are discussed and applied in the connected autonomous vehicle to improve energy consumption, vehicle safety, passenger feeling, and transportation efficiency.

The book series aims to introduce essential and recent achievements in the autonomous vehicle to researchers and engineers. The individual book volumes in the series are thematic. The goal of each volume is to give readers a comprehensive overview of how developing control algorithms can be used to improve vehicle performance and transportation efficiency. As a collection, the series provides valuable resources to a broad audience in academia, the engineering research community, industry, and anyone else who are looking to expand their knowledge of the autonomous vehicle.

More information about this series at http://www.springer.com/series/16707

Yueying Zhu

The Key Technologies for Powertrain System of Intelligent Vehicles Based on Switched Reluctance Motors

Springer

Yueying Zhu
Tianjin University of Science
and Technology
Tianjin, China

ISSN 2731-0027 ISSN 2731-0035 (electronic)
Recent Advancements in Connected Autonomous Vehicle Technologies
ISBN 978-981-16-4850-2 ISBN 978-981-16-4851-9 (eBook)
https://doi.org/10.1007/978-981-16-4851-9

This Springer imprint is published by the registered company Springer Nature Singapore Pte Ltd.
The registered company address is: 152 Beach Road, #21-01/04 Gateway East, Singapore 189721, Singapore

Contents

Chapter 1
Modeling of SRM Drive System for EV

1.1 Introduction

As the energy crisis and the environment problems are more and more serious, the key technologies on electric vehicles (EVs) are paid more attention to currently and in the future. Furthermore, the traction motors which are the core part of the EV need to be concerned deeply. The switched reluctance motor (SRM) has a greatly potential to be used in EVs because of its features of low maintenance cost, high efficiency, simple structure, and reliable operation. Therefore, it becomes one of the most competitive drive motors in the industrial applications of EVs, airplane, renewable energy harvesting, and so on [1, 2]. For SRMs, when one of the phases is excited without considering phase current direction, the adjacent rotors are attracted to reduce the reluctance of the magnetic path [3]. Thus, torque generation doesn't depend on the direction of the phase current. In addition, when the phases are energized in a certain order, the torque from each phase remains continue such that a steady total torque is output [4]. However, the modeling of the SRM drive system are one of the most important issues for the SRMs used in EVs.

In the past decade, many studies have been done, which includes motor ontology [5–8], excitation and winding modes [9–11], and control strategies [12–15]. In general, the reason that the magnetic circuit of SRM is usually designed to be saturated is to improve the electromechanical energy conversion capability of SRM. However, the electromagnetic characteristics of SRM are highly nonlinear due to the high saturation of the magnetic circuit, which could directly lead to the result that it is hard to acquire accurate non-linear electromagnetic mathematical model of SRM for forecasting the drive system performance [4]. Therefore, accurate modeling of SRM is of great significance to study its electromagnetic characteristics, improve output efficiency, and obtain good performance characteristics.

In [16], for the problem of the non-linear modeling, a model, presuming that the mutual flux does not depend on currents of all exciting phases, was presented, which is not appropriate in the case of magnetic saturation. Considering mutual

Y. Zhu, *The Key Technologies for Powertrain System of Intelligent Vehicles Based on Switched Reluctance Motors*, Recent Advancements in Connected Autonomous Vehicle Technologies 1, https://doi.org/10.1007/978-981-16-4851-9_1

coupling, a dynamic two-phase excitation model was built in [17]. Nevertheless, when the number of phase and winding mode change, the model can be not compatible due to loss of symmetry. Reference [18] compared and studied the dynamic performance and magnetic characteristics of the dual-channel switched reluctance machines (DCSRMs) under single and double excitation modes, the modeling of the DCSRMs and simulation for the drive system with the coupling effects were done. However, it was only suitable for DCSRMs. In [19], the coupling effects were discussed and considered into the dynamic model, which greatly decreased the harmonic content in the torque ripple. Based on single-phase nonlinear model, [20] proposed a new model with multi-phase excitation, and further verified the validity of the model. However, the accuracy of the mathematical model can be affected due to the changes of the equivalent magnetic resistance under the magnetic saturation. The studies mentioned above greatly developed the nonlinear modeling theory of SRM based on the coupling effects between each exciting phase. Nevertheless, for SRM, the total torque generated by coupling action is approximately computed by means of the torque sum of two phases during the period of simultaneous excitation, which could directly influence the accuracy of the model.

Focusing on the problem mentioned above, considering the excitation overlap, a new modeling method for an SRM with four phases is proposed to improve the model accuracy. Based on the theory of mutual inductance, a novel nonlinear mathematical model is also built, and the coupling effects between two excited phases on electromagnetic characteristics are analyzed and compared, in this chapter.

1.2 The Basic Theory of SRM

1.2.1 The Basic Structure of SRM

Differing from the principle and structure of most DC and AC drive motors, SRM has unique structure and operating principle. Figure 1.1 describes the schematic diagram of a typical four-phase 8/6 SRM.

As a kind of variable reluctance motor with double salient poles and no brush, SRMs have not permanent magnetic material in the rotor and winding around the rotor pole. The motor body is made of stacked silicon steel and independent phase windings are distributed on the stator salient poles. In general, the winding on the stator salient pole are connected in series with that on the opposite salient pole to form a phase of the SRM. The terminals on both sides of the winding are connected to the positive and negative poles of the power supply *Us* by switches S1 and S2 and diodes D1 and D2, as shown in Fig. 1.1.

1.2.2 The Basic Operating Principle of SRM

SRM follows the principle of reluctance minimum when operation, that is, the magnetic circuit forms a closed loop along the path with the minimum reluctance. For Fig. 1.1, when the two switches S1 and S2 of phase A are turned on, this phase can begin exciting and the magnetic flux lines distribute from the stator pole A to A′ through the adjacent rotor salient poles. In this process, the bending of magnetic flux lines can produce magnetic force, which makes the rotor rotate to the position that the centerlines of the stator pole and the adjacent rotor pole coincide, such that the magnetic resistance of magnetic circuit decreases, further resulting in the electromagnetic torque. If the excitation order is phase D, phase C, phase B, and phase A, the rotor can rotate clockwise, and when the windings of phase A, phase B, phase C, and phase D are energized for a period in the given order, the rotor can rotate one rotor pole distance.

As mentioned above, if the exciting overlap area isn't considered for the single-phase excitation mode, the excitation of each phase of SRM is energized in certain sequence, which represents that only one phase is turned on and excited each time, while the others are in closed state without phase current, as showed in Fig. 1.2. It can be seen that a long flux path can be generated during the excitation of each phase. Figure 1.3 shows the ideal phase diagram of the phase current of an 8/6 four-phase SRM during single-phase excitation. In this excitation mode, each phase separately turns on 15 degrees in order such that the interaction between each phase is very small, and the turn-on and turn-off of each phase are relatively easy to control.

In addition, to improve the motor dynamic performance, two or more phase windings of SRM are usually turned on at the same time when SRM works, and the operating principle of the motor does not change in this excitation mode. For 8/6 four-phase SRM, two-phases excitation mode is commonly used in industrial applications, which means two phases are turned on and excited at the same time, and the phase diagram of the phase current is illustrated in Fig. 1.4. As can be seen from the figure, the turn-on sequence of each phase is AB, BC, CD, and DA. In this excitation mode, each phase could separately turn on 30 degrees in order, and the

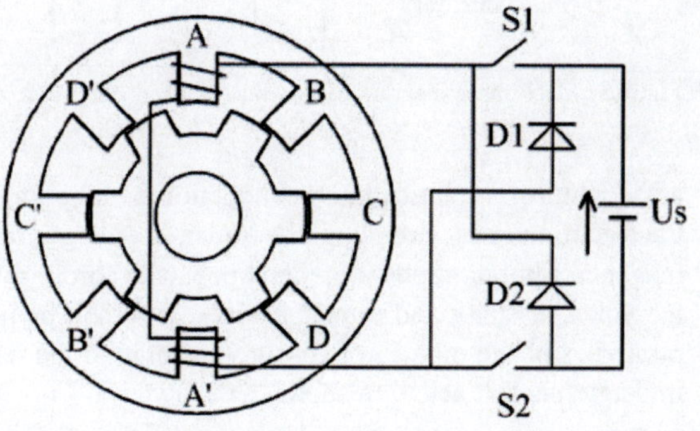

Fig. 1.1 Structure diagram of a typical four-phase 8/6 SRM

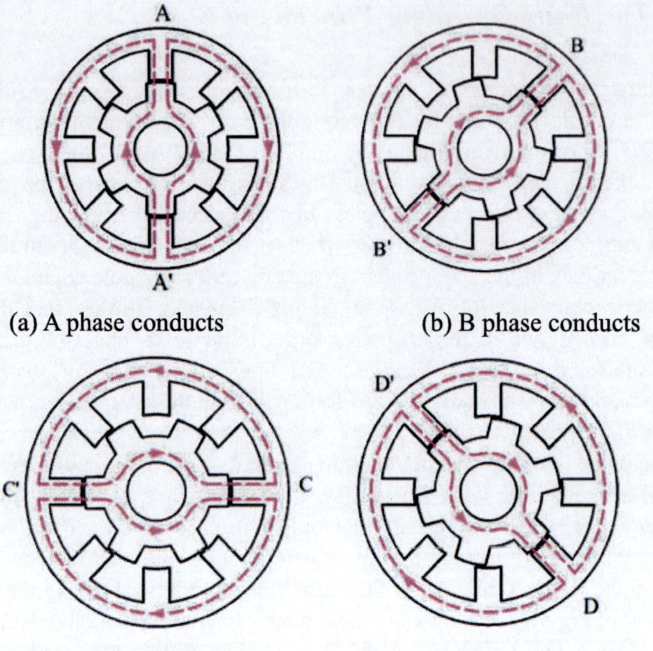

(a) A phase conducts (b) B phase conducts

(c) C phase conducts (d) D phase conducts

Fig. 1.2 Working principle of single-phase excitation mode for SRM

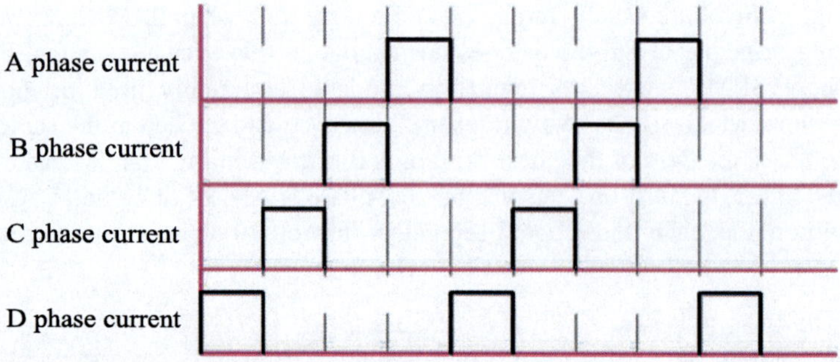

Fig. 1.3 Ideal phase diagram of the phase current for single-phase excitation mode

adjacent turn-on phases have conduction overlap angle of 15 degree. At this time, the electromagnetic coupling phenomenon emerges between the excited phases, and magnetic saturation phenomenon appears accordingly because the magnetic flux in the yokes of stator and rotor increases. It should be noted that, the electromagnetic properties of the motor in two-phase excitation mode are quite different from those in single-phase excitation mode, as showed in Fig. 1.5.

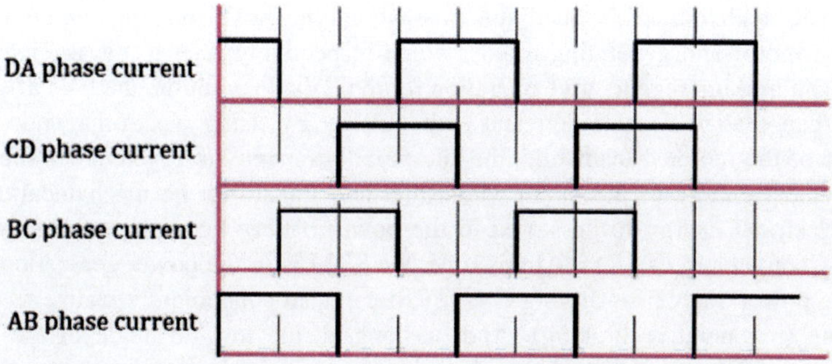

Fig. 1.4 Ideal phase diagram of the phase current for two-phase excitation mode

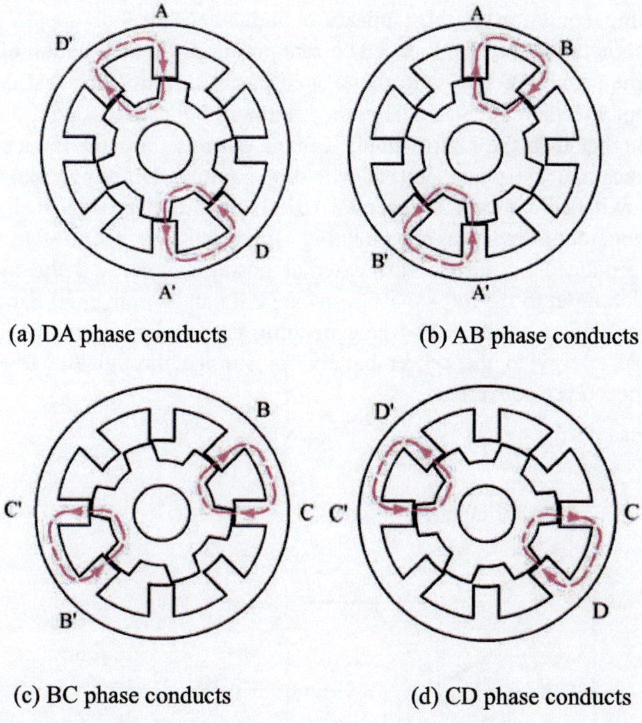

Fig. 1.5 Working principle of two-phase excitation mode for SRM

The EVs driven by SRM mainly have driving conditions, sliding conditions, and braking conditions when they drive on the road normally, which respectively correspond to the driving state and power generation state of the SRM drive system. The SRM drive system, as a kind of high-performance speed control system that can easily realize the four quadrants positive and negative rotation, can be used as driving motor and generating motor, which respectively refer to the driving operation and braking regenerative operation for the EVs. In addition, the EVs driven by SRM can realize deceleration and stop quickly by using electromagnetic brake torque of the motor. Meanwhile, the automobile transmission system can input the mechanical energy to the motor rotor shaft and transform the mechanical energy into electrical energy to be stored in the power battery because of the reversible energy conversion of SRM. At this time, the SRM is in the power generation state. As the power source for driving EVs, SRM can also conveniently realize feedback braking (regenerative braking). And, as long as the turn-on angle of the phase winding is changed, most of the phase current will be in the inductance decline stage ($\partial L/\partial \theta < 0$), as shown in Fig. 1.1. It is seen from Fig. 1.1 that the current waveforms in the braking state and the driving state are mutually symmetric, which also indicates the reversibility of the SRM drive system. And the phase current under braking condition mainly appears in inductance decline stage. When the phase switch is turned off, the induced current produced by the rotation of the rotor will be in the same direction with the excited phase current, such that the excited phase current will increase until it is in the minimum inductance area if the induced voltage is higher than the motor supply voltage. At this time, the induced voltage will disappear and the phase current will drop rapidly. Differing from the drive mode, the switched reluctance generator (SRG) operation process includes excitation and generating stages, as shown in Fig. 1.6. For the excitation state, the phase current is produced because of the external power source, and the mechanical energy, which input to the rotor shaft from external part, is converted into magnetic energy storage of the winding. For the generating state, the phase windings transmit the electrical energy to the power battery for storage through the freewheeling diodes of the power converter.

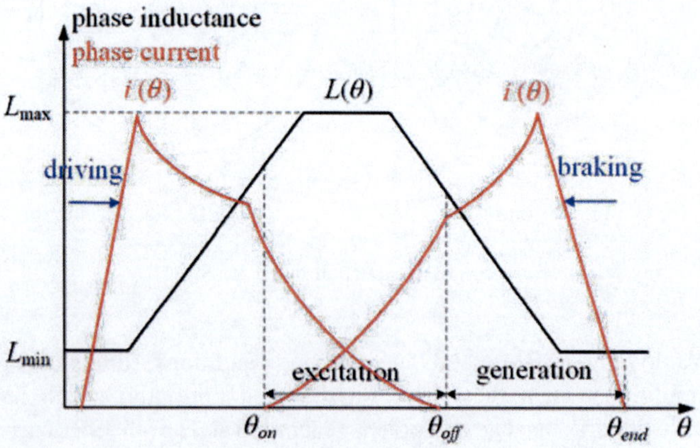

Fig. 1.6 Typical relationship between phase current and phase inductance

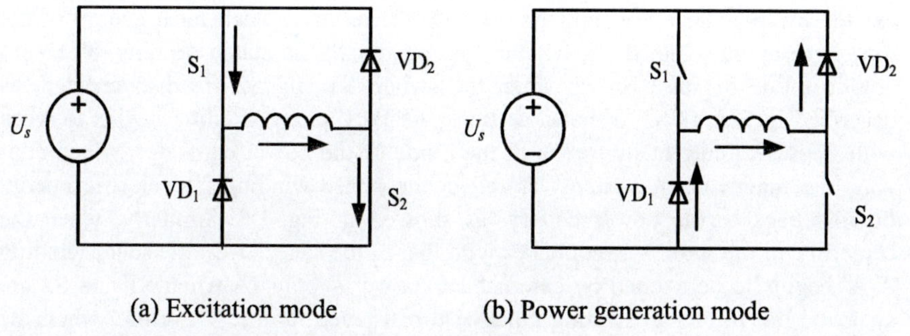

Fig. 1.7 Equivalent circuit model of one-phase winding

In general, the SRM drive system must meet three conditions to realize energy generating operation. The first one is that there must be external mechanical energy inputting to the rotor shaft of the motor. Then, the stator winding must be energized in proper given order to provide excitation. Finally, each phase must be controlled according to the rotor position information to realize the machine-electric energy conversion.

The one-phase winding equivalent circuit of the power converter is described in Fig. 1.7, where the excitation mode and generation mode respectively correspond to the excitation stage and generation stage of the phase winding. Combined with the structure of the SRM, the working principle of the generator is explained according to Fig. 1.8. Suppose the rotor axis of SRM rotates along the counter-clockwise because of the effect of external power supply until the rotor pole 1–1′ coincides with the stator pole A-A′ axis (the minimum reluctance position), the switch S1 and S2 are turned off at this time, as showed in Fig. 1.7a. The phase windings of the stator pole B-B′ can be supplied with electricity by the external DC power supply *Us* such that the B-B′ phase winding begins exciting. Then, the clockwise resistance torque of rotor is generated because of the magnetic field

Fig. 1.8 The structure of SRM

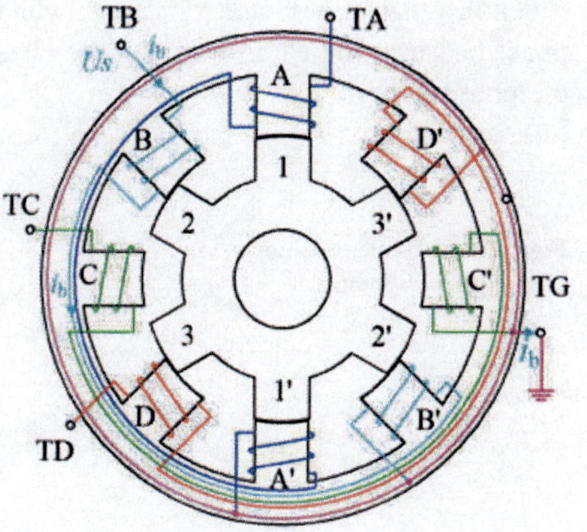

excited by this phase winding. As a result, the reduced mechanical energy of the rotor is converted into the magnetic field energy of the stator winding. The generating mode begins to work when the switch S1 and S2 are disconnected, as described in Fig. 1.7b. At the same time, the B-B′ phase winding begins to work with freewheel current by means of the diode of the power converter, which converts the magnetic field energy stored in the phase winding into electric energy flowing back to the power battery, as showed in Fig. 1.8. Similarly, when the centerline of the rotor 3–3′ coincides with that of the stator D-D′, the stator winding A-A′ begins to be excited by external DC power supply *Us* when S1 and S2 are switched on, further generating clockwise resistance torque. Therefore, when the stator windings are energized in the order of B-A-C-D, counterclockwise resistance torque will be generated in the rotor, and the magnetic energy stored in the magnetic field will be converted into electrical energy, thus achieving the electromechanical energy conversion between mechanical energy, magnetic energy, and electrical energy.

1.2.3 Electromagnetic Model of SRM

Taking four-phase 8/6 SRM as an example, the voltage, flux linkage, phase resistance, phase current, phase self-inductance, mutual inductance, and torque of *k*-phase are set as U_k, ψ_k, R_k, i_k, L_k, M_{k1}, M_{k2}, and T_k ($k = a, b, c, d$, representing four phases, respectively). The motor speed is w, and the rotor position angle is θ.

As shown in Fig. 1.9, according to Kirchhoff voltage law, the terminal voltage applied to the stator winding is equal to the sum of the phase resistance voltage and the induced voltage caused by changes of the flux linkage, as shown in Eq. (1.1).

$$U_k = R_k i_k + \frac{d\psi_k}{dt} \tag{1.1}$$

where U means column vector of the phase voltage, that is, $U = [u_a\ u_b\ u_c\ u_d]^{\mathrm{T}}$; R denotes matrix of phase resistance, which can be considered as the same for each phase because of the uniformity of the phase windings; I refers to column vectors of the phase current, that is, $I = [i_a\ i_b\ i_c\ i_d]^{\mathrm{T}}$; Ψ represents column vectors of the flux linkage, namely, $\Psi = [\psi_a\ \psi_b\ \psi_c\ \psi_d]^{\mathrm{T}}$; k is number of each phase, $k = a, b, c, d$.

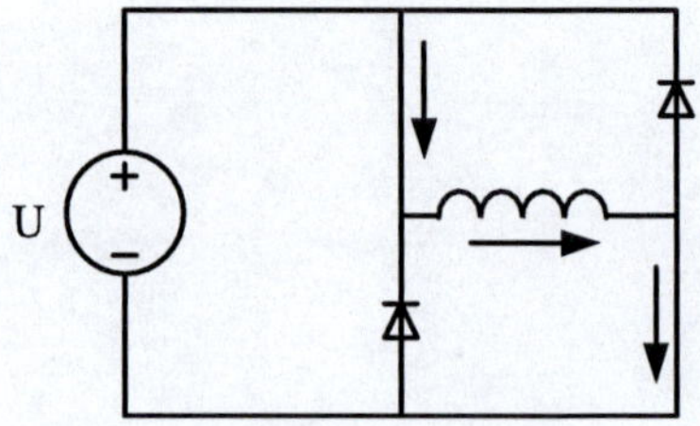

Fig. 1.9 Equivalent circuit of one phase winding

The flux linkage of each phase winding of SRM can be computed by inductance and current of the phase, which is expressed as Eq. (1.2).

$$\Psi = Li_k \tag{1.2}$$

where L is the phase inductance matrix, including self-inductance and mutual inductance, both of which are functions of phase current i_k and rotor position angle θ, which is calculated as Eq. (1.3).

$$L_k(\theta, i_k) = \begin{bmatrix} L_a & M_{ab} & M_{ac} & M_{ad} \\ M_{ba} & L_b & M_{bc} & M_{bd} \\ M_{ca} & M_{cb} & L_c & M_{cd} \\ M_{da} & M_{db} & M_{dc} & L_d \end{bmatrix} \tag{1.3}$$

Thus, the flux linkage ψ_k of each phase is also the function of the current i_k of each phase and rotor position angle θ, which is computed as Eq. (1.4).

$$\psi_k(\theta, i_k) = L_k(\theta, i_k)i_k \tag{1.4}$$

Due to the electromagnetic nonlinearity of SRM, the electromagnetic torque could be obtained according to the principle of virtual displacement, which can be expressed as Eq. (1.5).

$$T_{ek}(\theta, i_k) = \frac{\partial W_{ek}(\theta, i_k)}{\partial \theta}\Big|_{i_k = const} \tag{1.5}$$

where W_{ek} is the magnetic common energy of each phase, and it can be calculated as Eq. (1.6).

$$W_{ek}(\theta, i_k) = \int_0^i \psi_k(\theta, i_k)di \tag{1.6}$$

Therefore, the total output torque of the motor can be computed as Eq. (1.7).

$$T_{em}(\theta, i_k) = \sum_{k=1}^{4} T_{ek}(\theta, i_k) \tag{1.7}$$

According to the principle of mechanical dynamics, the mechanical equation of SRM can be described as

$$T_{\text{em}} = J\frac{dw}{dt} + Kw + T_L \tag{1.8}$$

where J is the inertia moment of the SRM rotor, K means the friction coefficient, and T_L represents the load torque of the motor.

1.3 Magnetic Pole Distribution of SRM

The winding of SRM is distributed on the stator pole and consisted by one or more coils. In the process of motor excitation, the winding on the two stator salient poles of each phase will form N or S poles. Therefore, for multiphase SRM, various distribution ways of winding will form different distribution of magnetic poles, which can directly affect the electromagnetic filed characteristics of each phase for SRM.

For four-phase 8/6 SRM, there are two common magnetic pole distributions, which are NNNNSSSS and NSNSSNSN respectively, as shown in Fig. 1.10. The remaining distribution methods of magnetic poles seldom appear in practical application and are beyond the scope of this chapter.

The two kinds of magnetic pole distribution can form different magnetic paths in the multi-phase excitation mode of the motor. Take the two-phase excitation mode as an example, Fig. 1.11 shows two typical magnetic circuit ways, long magnetic circuit (LMC) and short magnetic circuit (SMC), generated by these two magnetic pole distributions. It can be predicted that it could have a non-negligible impact on the electromagnetic characteristics of the motor and should be considered separately in electromagnetic characteristic analysis and dynamic modeling.

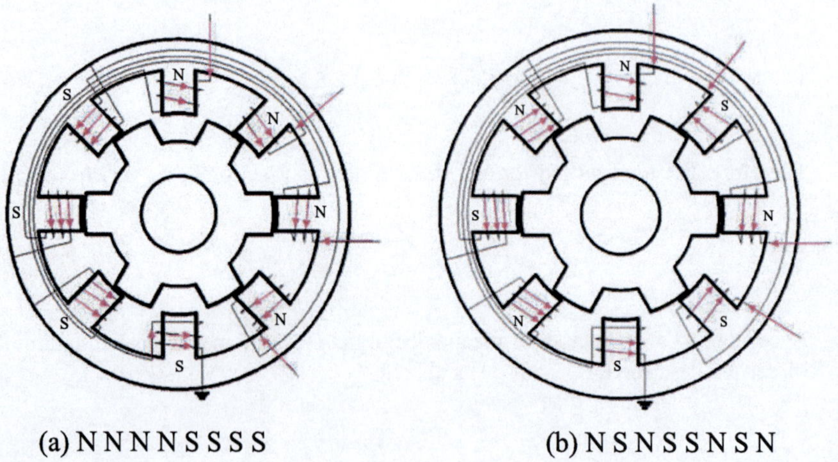

Fig. 1.10 Magnetic pole distribution of SRM

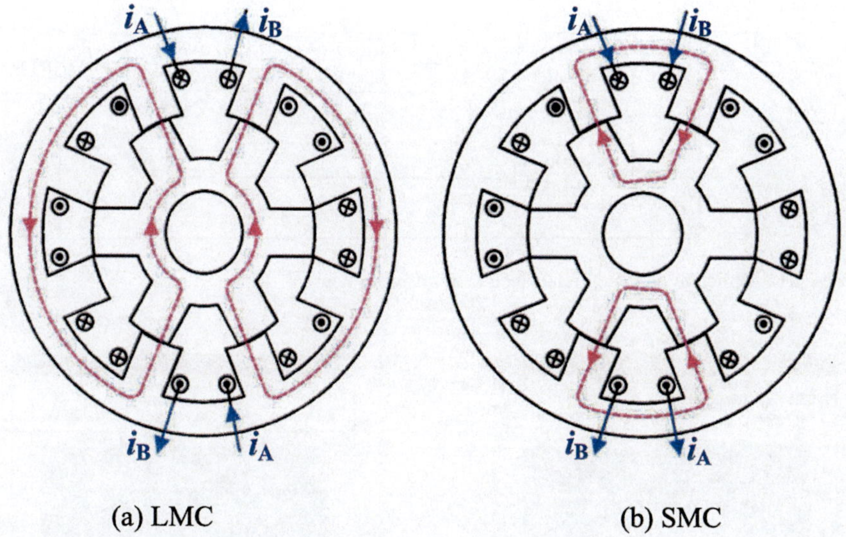

Fig. 1.11 Two typical flux circuits under two-phase excitation mode

1.4 The SRM Drive System

1.4.1 The Basic Composition of SRM Drive System

Typically, the SRM drive system (SRD) is mainly composed of six parts, SRM, power converter, current collector, angle controller, speed controller, and current controller, as shown in Fig. 1.12.

In SRD, SRM is the actuator of the whole drive system, and it is also the key to realize electromechanical energy conversion. The power converter is the energy provider of the whole drive system, which makes it has a high demand for its structure, because the ideal power converter should have some excellent characteristics, such as high efficiency, low cost, easy control, few main switching devices, and so on. In this chapter, ANSYS/Maxwell software is used for the electromagnetic finite element analysis (FEA) of the 8/6 four-phase SRM, as shown in Fig. 1.13, which provides three common main circuit forms of power converter, corresponding to asymmetric half-bridge type, H-bridge half-voltage type, and double-winding coupling type, showed in Fig. 1.13a–c respectively.

In Fig. 1.13a, the main circuit of asymmetric half-bridge type has some advantages, such as, independent control of each phase winding, clear supply voltage from the rated voltage, no limitations for the number of motor phases, and so on. However, there are also some disadvantages, including more main switching devices and poor economy. Considering that this main circuit is easy to control and the control mode is flexible, such that the asymmetric half-bridge type main circuit

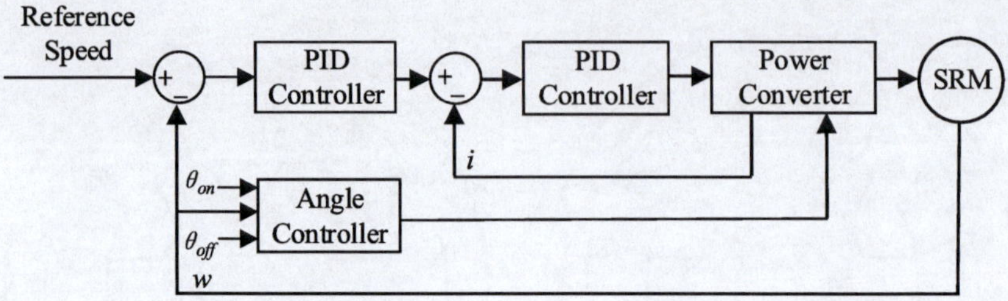

Fig. 1.12 Structure of switch reluctance motor drive system

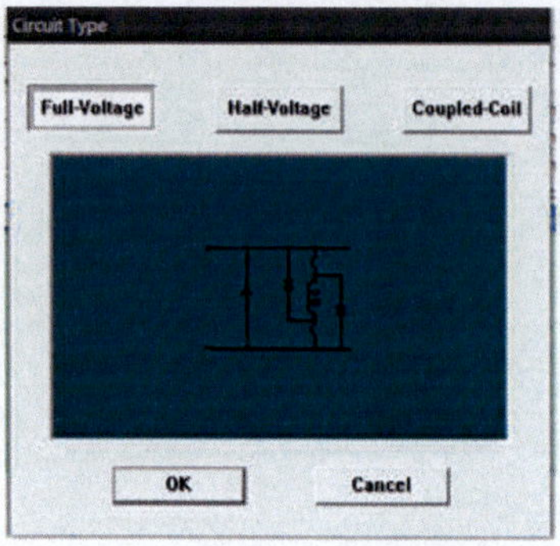

(a) asymmetric half-bridge type

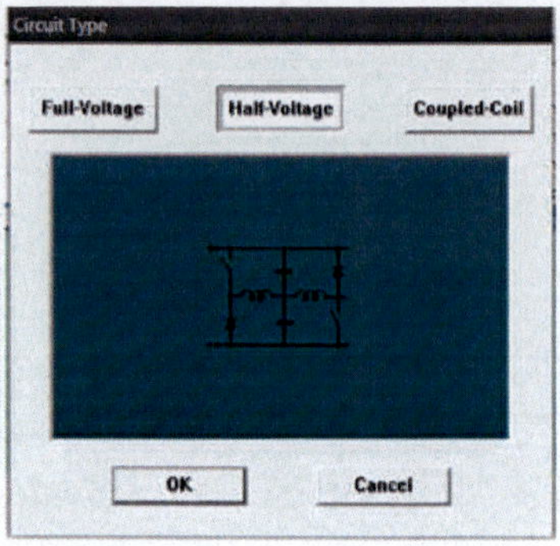

(b) H-bridge half-voltage type

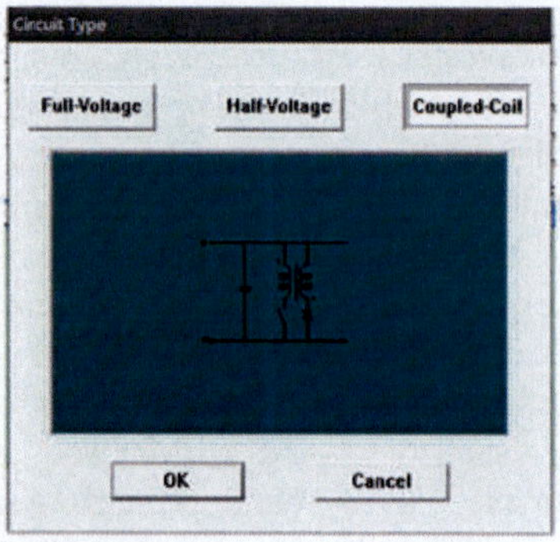

(c) double-winding coupling type

Fig. 1.13 Three common main circuit forms of power converter

is adopted in this chapter. The complete circuit diagram of the asymmetric half - bridge main circuit is showed in Fig. 1.14.

From Fig. 1.14, different states of switch VT1 and VT2 can cause three working states. When both VT1 and VT2 are turned on, the positive voltage is applied on

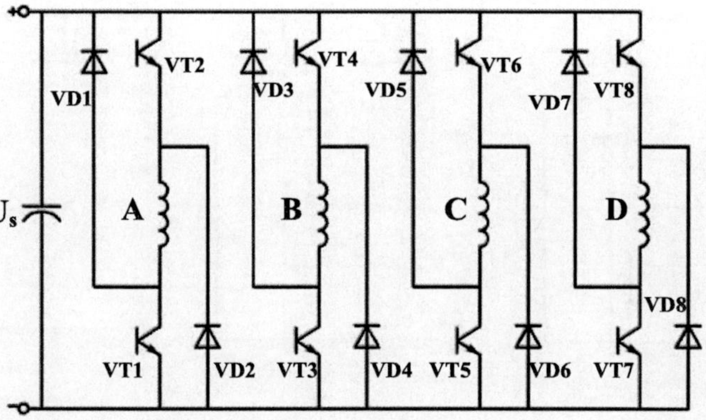

Fig. 1.14 Diagram of asymmetrical half-bridge main circuit

both sides of the corresponding phase winding, causing the increase of phase current, and this state can be denoted as "stage 1". When either VT1 or VT2 is turned on and the other is switched off, the terminal voltage of the phase winding is zero and the phase current decreases, such that the state can be denoted as "stage 0". When both VTI and VT2 are turned off, the current is in a freewheel current state and feeds back to the power supply, and this state is denoted as "stage −1". These three states are shown in Fig. 1.15a–c, respectively.

As shown in Fig. 1.13b, the H-bridge main circuit has less main switching devices, but only applies to the SRMs with four or more phase numbers. Furthermore, it has to works under two phase synchronous excitation mode. It can be seen that the terminal voltage on both sides of each phase winding is half of the supply voltage due to the circuit structure. In order to achieve the successful work, each one from both high-bridge switches and low-bridge switches should be turned on at the same time. It has some advantages, such as, few main switching devices, relatively simple the structure, good economy performance, and being suitable for the drive system with low control requirements, although the control way of this main circuit is not flexible enough. Its complete circuit diagram is shown in Fig. 1.16.

As shown in Fig. 1.13c, the double-winding coupling main circuit requires only one main switch for each phase, which can greatly simplify the structure of the drive circuit. However, it also requires the corresponding winding structure of SRM to be double-winding type, which directly limits the types of motors. Besides, the electromagnetic characteristics of double-winding SRM is greatly complicated such that it is hard to achieve high-performance control. Moreover, the nominal voltage of the switch device must be twice of the winding voltage, such that this main circuit is suitable for low supply voltage occasions. Its complete circuit diagram is shown in Fig. 1.17.

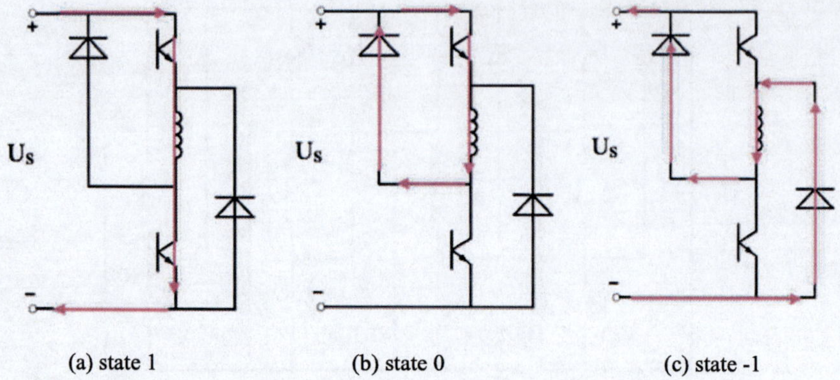

Fig. 1.15 Diagram of three circuit types

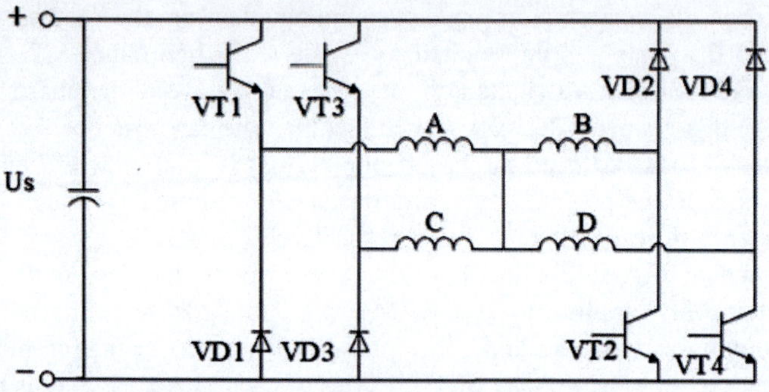

Fig. 1.16 H-bridge main circuit

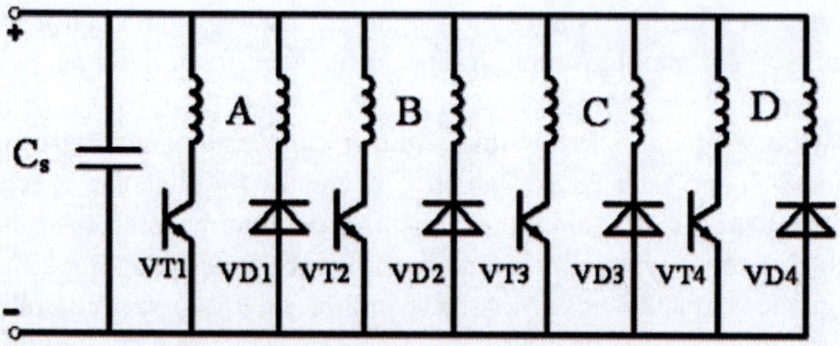

Fig. 1.17 Double-winding coupling main circuit

1.4.2 The Control Methods for SRM

The control method of SRM is the core to realize its speed regulation function, whose ability is to adjust the controlled variables of the drive system according to the different output requirements on speed and torque under various working conditions, such that it can meet the operating state of corresponding working conditions. For SRM, the commonly used controlled variables include the phase voltage applied to both ends of the winding, the turn-on and turn-off angles θ_{on} and θ_{off}. Considering the potential control variables, three common control modes are widely used, that is, angle position control (APC), current chopping control (CCC), and voltage PWM control (PWM).

In the operation process of SRM, the phase current waveform mainly depends on the turn-on angle θ_{on} and the turn-off angle θ_{off}. Therefore, to achieve the adjustment for the phase current, the two angles need to be controlled. Afterwards, the control of the output torque and speed are also achieved. In general, the turn-on angle mainly affects the peak value, phase width, and effective value of the phase current. Although the phase current peak value could not be affected by the turn-off angle, the phase width and the effective value of the phase current can be affected by the change of turn-off angle, such that the output torque and speed of the motor can also be affected. For APC, the waveform of phase current will change greatly in each combination between two angles, and it can be predicted that the small change of turn-on and turn-off angles can have a serious effect on the phase current. Therefore, it is difficult to achieve the accurately control for the phase current if only APC is employed for the drive system.

CCC, known as current PWM control, is also a kind of chopping control method with given upper and lower limits of phase current. Generally, in the actual operation process, the adjustment of the phase current is achieved by controlling the turn-off time after the upper phase current value is detected. The principle of the control method can be described as follows. when the rotor position angle is located in the excitation region of the motor, i.e., $\theta_{on} \leq \theta \leq \theta_{off}$, the corresponding main switching device is keeping on if the actual phase current is less than the maximum allowable value. Then, the phase current increases and gradually reaches the upper limit of the allowable chopping current. At this time, the switching device of the corresponding phase is disconnected, and then the phase current can decrease until the phase current reaches the lower limit value. By means of this method, the phase current can be controlled in the range of chopping current limits according to the turn-on and turn-off of the corresponding switching devices. As a result, the output torque and speed of the motor can be controlled. Compared to the APC, this control mode can directly limit the phase current amplitudes, and setup certain hysteresis thresholds according to different control requirements, which further improves the control effect for the phase current. It should be noted that CCC method is usually used in the operating condition of low speed. Because, the phase current change range is wide enough to obtain much high phase current at this time, which could directly result in circuit faults. Therefore, in order to protect the power circuit, CCC method is widely used when the SRM starts up.

The voltage PWM control method is also well known as PWM. The principle of this control method can be briefly described as follows. A PWM information can be obtained according to the working state of the motor. Then, the PWM signal is sent to the drive unit of the main switches, such that the power switches can be controlled. Based on this, the phase voltage can also be controlled by means of the pulse width modulation. In addition, the phase current is also regulated at the same time, although it is not controlled directly. Afterwards, the purpose of adjusting the motor output torque and speed is achieved. The control method can greatly adjust the amplitude and change rate of the phase current, which is conducive to forming a closed-loop control system combined with PID control, and also has good dynamic performance. However, this control method is relatively complex, and frequent switching of the main switch devices will lead to increase switching loss and reduce the controller lifetime.

1.5 Electromagnetic Characteristic Analysis Under Interactive Excitation

1.5.1 The Finite Element Model of SRM

The prototype studied in this chapter is an 8/6 four-phase SRM, whose physical structure is shown in Fig. 1.18.

The body of SRM has linear model, quasi linear model, and nonlinear model, and the dynamic characteristics of SRM cannot be well reflected by liner or quasi-linear model due to its high nonlinear characteristics of complex magnetic

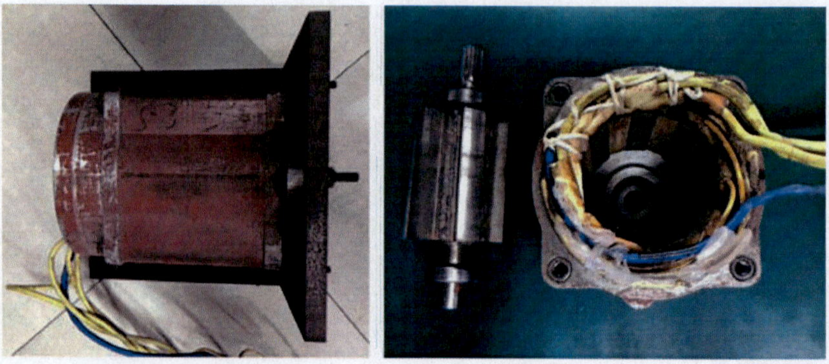

(a) Physical appearance (b) The structure of stator and rotor

Fig. 1.18 Prototype of SRM

flux, the nonlinear coupling of each phase, and high saturation characteristic of magnetic circuit of SRM. Therefore, this chapter chooses the method of nonlinear model to construct the model of SRM. In general, there are four typical modeling methods that include tabular lookup method, finite element analysis method (FEA), neural network method, and analytical method. Among them, the combination between FEA and tabular lookup method can accurately express the nonlinear characteristics of SRM, which is widely used in modelling SRM and also adopted in this chapter. Therefore, The FEA is used to calculate the characteristics of inductance and torque of the SRM, and then a lookup table is established to express the characteristics in this chapter.

In this chapter, rated voltage of the SRM is 72 V, rated torque is 25 N m, and rated power is 4 kW. As mentioned above, the inductance and torque characteristics in this chapter are obtained by FEA. The structural parameters of the SRM are listed in Table 1.1. According to the structural parameters, the finite element model of the SRM is established which is shown in Fig. 1.19.

1.5.2 *Magnetic Field Characteristics Under Interactive Excitation*

Magnetic flux lines, magnetic density distribution, inductance, mutual inductance, and electromagnetic torque can be obtained by the FEA. Considering potential overlap excitation, the magnetic flux lines and flux density distributions of the SRM under four special rotor position angles are shown in Figs. 1.20, 1.21, 1.22 and 1.23, where the phase current of 60 A is employed and two kinds of magnetic circuit (long magnetic circuit and short magnetic circuit) can be obtained. It can be seen that when the stator and rotor are in different relative positions, the magnetic flux lines and flux densities distribution in the yoke and air gap are obviously different. When the rotor of the motor is located in the position where the coincident area of the salient pole of stator and rotor is smaller, the reluctance and flux leakage

Table 1.1 Structure parameters of the SRM

Structural variables	Parameter values
Stator outer diameter	180 mm
Stator intermediate diameter	150 mm
Stator inner diameter	105 mm
Rotor outer diameter	104 mm
Rotor intermediate diameter	75 mm
Rotor inner diameter	36 mm
Iron core stack length	125 mm
Pole arc angle of stator	22 degrees
Pole arc angle of rotor	23 degrees
Air gap	0.5 mm

Fig. 1.19 Finite element model of the SRM

decrease accordingly, such that the saturation phenomenon becomes more serious. When the salient poles of stator and rotor are aligned, it is the minimum resistance position of the condition with two excited phases, which is defined as 37.5 degrees in this chapter.

Thus, it can be seen that when two phases are interactively excited in the single-phase excitation mode, the magnetic flux line and magnetic density distributions inside the motor are already different from those under the classic single-phase excitation mode and more similar to those under the two-phase excitation mode, and will deviate a certain extent to the side with large coincidence area between the stator and the rotor poles. In the long magnetic circuit (LMC), the magnetic poles of the stators from two excited interactive phases are the same, such that they repel each other and the magnetic flux lines have completely opposite directions. Therefore, the magnetic flux lines have a relatively long path and a wide distribution range within the motor. However, in the short magnetic circuit (SMC), the magnetic poles are attracted to each other and the magnetic flux lines have the same direction, such that the magnetic flux paths are relatively short and the magnetic field distribution is concentrated, compared to the LMC.

It can be seen from Fig. 1.20 to Fig. 1.23 that the magnetic flux length in the stator yoke and rotor yoke under LMC are three times and two times that of the SMC, respectively. Through the analysis of LMC, when the rotor position angle is 0 and 22.5 degrees, the stator and rotor salient poles overlap area of one phase is relatively large and the magnetic induction intensity is relatively high, such that the magnetic flux distribution is more intensive in this phase, compared to the other phase. In the stator yoke between the two phases, the weaker side of the magnetic field will be offset by the stronger side, and the magnetic density cloud image will

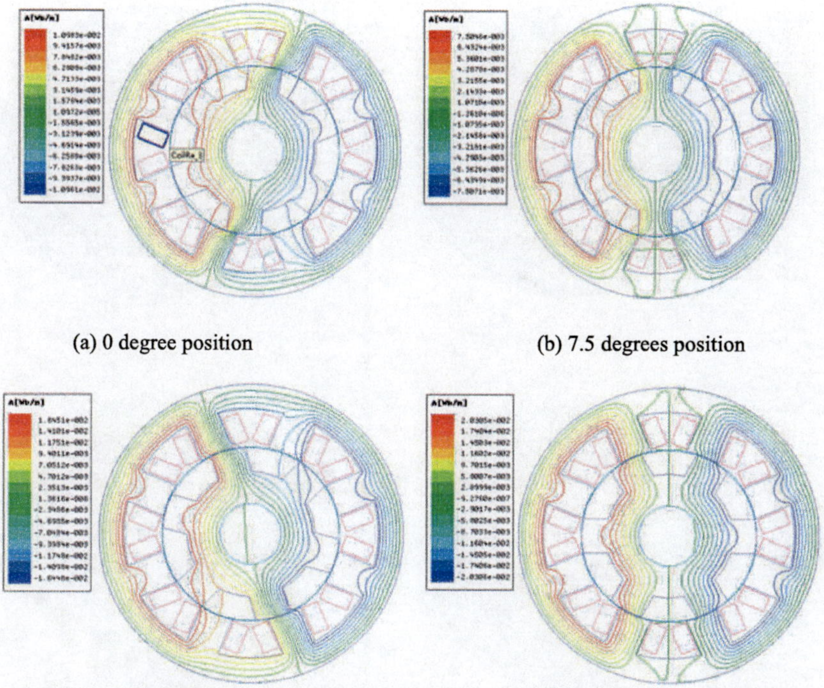

(a) 0 degree position (b) 7.5 degrees position

(c) 22.5 degrees position (d) 37.5 degrees position

Fig. 1.20 Magnetic flux lines distribution of LMC at four typical positions

darken at the corresponding position. At the rotor position of 7.5 degrees, the coincidence area of the salient poles of stator and rotor is the smallest, the magnetic resistance is the largest, and the magnetic induction intensity is the lowest, such that the magnetic flux distribution is the sparsest and the brightness of the magnetic density cloud image is the darkest. When the rotor position is 37.5 degrees, the coincidence area between stator and rotor salient poles is the largest, and the magnetic induction intensity is also the highest, such that the magnetic flux distribution is the densest and the magnetic density cloud image brightness is also the highest.

Differing from the LMC, at the rotor position of 0 and 22.5 degrees for SMC, the magnetic flux distribution focuses on the area nearby the two magnetic poles because of the two attracted magnetic poles. In addition, only a small amount of magnetic flux line overflow to other yoke parts. Compared to the LMC, the offset phenomenon in the yoke caused by the different magnetic flux directions does not appear in SMC. When the position angle is 7.5 degrees, the coincidence area of the stator and rotor salient poles is the minimum, such that the magnetic induction intensity is also the minimum, the magnetic flux lines are concentrated near the two magnetic poles and almost do not escape, and the brightness of the magnetic density

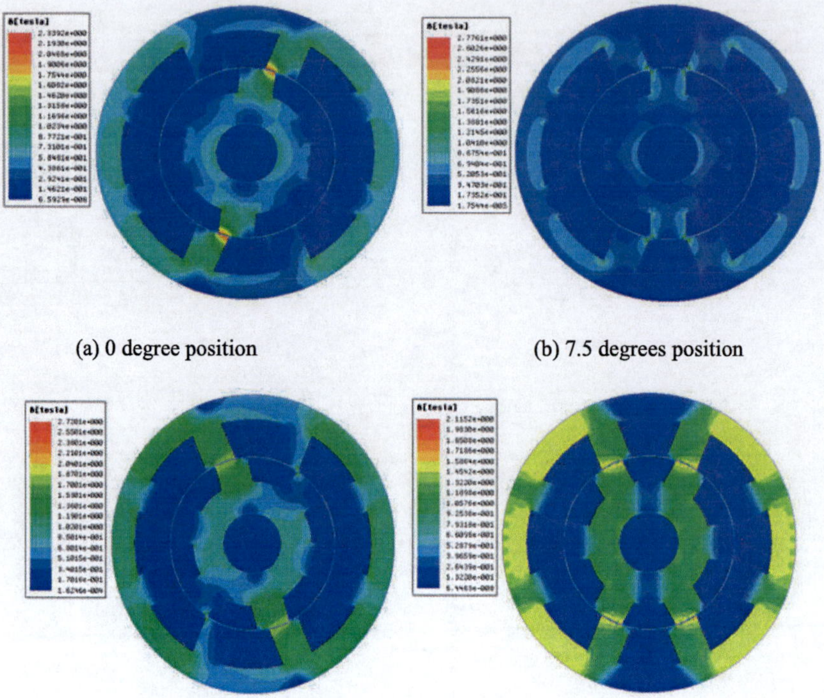

(a) 0 degree position (b) 7.5 degrees position

(c) 22.5 degrees position (d) 37.5 degrees position

Fig. 1.21 Magnetic flux density cloud map of LMC at four typical positions

distribution cloud image is the lowest. At the rotor position of 37.5 degrees, the coincidence area of stator and rotor salient poles is the largest and the magnetic induction intensity is also the largest. At this time, the magnetic flux lines are the most concentrated and the magnetic density distribution is also the largest, and the magnetic density cloud image has the maximum brightness. By comparing the magnetic fields of LMC and SMC, it can be seen that the magnetic field distributions of the two magnetic circuits are more different when the coincidence area of stator and rotor salient poles is larger.

1.5.3 Flux Linkage Characteristics Under Interactive Excitation

The flux characteristics of the four-phase 8/6 SRM in the single-phase excitation mode are computed by FEA and shown in Fig. 1.24, where the interactive excitation between two excited phases is considered.

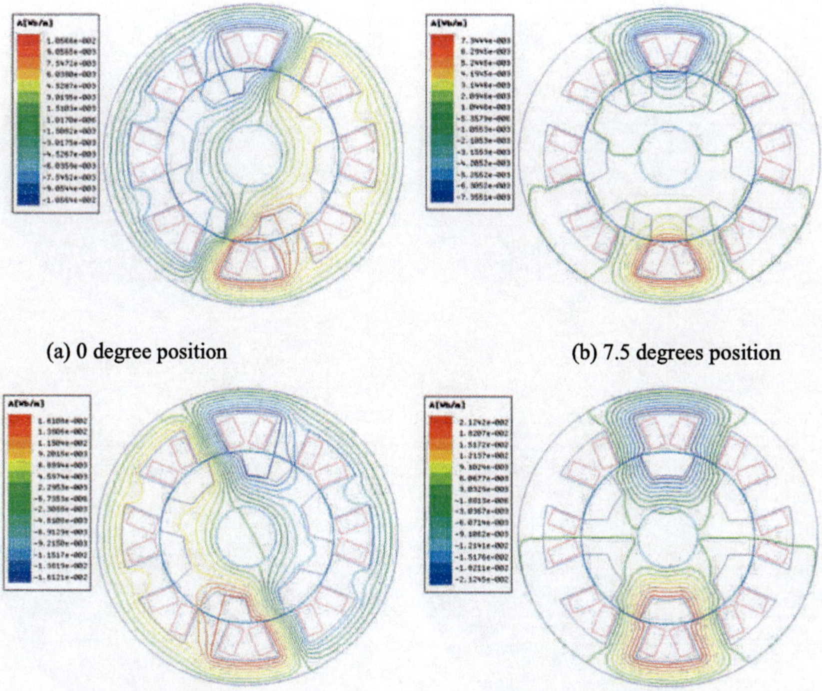

(a) 0 degree position

(b) 7.5 degrees position

(c) 22.5 degrees position

(d) 37.5 degrees position

Fig. 1.22 Magnetic field lines distribution of SMC at four typical positions

According to the analysis of the magnetic flux characteristics under two different magnetic circuits as shown in Fig. 1.24, when the rotor position angle changes from 0 to 60 A, the overall trend of the magnetic flux increases with the increase of the phase current. When the phase current is less than 30 A, the magnetic field of the motor is not saturated, such that the magnetic flux increases linearly with the increase of the phase current. However, when the phase current is more than 30 A, the magnetic field of the yoke of the motor is saturated, the flux linkage is nonlinear with the increase of the phase current, and the growth trend slows down. When the phase current of the SRM is constant, the magnetic flux increases first and then decreases with the increase of rotor position angle, reaching the maximum value until the rotor position angle is 30 degree, and the change range of the magnetic flux is small near the reluctance maximum and minimum positions. When the phase current and rotor position angle are constant, the magnetic flux value of SMC is slightly larger than that of LMC.

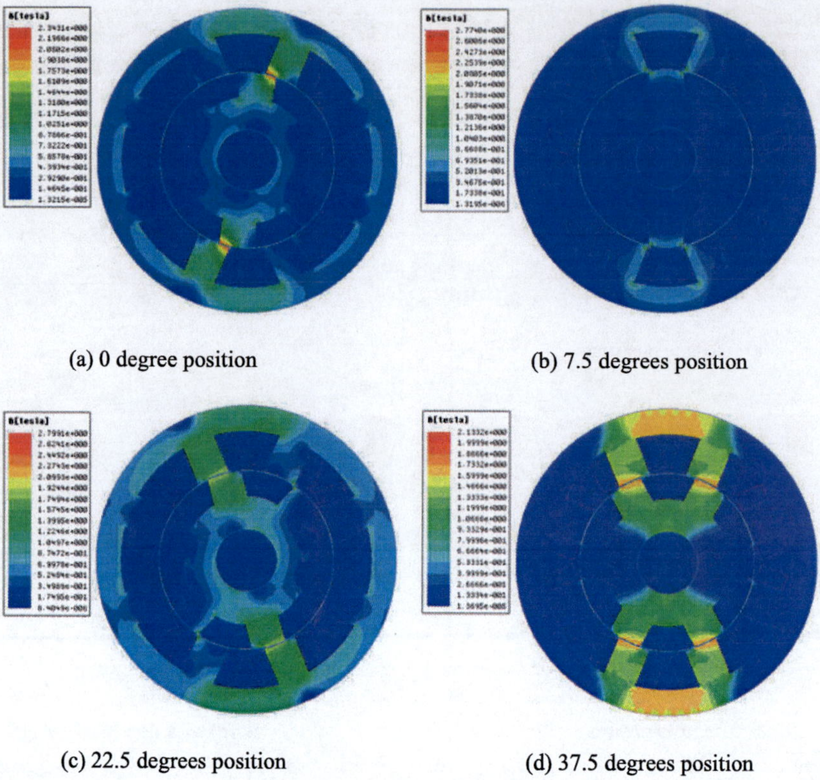

Fig. 1.23 Magnetic flux density cloud map of SMC at four typical positions

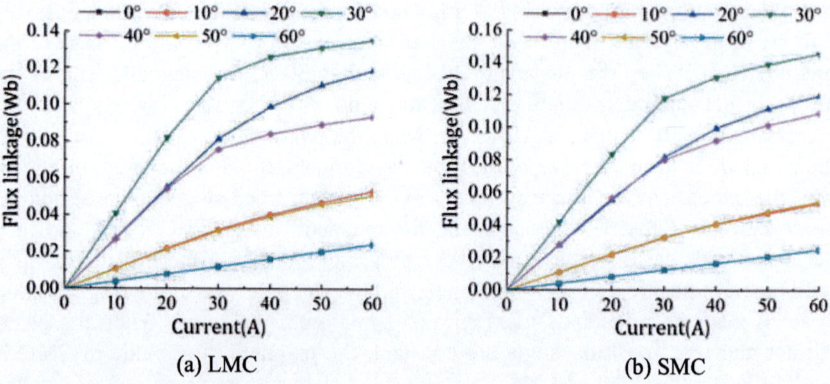

Fig. 1.24 Flux linkage characteristics of two different magnetic circuits

1.5.4 Mutual Inductance Characteristics Under Interactive Excitation

The mutual inductance of two interactive excited phases is a function of the rotor position angle and the phase current, which can represent the mutual coupling effect of the two interactive excited phases. The curves of mutual inductances in LMC and SMC are described in Fig. 1.25, where the phase currents are determined in the range of 10–60 A with a step of 10 A and the rotor positions are selected as 0–60 degrees.

It can be seen from Fig. 1.25a that mutual inductance under the magnetic circuit of LMC is negative. This is because the magnetic field in the yoke of the stator and rotor is more easily saturated and the magnetic path is relatively long, and the magnetic potential drop in the yoke is larger, such that the actual total magnetic flux under the LMC is less than the sum of the magnetic flux values of the two excited phases. Moreover, under the SMC, the magnetic field in the yoke is saturated and the magnetic potential drop in the stator and rotor yoke is relatively large when the currents of phase A and phase B are high enough, i.e., 30–60 A. As a result, the actual total flux is also less than the sum of the magnetic flux generated by the two excited phases respectively. As shown in Fig. 1.25b, mutual inductance is also negative in this case, however, when both phase currents are small enough (10 and 20 A), the magnetic field is not saturated, the magnetic path is relatively short, and the magnetic potential drop is small, which causes that the total magnetic flux is higher than the sum of the magnetic flux generated by the two excited phases respectively. Therefore, mutual inductance in the SMC is positive. In addition, when the rotor positions and the phase currents are both the same, the magnetic saturation in the two magnetic circuit modes is basically the same, and the absolute value of mutual inductance in the SMC is always smaller than that in the LMC.

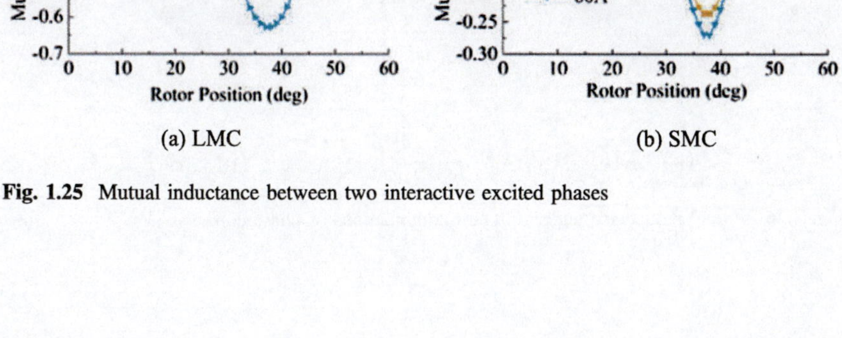

Fig. 1.25 Mutual inductance between two interactive excited phases

Under both the two magnetic circuits, the absolute value of mutual inductance reaches the maximum at the alignment position where the center lines of the stator and rotor slot coincide.

The curves that the mutual inductance varies with different combinations of the two-phase currents in alignment position of two interactive excited phases can be obtained by further analyzing the results from FEA, as shown in Fig. 1.26. It can be seen from the figure that the absolute value of mutual inductance under LMC increases correspondingly with the increase of the both phase currents, and reaches the maximum absolute value when both phase currents are 60 A. However, for the SMC, when both phase currents are low enough, i.e., 10 and 20 A, the mutual inductance is positive due to the unsaturated magnetic circuit and the large flux linkage. Nonetheless, with the increase of the both phase currents, the magnetic circuit gradually reaches saturation, the flux linkage gradually decreases, and the mutual inductance becomes negative. In addition, when the phase current increases to 60 A, the absolute value of mutual inductance also reaches the maximum.

1.5.5 *Electromagnetic Torque Characteristics Under Interactive Excitation*

In the static finite element analysis of SRM, the calculation of torque is also a very important and indispensable. In the environment of Maxwell, the virtual displacement method is used to calculate the electromagnetic torque, also known as the virtual work principle, that is, the work done by the electromagnetic torque is equal to the change of the magnetic field energy at this stage. The two-dimensional curves of electromagnetic torque are shown in Fig. 1.27, where the phase currents are

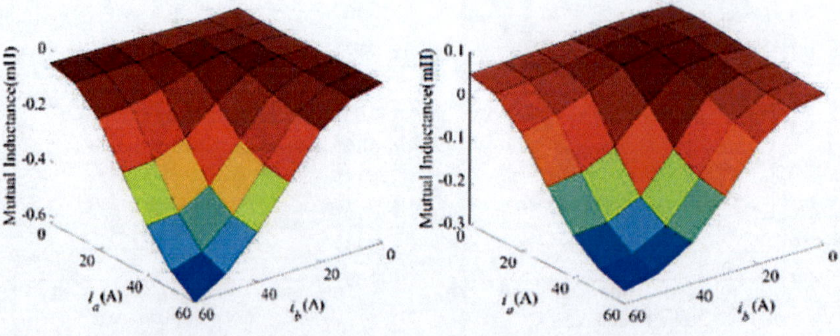

(a) LMC (b) SMC

Fig. 1.26 Mutual inductance under different combinations of phase currents determined in the range of 0–60 A with a step of 10 A and the rotor positions are selected as 0–60 degrees.

As for Fig. 1.27, it can be seen that the output torque of SRM is positive in the range of rotor position angle from 0 to 30 degrees because the rotor is in the inductance rising zone of the excited phase at this time, while the output torque is negative in the range of rotor position angles from 30 to 60 degrees thanks to the inductance drop zone of the excited phase. At the maximum reluctance position 0 degree and the minimum reluctance position 30 degrees, the output torque value of the SRM becomes zero, because the center line of the salient pole of the phase stator aligns that of the rotor groove at the maximum reluctance position, while the center line of the stator salient pole of the phase aligns that of the rotor salient pole at the position of the minimum reluctance. Thus, these two positions cannot deflect the magnetic flux line such that the magnetic forces at both positions are zero.

The torque–angle characteristics between two adjacent excitation phases are also dependent on the rotor position and the phase current of the two interactive excited phases. The torques of phase A and phase B under single-phase excitation mode, T_A and T_B, and the torques under LMC and SMC considered the coincidence of interactive excitation between both phases, T_{AB}, are computed by FEA and shown in Fig. 1.28, where three phase current conditions of 20, 40 and 60 A are set for comparison. It should be noted that the $T_A + T_B$ refers to the sum of separated phase torque.

Under the phase current of 20 A in both excited phases, through comparing the torque generated in the two magnetic circuits of LMC or SMC with the that generated in the single-phase excitation modes of phase A and phase B, which are as shown in Fig. 1.28a, it can be seen that the three curves, actual torque in LMC, actual torque in SMC, and summing torque from two phases, are very close to each other, but slightly different in value. This is because when the phase current is 20 A, the mutual inductance value of LMC and SMC is very small, and the influence on the torque is also very small, such that these three curves are very close. It can also

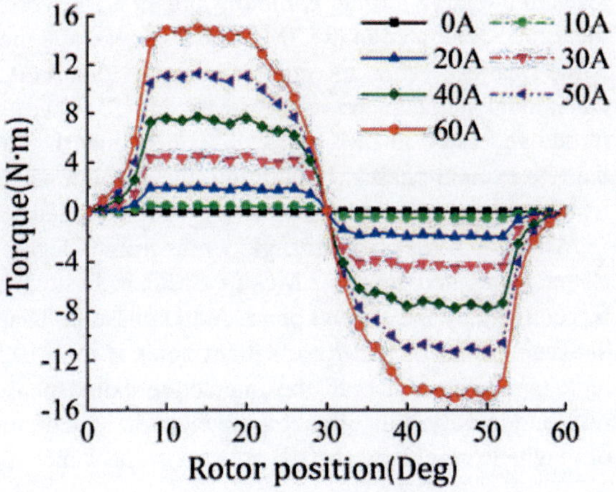

Fig. 1.27 Torque curves under single-phase excitation mode

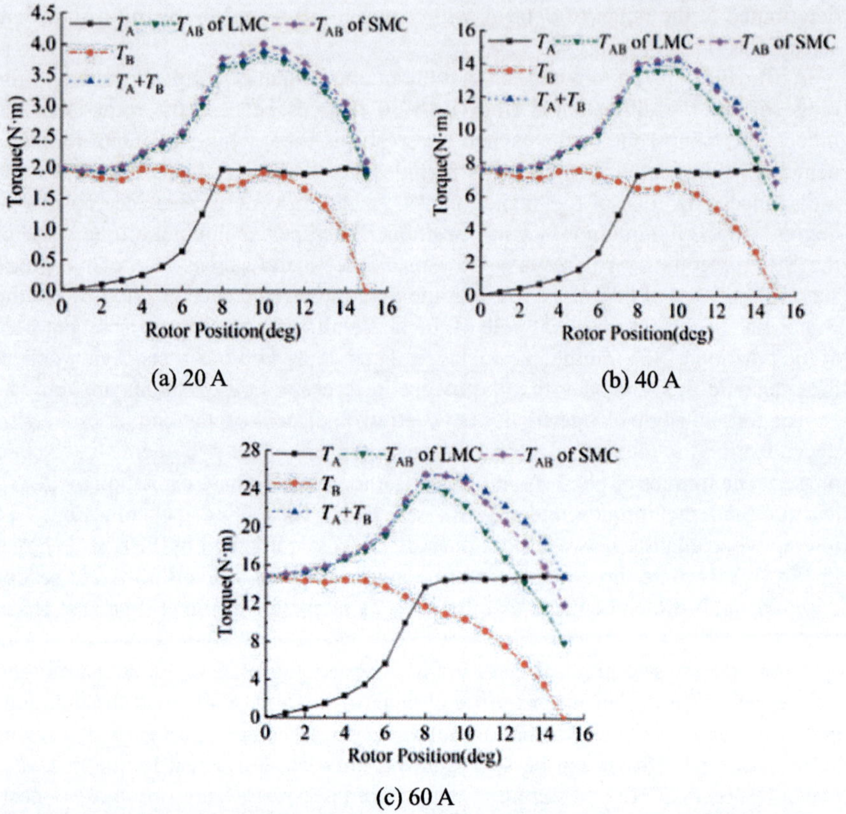

Fig. 1.28 Torque characteristics under different phase currents

be found from Fig. 1.26b that mutual inductance is positive under the SMC, which leads to the large torque, while the situation is reversed under the LMC. Therefore, the torque generated in the SMC is the largest, and the torque generated in the LMC is the smallest. When the current is set to 40 and 60 A, as shown in Fig. 1.28b, c, the mutual inductances in the LMC and SMC are negative. Therefore, the total torque generated in LMC or SMC is less than the sum of the torque generated by the two excited phases. In addition, the absolute value of mutual inductance in the LMC is more than that in the SMC, which can lead to smaller torque in the LMC.

When the rotor position angle varies from 0 degree to around 8 degrees, the total torque generated in the LMC or SMC is basically the same as that generated respectively by the excited phase A and phase B. That is because the overlap region between the stator and rotor salient poles is too much small, such that the mutual inductance between both the interactive exited phases is also very small, further resulting in very little effect on the electromagnetic torque. Furthermore, the values of magnetic co-energy (MCE) are almost the same when the total MCE in the LMC or SMC and the sum of the MCE respectively generated by the phase A and phase B are compared. However, as the rotor position angle rises to the range of 8–15 degrees, the discrepancy values among the three curves shown in Fig. 1.22 increases correspondingly. It can be observed after analysis that the overlap region between stator pole and rotor pole becomes large when the rotor position angle near to 15 degrees. As a result, the absolute value of mutual inductance can become higher and the effects of mutual inductance on the electromagnetic torque can be greater. Therefore, the three curves diverge more and more when the rotor position angle increases from 8 to 15 degrees.

In addition, to more distinctly compare the numerical differences of the three torques when the rotor is at different positions, the values of each torque value and their differences are listed in Table 1.2, where *TT*, *TS*, and *TL* refer to the sum of the torque from phase A and phase B, the torque generated in the SMC, and the torque generated in the LMC, respectively. Besides, it should be noted that the ΔTS and ΔTL are defined and computed by subtracting summing torque *TT* from the SMC torque *TS* and the LMC torque *TL*, respectively, which can be used to directly show the differences among the three torque values. From Table 1.2, the follows can be also observed that when the rotor position angle is small, the differences among the three torques are very small, while when the rotor position angle becomes large, the differences of the three torques become larger and larger. Furthermore, it can be seen that the maximum differences between *TT* and *TL*, and *TT* and *TS*, can reach 7.0 and 3.0 N m with difference ratios up to −46.98% and −20.13%, respectively. Thus, such large difference values can directly prove that the torque–angle characteristics of the SRM can be changed because of the interactive excitation between the two excited phases, which should be fully considered in the process of modeling the motor to improve the model accuracy.

As shown in Fig. 1.29 where the rotor position angles are 7.5, 22.5, and 37.5 degrees, the influences of the various combinations between two phase currents of phase A and phase B on torque–angle characteristics are computed and obtained under three different rotor position angles in SMC in this chapter, in order to further study the effects of the interactive excitation. As can be seen from Fig. 1.29a, the reason why the torque is zero when the two excited phases are the same is that the

Table 1.2 Torque difference data of two interactive excited phases

Position angle	0 degree	3 degrees	6 degrees	9 degrees	12 degrees	15 degrees
TT	14.9 N m	15.7 N m	19.5 N m	25.5 N m	22.3 N m	14.9 N m
TS	14.9 N m	15.9 N m	19.7 N m	25.5 N m	20.6 N m	11.9 N m
TL	14.8 N m	15.6 N m	19.1 N m	23.7 N m	16.9 N m	7.9 N m
ΔTS	0.0 N m	0.2 N m	0.2 N m	0.0 N m	−1.7 N m	-3.0 N m
ΔTL	−0.1 N m	−0.1 N m	−0.4 N m	−1.8 N m	−5.4 N m	-7.0 N m
$\Delta TS/TT$	0	1.27%	1.03%	0	−7.62%	−20.13%
$\Delta TL/TT$	−0.67%	−0.64%	−2.05%	−7.06%	−24.22%	−46.98%

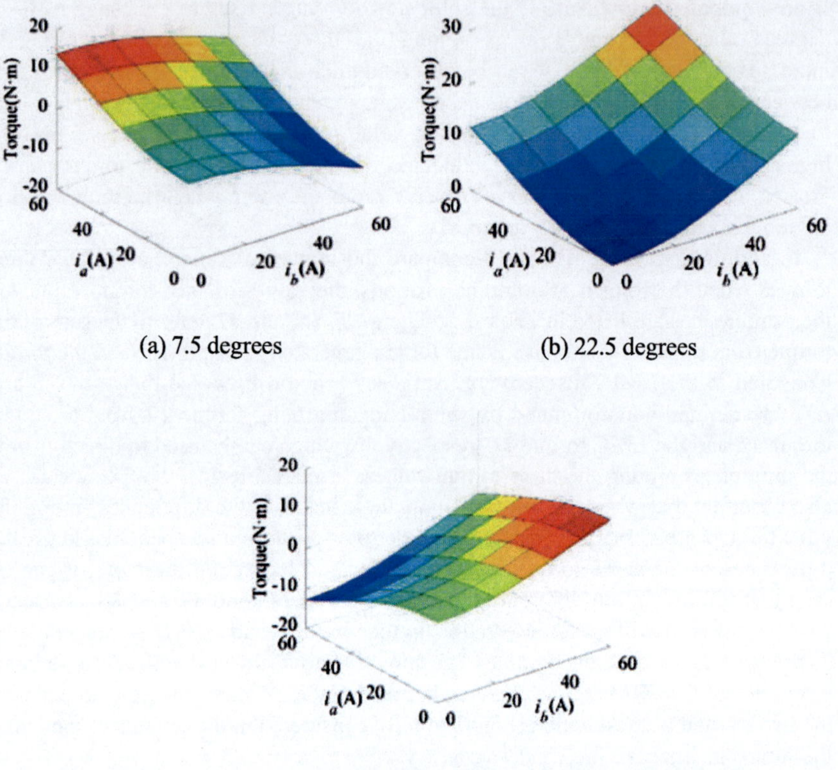

(c) 37.5 degrees

Fig. 1.29 Torque characteristics under different combinations of both currents

salient poles of the stator and rotor are not aligned when the rotor position angle is 7.5 degrees under the single-phase excitation mode. However, when the current of one phase is higher than that of the other phase, the torque may become positive or negative accordingly. As for Fig. 1.29b, it can be seen that when the rotor position angle is 22.5 degrees, a positive torque value can be obtained and it can gradually increase according to the rise of the phase current since there are both with increasing inductance for the phase A and phase B such that the positive torques for both phases are generated at this time. As shown in Fig. 1.29c, where the rotor position angle is fixed to 37.5 degrees, it can be seen that the situation is completely opposite after compared to the results from condition of 7.5 degrees.

In addition, the torque characteristics in LMC should be also computed and obtained. After compared the results from LMC and SMC, an obvious conclusion can be obtained that the torque changing trend between the two circuits are basically the same. As a result, the results from LMC would not be described in this chapter.

1.6 Modelling the SRM Drive System

1.6.1 The Principle of Overlap Between Excited Phases

Taking the four-phase 8/6 SRM under the single-phase excitation mode as an example, in the actual operation process, the excitation interval angle of each phase is usually greater than 15 degrees, which can effectively improve the performance of SRM and greatly reduce the torque ripple. When the main switch is turned off, then the power circuit is shut off, but there is still current in the circuit because of the freewheel current. Therefore, even under the single-phase excitation mode, the power circuit of the two adjacent excitation phases will flow simultaneously current at the same time for a time, which is much similar to the simultaneous excitation of two phases. However, it should be noted that the current values of the two phases cannot keep the same in the operation process, which can result in different magnetic fields generated by both phases. In this case, since the coupling effect between the two excitation phases cannot be ignored, the total torque is not equal to the sum of the torque generated by the two phases separately. The diagram of phase current overlap is showed in Fig. 1.30 under the single-phase excitation mode, where the turn-on and turn-off angles are set to 3 and 22 degrees. By this time, the original magnetic field of the motor has been changed, and the coupling effect between phases magnetic field should not be ignored. Therefore, the establishment of SRM dynamic model considering the mutual inductance effect of interaction excitation between two excited phases is of great significance for accurate analysis of electromagnetic characteristics and prediction of motor performance.

As for 8/6 four-phase SRM, the magnetic field distribution of NSNSSNSN is widely employed for the SRM with single-phase excitation mode. The typical power circuit distribution diagram under the single-phase excitation mode is showed in Fig. 1.31, where all phases are linked to the power supply U_s through combinations between main switches and diodes. Take the phase A as an example, there may be two kinds of phase currents in the power converter when the SRM works, which are divided into the excitation current i_{A1} and the continuation current i_{A2}, respectively. Furthermore, when phase B is excited with the excitation interval angle greater than 15 degrees, there could still have i_{A1} or i_{A2} in the power circuit.

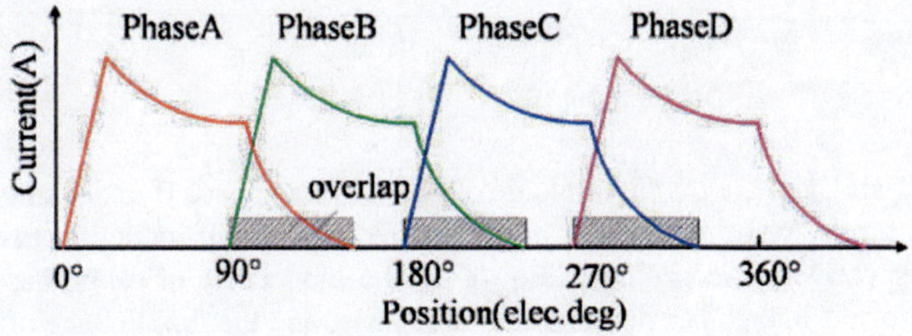

Fig. 1.30 Current overlap of two interactive excited phases

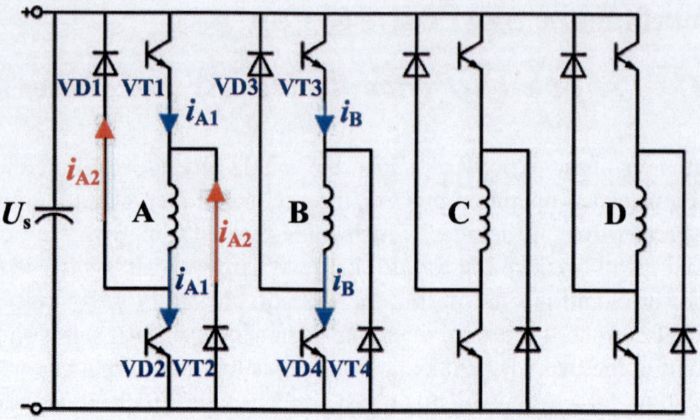

Fig. 1.31 Power converter circuit of two-phase overlap under single-phase excitation mode

As a result, two phases are excited at the same time, even though the SRM works under single-phase excitation mode. Therefore, the magnetic fields excited by phase A and phase B exist at the same time such that they could interact with each other, which will change the original magnetic field of the motor in the single-phase excitation mode.

1.6.2 The Mathematical Model of the SRM

Under single-phase excitation mode, the electromagnetic characteristics of the motor will be changed because of interactive excitation between two excited phases, and the original mathematical model is no longer applicable by this time. Phase A and phase B are taken as examples in this chapter, considering the mutual effect of interactive excitation between two excited phases, the mathematical model of SRM is established, and the data required for modeling are all from the FEA method.

When the phase A and phase B of the SRM are excited separately, the flux linkages for each phase can be expressed as Eqs. (1.9) and (1.10).

$$\psi_{AA} = L_A i_A \tag{1.9}$$

$$\psi_{BB} = L_B i_B \tag{1.10}$$

where ψ_{AA} and ψ_{BB} are the flux linkage of phase A and phase B under single-phase excitation mode, respectively; L_A and L_B refer to the self-induction of the two phases, respectively. According to Eq. (1.3), the inductance of each phase can be divided into self-inductance and mutual inductance, and the flux linkage of phase A and phase B will be affected by the phase currents of each phase because of the interactive excitation between two excited phases. Thus, the total flux linkage of phase A and phase B can be calculated as Eq. (1.11).

$$\psi = \psi_{AA} + \psi_{BB} + \psi_{m} \tag{1.11}$$

where ψ is the total flux linkage of the interactive excitation between two excited phases, and ψ_m is the mutual-inductance flux linkage of phase A and phase B. At this time, the flux linkage of phase A should be added the effects of the mutual inductance, other than the product of phase current and self-inductance of phase A. It should be noted that the mutual-inductance flux linkage of phase A and phase B can also be signed as ψ_{AB} and ψ_{BA}, such that the following Eq. (1.12) are expressed.

$$\psi_{m} = \psi_{AB} = \psi_{BA} \tag{1.12}$$

Therefore, for phase A, the expression of the flux linkage can be described as

$$\psi_{A} = \psi_{AA} + \frac{1}{2}\psi_{AB} \tag{1.13}$$

In short, the A-phase current can be obtained by integral calculation as follows.

$$i_{A} = \int \frac{U_{A} - R_{A}i_{A} - i_{A}\frac{dL_{A}(\theta,i_{A})}{dt} - \frac{1}{2}i_{B}\frac{dM_{AB}(\theta,i_{A},i_{B})}{dt}}{L_{A}(\theta,i_{A})}dt \tag{1.14}$$

According to the electromechanical energy conversion principle, the magnetic common energy can be calculated as

$$W_{c} = \int \psi di \tag{1.15}$$

And the magnetic common energy can be expressed as

$$W_{c}(i_{A}, i_{B}, \theta) = \int_{0}^{i_{A}} L_{A}(i_{A}, \theta)i_{A}di_{A} + \int_{0}^{i_{B}} L_{B}(i_{B}, \theta)i_{B}di_{B} + \int_{0}^{i_{A}} M_{AB}(i_{A}, i_{B}, \theta)i_{B}di_{A} \tag{1.16}$$

According to Eqs. (1.16) and (1.5), the output torque of SRM in the case of interactive excitation under single-phase excitation mode can be written as

$$\begin{aligned} T_{\text{total}} &= \frac{\partial W_{c}(i_{A}, i_{B}, \theta)}{\partial \theta} = \frac{1}{2}i_{A}^{2}\frac{\partial L_{A}(i_{A}, \theta)}{\partial \theta} + \frac{1}{2}i_{B}^{2}\frac{\partial L_{B}(i_{B}, \theta)}{\partial \theta} + i_{A}i_{B}\frac{\partial M_{AB}(i_{A}, i_{B}, \theta)}{\partial \theta} \\ &= T_{A} + T_{B} + T_{AB} \end{aligned} \tag{1.17}$$

where T_A and T_B are the torque generated by phase A and phase B under single-phase excitation mode, and T_{AB} refers to the torque generated by the coupling of phases under the interactive excitation between two excited phases. According to the mathematical model established above, considering the mutual effect of interactive excitation between two excited phases, the schematic diagram of the ontology model of SRM under single-phase excitation mode can be shown in Fig. 1.32.

1.7 Dynamic Simulation Analysis

According to the ontology model established in Fig. 1.32 and the electromagnetic characteristics obtained by FEA, the dynamic model of the SRD based on the four-phase 8/6 SRM is established under the Matlab/Simulink environment. The simulation results are obtained under various combinations of two speeds (800 and 1500 r/min) and two loads (1 and 10 N m). By contrast with the simulation results of the traditional dynamic model without considering the overlap of excitation, the accuracy of the SRM model considering the mutual effect of interactive excitation between two excited phases was compared and analyzed. It should be noted that, the model considering the mutual effect of interactive excitation between two excited phases is marked as M1, and the model without considering the mutual effect of interactive excitation between two excited phases is marked as M2.

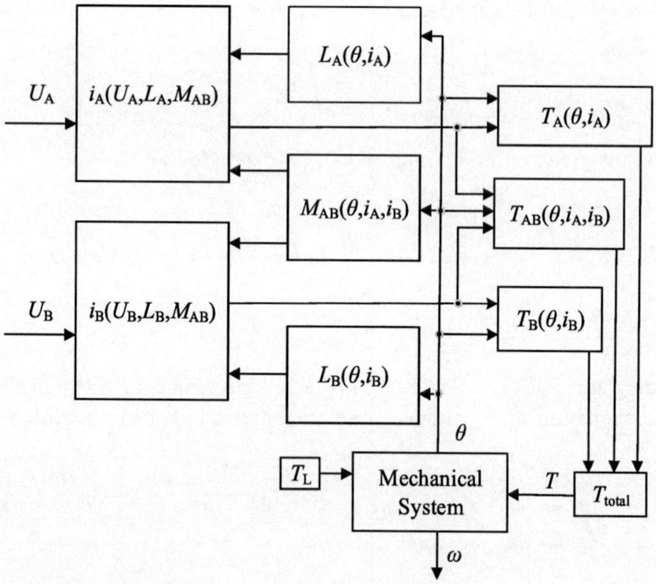

Fig. 1.32 Ontology model diagram of the SRM

1.7.1 Phase Current Results

The current curves under M1 and M2 are obtained through simulation under two working conditions (800 r/min with 1 N m, and 1500 r/min with 10 N m), as shown in Fig. 1.33, where the differences of the phase current between M1 and M2 are clearly illustrated. It can be seen that the maximum difference value can be up to nearly 0.5 A between M1 and M2. Therefore, it can be observed that the effects of interactive excitation between two excited phases is high enough such that the phase current error is too large. Besides, with the increase of the load torque, the differences of the phase currents between M1 and M2 could become larger.

1.7.2 Dynamic Torque Results

The dynamic torque curves under M1 and M2 are also obtained through simulation under the same two working conditions, as shown in Fig. 1.34, where the operation speeds are 800 and 1500 r/min with two different load torques. It can be clearly seen that the torques produced by the two models are obviously different, because the significant influence of interactive excitation between excited phases is precisely accounted in M1. As shown in Fig. 1.35, the ratio of torque difference between M1 and M2 is computed, which can directly show the torque differences in value. It can be seen from Fig. 1.35, when the rotation speeds are 800 and 1500 r/min, and the load is 1 N m, the torque difference ratio can be 15% and 8%, respectively. Meanwhile, the ratio values can reach 3.6 and 2.9% with the increase of the load torque to 10 N m. In total, the interactive excitation between the two excited phases should be taken more attention when a more accurate SRM model is required in the simulation work.

Moreover, in order to clearly know the differences between the two magnetic circuits, that is, LMC and SMC mentioned above, LMC and SMC are marked in the

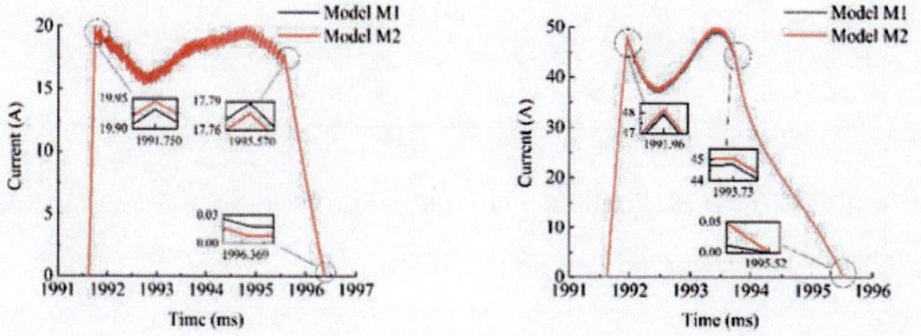

(a) speed of 800 r/min and load of 1 N·m (b) speed of 1500 r/min and load of 10 N·m

Fig. 1.33 Phase current simulation comparison

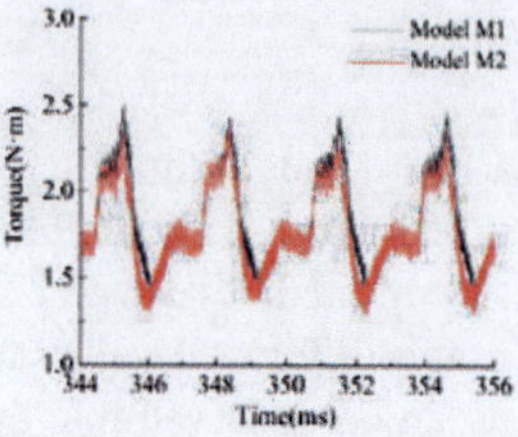

(a) speed of 800 r/min and load of 1 N·m

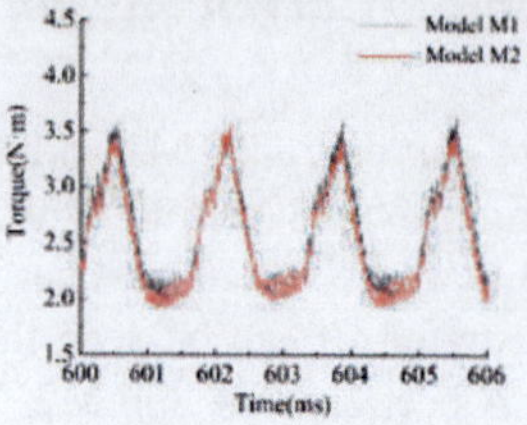

(b) speed of 1500 r/min and load of 1 N·m

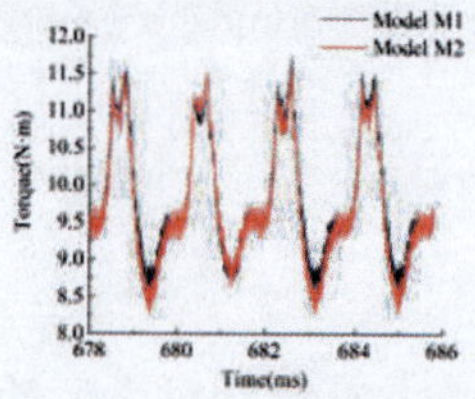

(c) speed of 800 r/min and load of 10 N·m

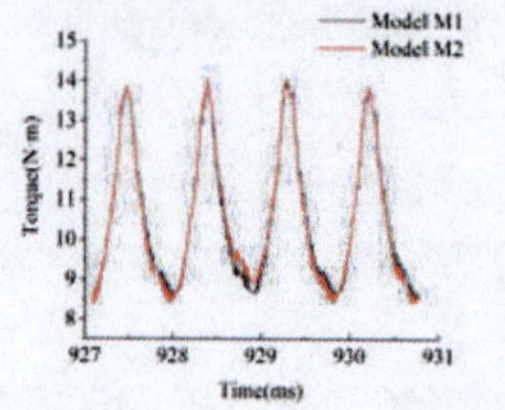

(d) speed of 1500 r/min and load of10 N·m

Fig. 1.34 Torque simulation comparison

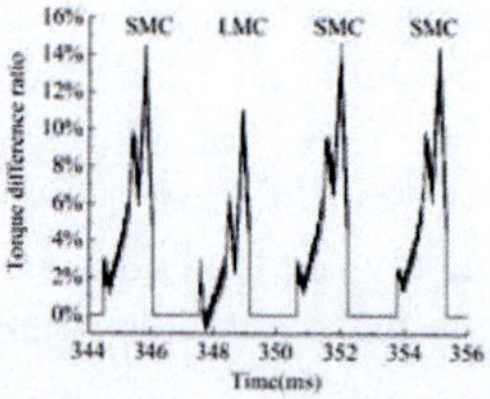

(a) speed of 800 r/min and load of 1 N·m

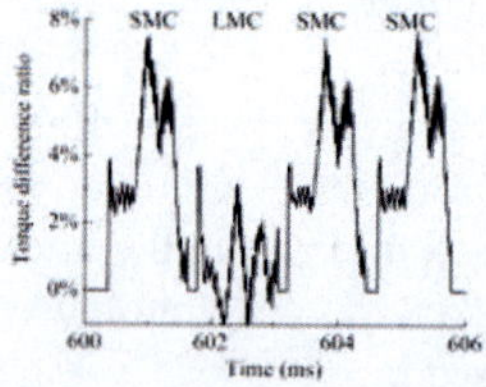

(b) speed of 1500 r/min and load of 1 N·m

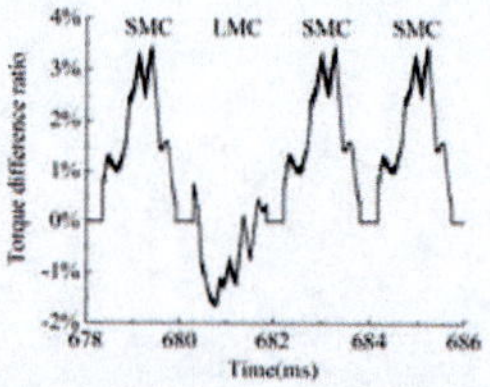

(c) speed of 800 r/min and load of 10 N·m

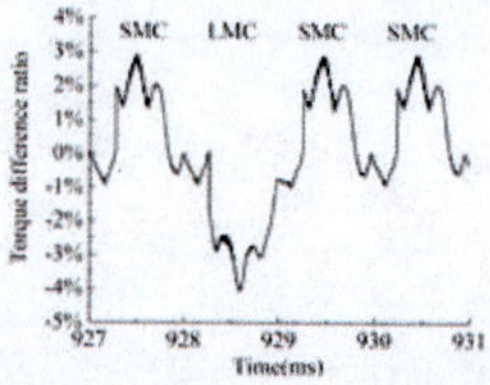

(d) speed of 1500 r/min and load of 10 N·m

Fig. 1.35 Torque difference ratio between M1 and M2

Fig. 1.35, where there are three SMC and one LMC in one electric cycle according to the magnetic field characteristics. It can be obviously seen that the difference radio between LMC and SMC is completely different. As for SMC, the maximum difference ratios are around 14.5%, 7.5%, 3.5%, and 3%, when the conditions are 800 r/min with 1 N m, 1500 r/min with 1 N m, 800 r/min with 10 N m, and 1500 r/min with 10 N m, respectively. Compared to the results from SMC, the ratios in LMC are 11, 3, −1.5, and −4% when the conditions are set the same. Meanwhile, it can also be seen that the average difference ratio between M1 and M2 under low load torque is higher than that under high load torque, which is suitable for both LMC and SMC conditions. Therefore, after amazing the results, the torques obtained in LMC and SMC are greatly different from each other, and even different from the torque sum of the two excited phases. As a result, the magnetic circuits are also an important factor when the SRM is modelled, even though the interactive excitation between the two excited phases should be taken more attention.

References

1. Xu B, Xiang C, Qin Y, Ding P, and Dong M. Semi-active vibration control for in-Wheel switched reluctance motor driven electric vehicle with dynamic vibration absorbing structures: concept and validation. IEEE Access, 2018, 6: 60274–60285.
2. Boldea I, Tutelea LN, Parsa L, and Dorrell D. Automotive electric propulsion systems with reduced or no permanent magnets: an overview. IEEE Trans. Ind. Electron, 2014, 61(10): 5696–5711.
3. Azer P, Bilgin B, and Emadi A. Mutually coupled switched reluctance motor: fundamentals, control, modeling, state of the art review and future trends. IEEE Access, 2019, 7: 100099–100112.
4. Qu B and Song J. Mutual inductance of switched reluctance motor and its effect on torque. Electric Machines and Control, 2009, 13(3): 332–336.
5. Zhao L, Deng Z and Cao X. Mathematical model of single-winding bearingless switched reluctance motor considering two-phase coupling. In proc. IECON, 2013: 2971–2976.
6. Kuai S, Zhao S, Heng F, and Cui X. Position sensorless technology of switched reluctance motor drives including mutual inductance. IET Electr. Power Appl, 2017, 11(6): 1085–1094.
7. Zhu Y, Wei W, Yang C, and Zhang Y. Multi-objective optimisation design of two-phase excitation switched reluctance motor for electric vehicles. IET Electr. Power Appl, 2018, 12 (7): 929–937.
8. Bogusz P, Korkosz M, Powrozek A and Prokop J. Motor-generator operation modeling of switched reluctance drive. Przeglad Elektrotechniczny, 2012, 88(12A): 56–61.
9. Zhu Y, Yang C, Yue Y, Zhao C, and Zhang Y. Development and Analysis of a Two-Phase Excitation Switched Reluctance Motor with Novel Winding Distribution Used in Electric Vehicles. J. Electr. Eng. Technol, 2018, 13(6): 2364–2375.
10. Ahn JW, Oh SG, Moon JW, and Hwang YM. A three-hase switched reluctance motor with two-phase excitation. IEEE Trans. Ind. Appl, 1999, 35(5): 1067–1074.
11. Ma C, Qu L and Tang Z. Torque ripple reduction for mutually coupled switched reluctance motor by bipolar excitations. In proc. IEMDC, 2013: 1211–1217.
12. Husain T, Elrayyah A, Sozer Y, and Husain I. Unified control for switched reluctance motor for wide speed operation. IEEE Trans. Ind. Electron, 2019, 66(5): 3401–3411.

13. Alrifai M, Zribi M, Rayan M, and Krishnan R. Speed control of switched reluctance motors taking into account mutual inductance and magnetic saturation effects. Energy Convers. Manag, 2010, 51(6): 1287–1297.
14. Chang X, Yang C, and Xiong J. Quantized fuzzy output feedback H_{∞} control for nonlinear systems with adjustment of dynamic parameters. IEEE Trans. Syst. Man Cybern. Syst, 2019, 49(10): 2005–2015.
15. Chang X and Yang G. Nonfragile H_{∞} filter design for T–S fuzzy systems in standard form. IEEE Trans. Ind. Electron, 2014, 61(7): 3448–3458.
16. Panda D and Ramanarayanan V. Mutual coupling and its effect on steady-state performance and position estimation of even and odd number phase switched reluctance motor drive. IEEE Trans. Magn, 2007, 43(8): 3445–3456.
17. Jain AK and Mohan N. Dynamic modeling, experimental characterization, and verification for SRM operation with simultaneous two-phase excitation. IEEE Trans. Ind. Electron, 2006, 53 (4): 1238–1249.
18. Cui X, Sun J and Gu C. Comparative study of performance for different winding connections of dual-channel switched reluctance machine using frozen permeability. In Proc. IEEE Southern Power Electron. Conf. (SPEC), 2018: 1–6.
19. Weiss CP, Schoeler S, and Doncker RW. Direct instantaneous torque control for switched reluctance machines considering mutual coupling. J. Eng, 2019, 2019(17): 3701–3704.
20. Mihic DS, Terzic MV, and Vukosavic SN. A new nonlinear analytical model of the SRM with included multiple coupling. IEEE Trans. Energy Convers, 2017, 32(4): 1322–1334.

Chapter 2
Electromagnetic Analysis for SRM

2.1 Introduction

As we know, SRMs are considered to be very competitive candidates for the next generation drive system and they have already been successfully applied in electric vehicles, mining machinery, aircraft and other fields, because of their simple and cost-effective construction, mechanical and thermal robustness, wide speed range as well as reliability [1, 2]. In order to improve SRMs performance, many studies have been done in terms of sensor-less drive [3, 4], advanced control methods [5], direct torque control [6], bipolar excitation [7]. In addition, the studies of redesign machine topology are also carried out [8, 9], including two-phase excitation SRM for a three-phase SRM [10] and two-phase SRMs with e-core structures [11, 12]. Although the redesign of new topology for SRMs has good improvements for the motor performance, it should be carried out under consideration combining with the redesign of winding distributions.

Furthermore, the choice of winding configuration in multi-phase machines has been investigated to achieve a better torque performance and a more flexible control for SRMs [13, 14]. In addition, Single and multi-phase excitation modes under various winding distributions are also studied and compared [15]. However, the alternative winding configurations in these literatures could cause asymmetric instantaneous torque waveforms, high torque ripples, torque density and low efficiency, when the winding configurations are applied for four-phase SRMs. Therefore, to address this problem, further studies on a new winding connection are required.

In view of this, we propose an 8/6 SRM working under double-phase excitation with a new winding to maintain a continuous short flux path operation of the SRM. Winding configurations of SRMs are the only source of magneto-motive force (MMF) such that it has a momentous influence on machine performance. It is of great significance to investigate different winding structures of SRMs. As a result, two typical concentrated windings: (1) single excitation mode; and (2) double

Y. Zhu, *The Key Technologies for Powertrain System of Intelligent Vehicles Based on Switched Reluctance Motors*, Recent Advancements in Connected Autonomous Vehicle Technologies 1, https://doi.org/10.1007/978-981-16-4851-9_2

excitation mode windings, in this chapter, are comprehensively studied and compared to clearly show the benefits of the proposed SRM with two-phase excitation mode. Meanwhile, the reason and source of the difference among traditional single-phase excitation SRM (SESRM), double-phase excitation SRM with traditional winding (HESRM) and DESRM are also revealed in this chapter such that the essential performance difference in static and dynamic conditions can be understood more clearly.

2.2 Various Exciting Winding Distributions

The electromagnetic characteristics of SRM are the basis for simulating dynamic performance and developing high-performance control system. Due to the non-linear flux linkage characteristic of SRM, the flux linkage is a non-linear function of the rotor position angle and phase current, which is quite difficult to calculate by means of the mathematical equations. At present, there are two main methods to obtain the flux linkage of SRM. The first method to obtain it is directly experimental measurements or mathematical calculations based on the measured inductance or current and voltage. The second one is to use the method of the finite element analysis (FEA) to perform numerical simulation calculation based on the detailed structure parameters of the SRM. In this chapter, the electromagnetic analysis results are all obtained by FEA because of the high accuracy and low time cost of FEA.

On the other hand, the winding distributions have a great influence on the performance of SMR because of different excitation modes, which can greatly increase the average torque and reduce the torque ripple. Therefore, focusing on four-phase SRMs with different winding distributions, the electromagnetic field performances are computed by FEA and the results are compared in this chapter.

The principle of minimum reluctance is that magnetic flux always tends to close along the path of minimum reluctance. In the single-phase excitation mode, the position where the salient polar axes of the stator and rotor overlap is the minimum reluctance position. The position where the salient polar axis of the rotor coincides with the interpolar axis of the adjacent stator poles has the maximum reluctance. As shown in Fig. 2.1, the magnetic resistance is maximum when the phase A is excited if the position is Fig. 2.1a, resulting in the electromagnetic tangential force that forces the rotor to the position with minimum magnetic resistance. However, the magnetic resistance when phase A is excited under the position of Fig. 2.1b can be minimum. According to the minimum reluctance principle, the SRM can rotate in the counterclockwise direction with the phase order of A-D-C-B when each phase is excited as corresponding sequences. It should be noted that the SRM can rotate in the opposite direction as long as the excitation order is changed to A-B-C-D.

The control method in single-phase excitation mode is relatively simple, and its electromagnetic field coupling is less. In the two-phase synchronous excitation mode, the electromagnetic torque characteristics are quite different from those of

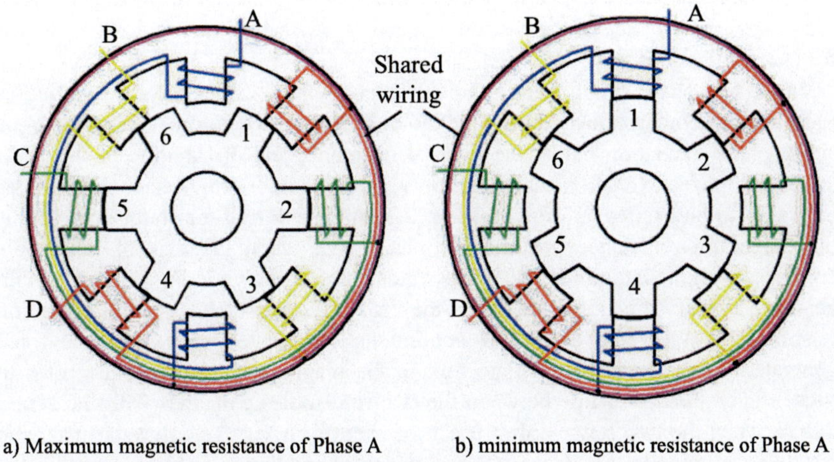

Fig. 2.1 SRM under single-phase excitation mode

single-phase excitation, such that the electromagnetic coupling effects between phases need to be considered. The two-phase excitation mode is beneficial to increase the average torque of the motor and reduce torque ripple, which can improve torque output and motor vibration. Figure 2.2 is the ideal waveforms of electromagnetic torque obtained under the two excitation modes. It can be seen from the figure that the output electromagnetic torque in the two-phase excitation mode is significantly higher than that in the single-phase excitation mode. Therefore, the SRM working under two-phase excitation mode can help improve

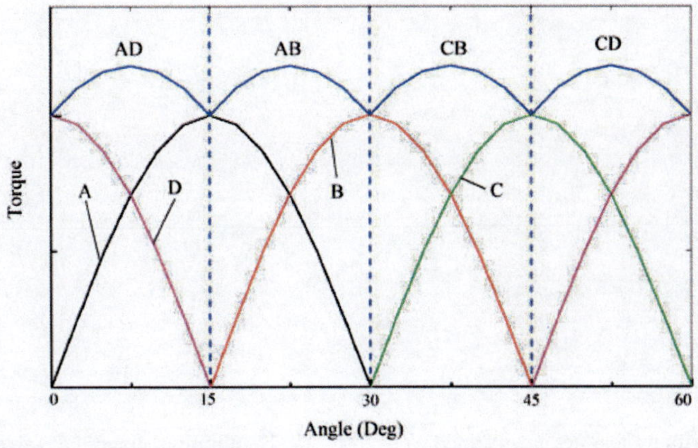

Fig. 2.2 The ideal torque waveforms

the dynamic characteristics of the motor, which is of great significance to improve the power performance of the electric vehicles (EVs), when the SRM is considered for EV applications.

In the two-phase simultaneous excitation mode, there can be always two phases to be excited synchronously during the motor operation. Therefore, compared to the single-phase excitation mode, the internal magnetic field distribution of the SRM under two-phase excitation mode can be very different. Taking phase A and D as examples to be excited as shown in Fig. 2.3, the position of the stator and rotor is the maximum reluctance position for phase AD, when the axis of rotor pole coincides with the interpolar axis of the stator poles as shown in Fig. 2.3a. It should be noted that the situation for the rotor is unstable at this position relationship. When the current is continuously energized, the tangential force generated by the twisted magnetic flux in the magnetic field drives the rotor to rotate. Once the center line between the two rotor poles coincides with the center line between the two stator poles, it can be seen from Fig. 2.3b that the magnetic resistance is minimal for phase AD and the rotor position is stable.

Although the operating principle of SRM has nothing to do with the direction of current, it is closely related to the winding direction and connection mode of the coil. The winding distribution of the traditional single-phase excitation SRM is shown in Fig. 2.1, in which the way of connection is mostly in series. When the two-phase synchronous excitation mode is performed in this winding distribution with this series connection mode, the magnetic flux is in the short magnetic circuit state when AB, CB, and CD are excited and is in the state of long magnetic circuit when AD is excited. As a result, the magnetic field is not steady during each period because of the differences between the short and long magnetic circuits. Therefore, an extra torque ripple can be generated when the SRM with traditional winding

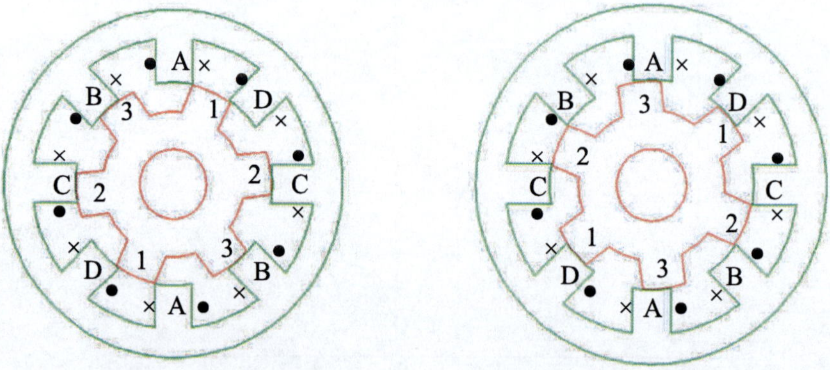

a) maximum magnetic resistance b) minimum magnetic resistance

Fig. 2.3 Two-phase excitation mode

distribution works under two-phase excitation mode, which could greatly reduce the torque performance of the motor.

As shown in Fig. 2.4, the torque curves in the long magnetic circuit and short magnetic circuit modes are analyzed by FEA and compared in different rotor positions. It can be seen that there is a great difference between the motor static torques generated by the long magnetic circuit (LMC) and the short magnetic circuit (SMC) with the same exciting current. The magnitude of the difference in value varies with the current, and the difference trend is basically the same. It can be observed that the SRM working LMC and SMC can directly cause the asymmetry of the excitation torque of each phase and increase the torque ripple, when the traditional winding distribution is used for two-phase excitation mode. Furthermore, the static torque in the long magnetic circuit mode is much smaller than that in the short magnetic circuit, especially when the rotor position angle is large enough, which can greatly reduce the output torque of the motor.

In order to improve the average torque and reduce the torque ripple of the four-phase SRM for electric vehicles, a symmetrical winding distribution based on two-phase synchronous excitation mode is developed, which is connected in series and can always work in short magnetic circuit. Differing from the traditional winding distribution, the winding direction diagram is shown in Fig. 2.5, where only the winding directions of phase A and B are illustrated, because the winding distribution of phase C and D are completely the same with phase A and B.

In this winding distribution, the magnetic poles of the adjacent stator poles of the two synchronous excited phases are N and S, respectively. During the operation of the motor in a whole period, the distribution of magnetic poles is the same and all are short magnetic circuit excitation. Therefore, compared to the traditional winding distribution, the SRM working under the new winding distribution not only reduces the torque ripple, but also improves the static torque of the motor.

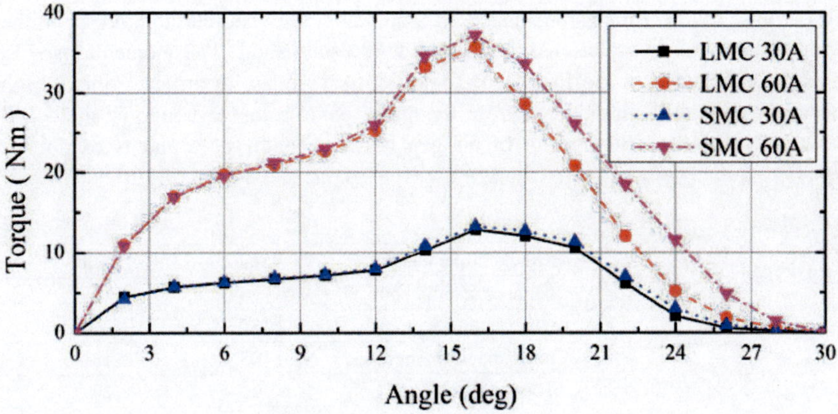

Fig. 2.4 Torque comparison in different magnetic circuits

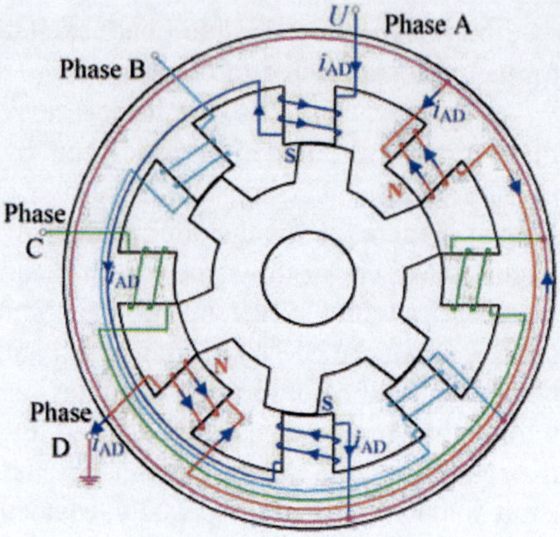

Fig. 2.5 Winding distribution for two-phase excitation mode

2.3 Analysis of Electromagnetic Parameters for SRM

2.3.1 Structural Dimensions of the Developed SRM

Based on the winding distribution developed in Sect. 2.1, a four-phase 8/6 SRM working under two-phase excitation mode is manufactured. The main structural dimensions of the prototype are shown in Table 2.1. In the next, the electromagnetic filed analysis for the prototype, including magnetic flux linkage, inductance, and static torque, is carried out by means of two-dimensional FEA in the software environment of ANSYS.

2.3.2 Analysis for Flux Linkage

According to the dimensions listed in Table 2.1, the geometric model for the prototype of the two-phase excitation mode is established. The elements used to analyze is selected in the form of triangle with six nodes. In order to obtain high accuracy results, distinguishing from the stator yoke, rotor yoke, motor shaft, and winding area, the grid density of air gap between stator pole and rotor pole is

Table 2.1 Configuration size of the SRM

Structural parameters	Stator	Rotor
Number of pole pairs	8	6
Outer diameter (mm)	180	104
Inner diameter (mm)	105	18
Tooth height (mm)	45	29
Tooth width (mm)	20	22

enhanced, because the magnetic flux density in air gap is greatly significant for the generate torque and force. As shown in Fig. 2.6, the mesh for the entire motor and air gap region is established, respectively.

The magnetic field characteristics of the SRMs with single-phase excitation and two-phase excitation are completely different. When two of the phases are simultaneously excited, they produce magnetic fields that are coupled to each other. As a result, the flux linkage of each phase and saturation situation in the yoke of stator and rotor are quite different from those of the single-phase excitation mode. Compared to the single-phase excitation, the mutual inductance between the two excited phases has a great influence and cannot be ignored. Therefore, the flux linkage characteristics and inductance characteristics of the SRM with the two-phase excitation mode need to be deeply analyzed. In this chapter, the magnetic field characteristics of the SRMs working under the single-phase excitation and the two-phase excitation modes are analyzed by FEA, and the variation principles of electromagnetic parameters are conducted and concluded by considering excitation mode, rotor position, and phase current.

It is very necessary to determine the rotor position angle before the analysis process, because the flux linkage and inductance characteristics of SRM have a nonlinear relationship with the rotor position angle and phase current of motor. As for the SRM with single-phase excitation mode, the rotor position angle is defined to be 30 degrees when the stator pole center line coincides with the rotor pole center line, which is also the minimum reluctance position for the excited phase. On the contrary, under the maximum reluctance position, the position angle is defined to be 0 degree, i.e., when the interpolar center line of the two adjacent rotors coincides with the stator pole center line. Figure 2.7 is the schematic diagram of the minimum reluctance position for the two excitation modes.

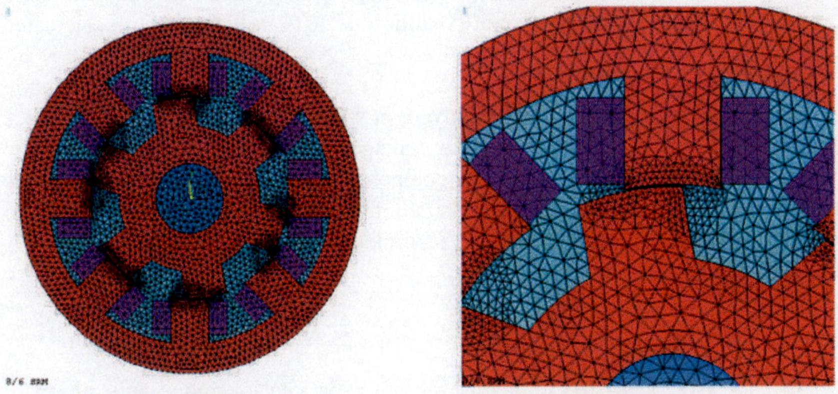

a) Entire motor b) Air gap region

Fig. 2.6 The mesh for the SRM

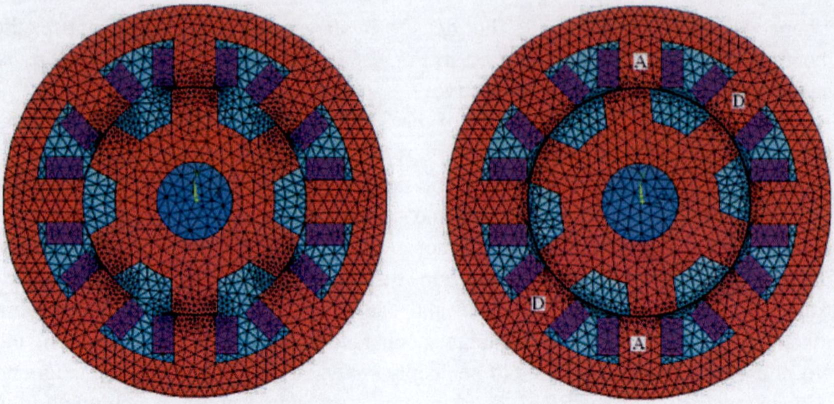

a) Single-phase excitation mode b) Phase AD excitation mode

Fig. 2.7 The minimal reluctance position

Regardless of single-phase excitation or two-phase excitation, the mechanical angles during one exciting cycle for the SRM are 60 degrees. According to the structural symmetry of the SRM, it is better to only calculate in the range of 0–30 degrees, which can save time on basis of keeping accuracy. After comprehensively considering calculation accuracy and efficiency, the calculation step is selected as 2 degrees. In addition, the phase current is selected from 5 to 70 A with the step length of 5 A, according to the performance characteristics of the motor.

The two-dimension static electromagnetic fields of the SRMs are calculated under the combinations of each rotor position and each phase current, and the internal magnetic field distribution characteristics of the SRMs under various conditions are obtained. Figures 2.8 and 2.9 respectively show the internal magnetic field distribution of the SRMs with single-phase excitation and two-phase excitation under the exciting current condition of 50 A, which are both calculated by FEA.

The axial magnetic field of the SRM can be considered as uniform distribution when the two-dimension FEA is employed, in which the end effect of the motor is ignored. As a result, the calculated flux linkage has slight difference between the results from FEA and experiment. To compensate the end effect, the flux linkage calculated by two-dimensional FEA can be multiplied by a correction coefficient to achieve the higher calculation accurac. The calculation method of this coefficient K_θ is determined as

$$K_\theta = K_{\max} + (K_{\min} - K_{\max})(\theta_{\max} - \theta)/(\theta_{\max} - \theta_{\min}) \tag{2.1}$$

where $\theta_{\max}$ and $\theta_{\min}$ mean rotor position angles at the maximum and minimum inductance, respectively.

a) Magnetic flux distribution b) Magnetic intensity vector

Fig. 2.8 Magnetic field distribution in single-phase excitation

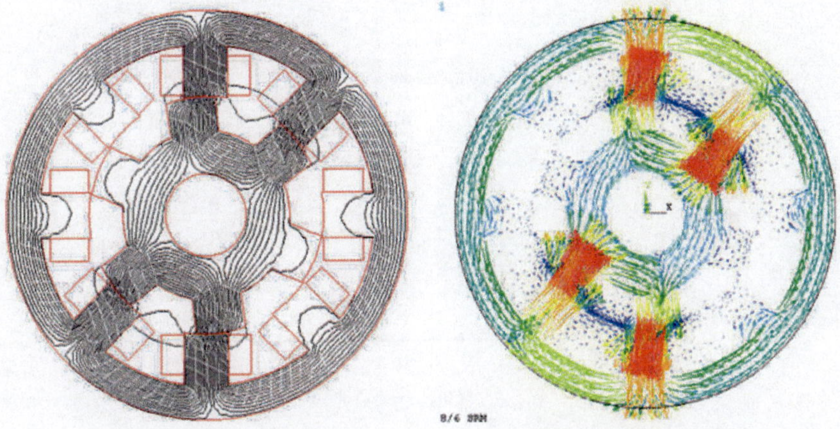

a) Magnetic flux distribution b) Magnetic intensity vector

Fig. 2.9 Magnetic field distribution in two-phase excitation

By combining the results obtained from the two-dimensional FEA with the correction by Eq. (2.1), the flux linkage values under conditions of single-phase excitation and two-phase excitation can be calculated. Then, the flux linkage data with respect to various position angles and exciting phase currents are processed to obtain the flux linkage characteristic curves as shown in Figs. 2.10 and 2.11, which are for single-phase excitation and two-phase excitation, respectively.

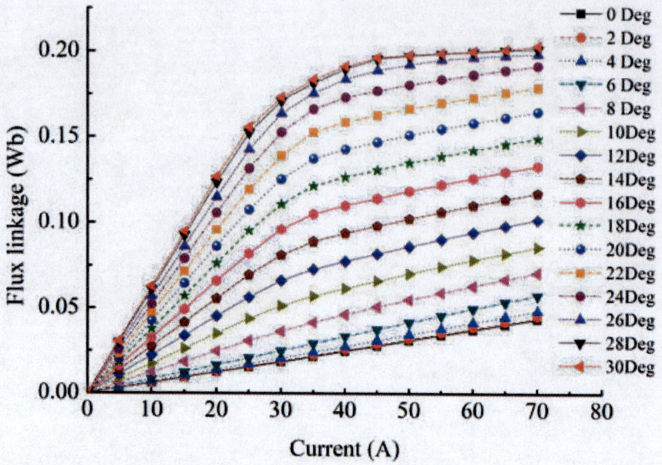

Fig. 2.10 Flux linkage in single-phase excitation

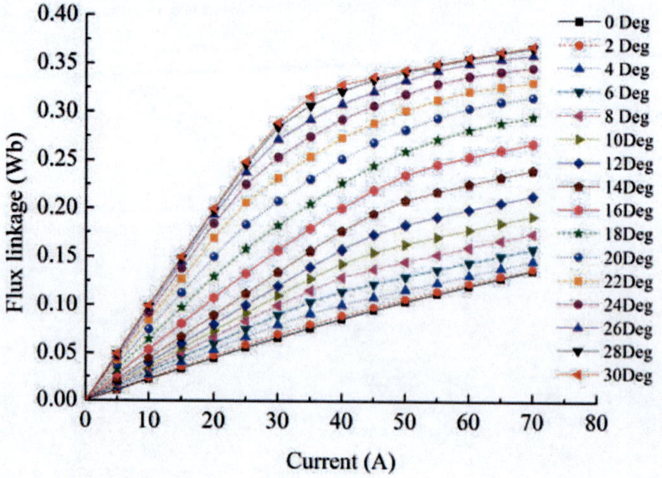

Fig. 2.11 Flux linkage in two-phase excitation

It can be seen from Fig. 2.10 that the flux linkage curve is almost a straight line when the rotor position angle is or is close to 0 degree, which means the flux linkage is a linear function of the current at this situation. However, as the rotor position angles increase, the flux linkage gradually displays nonlinear characteristics with respect to phase currents. On the other hand, the flux linkage changes faster as the current increases when the phase current with fixed position angle is low enough, while the flux linkage changes more and more slowly when the phase current is high enough due to the influence of magnetic circuit saturation.

Figure 2.11 is the total flux linkage generated by the two excited phases of the motor in the two-phase excitation mode. Similar to the flux linkage in the single-phase excitation, the flux linkage curve of two-phase excitation is also a straight line when the reluctance is or is close to minimum, i.e., it is a linear function with respect to phase current. With the increase of the rotor position angle, the flux linkage gradually becomes nonlinear. Influenced by the saturation of the magnetic circuit, the flux linkage increases faster when the current is at low level, while the increasing speed becomes smaller and smaller when the current reaches a certain level. According to the result comparison, it can be obviously observed that the flux linkage from two-phase excitation is approximately twice that of the single-phase excitation.

2.3.3 Analysis for Inductance

Considering the motor performance, inductance is one of the most important electromagnetic parameters. The characteristics of flux linkage of each phase are influenced by the phase current, self-inductance, and mutual inductance of each phase or phases. Under the condition of single-phase excitation, the value of mutual inductance is relatively small enough compared to the value of self-inductance. As a result, in the case of ignoring the effects from mutual inductance, the inductance is expressed as

$$\psi_k = L_k(\theta_i, i_k)i_k \tag{2.2}$$

where L_k is the self-inductance of k-phase winding, θ_k represents the rotor position angle, and i_k means the phase current of k-phase winding.

Currently, there are some methods to obtain the inductance values of the motor. The bridge method with an LCR measuring instrument is frequently used one among many inductance measurement methods. However, one of the key disadvantages of this method is that the measurement for inductance is carried out under a very small AC current. As a result, this method is not available for SRMs, because the inductance of SRM is a function as phase current, i.e., the inductances should be different under various phase currents. Many current researches have studied the self-inductance and mutual inductance characteristics of SRM [16], but the researches focus on SRM with the single-phase winding excitation mode, which cannot be used for the SRM with two-phase winding excitation mode. Because, the flux linkage, magnetic field distribution, and magnetic field saturation degree between the two excitation modes are completely different.

In this section, the characteristics of self-inductance and mutual inductance of the SRM with single-phase excitation are computed by FEA firstly as shown in Figs. 2.12 and 2.13. Figure 2.12 is the nonlinear curves of self-inductance with respect to rotor position angles and phase currents. Similarly, the mutual

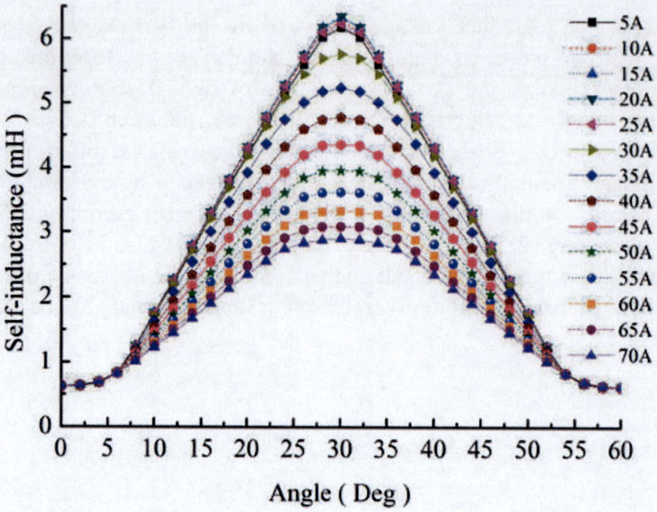

Fig. 2.12 Self-inductance in single-phase excitation

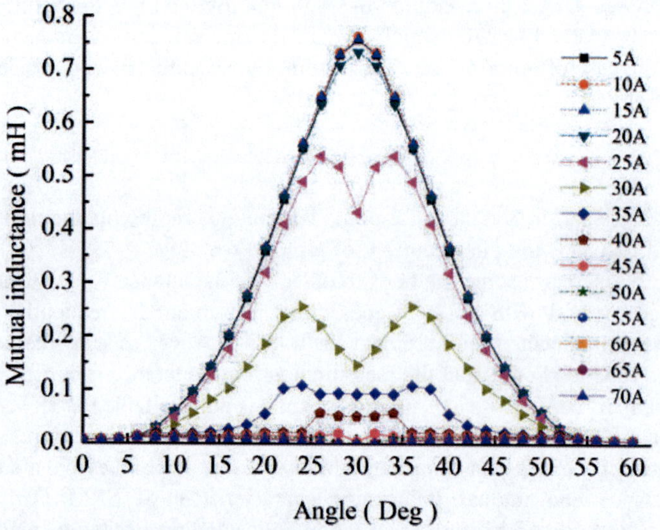

Fig. 2.13 Mutual inductance in single-phase excitation

inductances with respect to rotor position angles and phase currents are described in Fig. 2.13.

It can be seen from Fig. 2.12 that the self-inductance keeps the minimum when the center line of the stator salient pole coincides with the interpolar center line the two adjacent rotor poles because of the maximum reluctance. Nevertheless, the self-inductance can reach the maximum when the center lines of the stator pole and rotor pole overlap with the minimum reluctance. It should also be observed that, because the rotor pole arc is slightly larger than the stator pole arc in general, the self-inductance maintains its maximum value within this certain overlapped angle, where the surfaces of the stator pole and rotor pole fully overlap. In addition, in the current range of 0–20 A, the curves of self-inductance varying with the rotor position angles are basically the same, i.e., self-inductance has nothing to do with phase current when the phase current is low enough. As the current increases, the self-inductance can gradually decrease due to the saturation of the magnetic circuit.

After analyzing the mutual inductance results varying as phase currents and rotor position angles shown in Fig. 2.13, the followings can be observed obviously. Firstly, in the state of single-phase excitation, the mutual inductance between the two windings of the opposite stator poles is very small, and its maximum mutual inductance value is only about one-tenth of the self-inductance. Secondly, within the range from 0 to 20 A, the mutual inductance values are relatively large and basically the same under the same phase current and rotor position angle. Namely, the mutual inductance is only affected by rotor position angle rather than the phase current when the phase current is low enough. As the phase current increases, the mutual inductance under the same rotor position angle decreases correspondingly. When the phase current reaches a high level, the mutual inductance becomes low enough. Finally, within the current range from 25 to 40 A, the changing trend of the mutual inductance is nonlinear, and the fluctuation of the mutual inductance curve is mainly concentrated at the position where the center line of the stator pole coincides with that of the rotor pole. When the two center lines are completely coincident, the mutual inductance sharply becomes smaller than that in the nearby areas.

In order to visually compare the differences between self-inductance and mutual inductance of the SRM with single-phase excitation, the data computed under two rotor position angles, 30 and 20 degrees, are selected and arranged in a bar chart as shown in Figs. 2.14 and 2.15. It can be obviously seen that the influence of mutual inductance on the self-inductance is small enough such that it can be ignored in the entire phase current range, although the mutual inductance in low current is relatively a little high.

It has been mentioned that in the single-phase excitation, the influence of mutual inductance can be ignored because of the relative long distance between the two energized stator poles. In order to verify this conclusion, the data obtained from the FEA are selected and quantitatively compared. As shown in Fig. 2.16, representative self-inductance and mutual inductance data are computed under various rotor position angles, including 30, 24, 20, 16, 10, and 0 degrees, and the divisors obtained from dividing self-inductance by mutual inductance are calculated. After analyzing the ratio data, the followings can be observed. (1) When the phase current is small, the ratio between self-inductance and mutual inductance is relatively large with the maximum value of around 12%, especially at the area of the minimum reluctance. As a result, the mutual inductance should be considered because of the higher ratio value when the phase current is small and the surfaces of the stator and

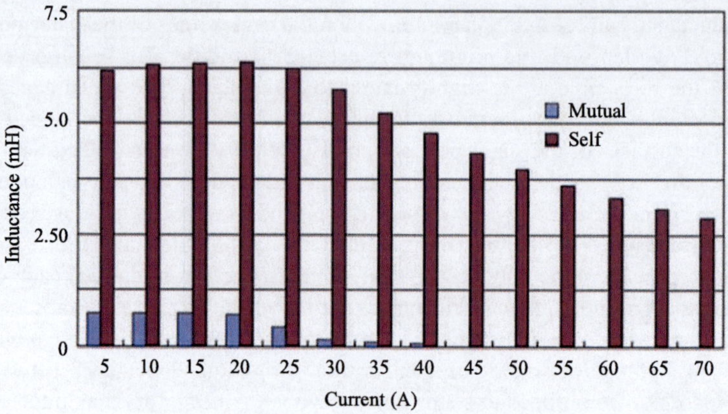

Fig. 2.14 Self-inductance and mutual inductance in 30 degrees

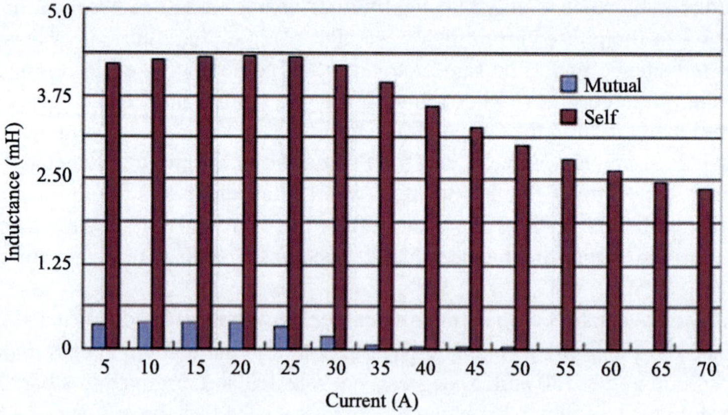

Fig. 2.15 Self-inductance and mutual inductance in 20 degrees

rotor poles overlap. (2) When the current is large enough, the ratio of mutual inductance to self-inductance gradually decreases and finally remains at about 1%. In this case, the value of mutual inductance is very small such that it can be ignored. (3) The ratio values become smaller and smaller as the rotor position angle decreases.

On the other hand, in the case of SRM with two-phase simultaneous excitation, the magnetic field distribution has a larger change compared to the single-phase excitation. The magnetic fields generated by two phases are coupled with each other, and the flux linkage of each phase is greatly different from that of single-phase excitation. Therefore, it can be predicted that the mutual inductance and self-inductance between the two excitation modes are completely different.

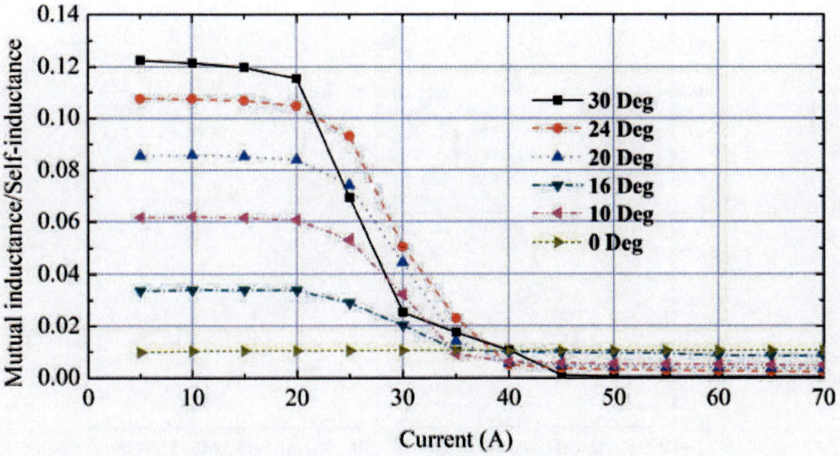

Fig. 2.16 Mutual inductance/self-inductance in single-phase excitation

In order to study the magnetic field distribution law of the SRM with two-phase excitation, the characteristics of self-inductance and mutual inductance of the SRM under two-phase excitation are analyzed. It should be noted that the characteristics of the self-inductance and mutual inductance of phase A and D are studied and the mutual inductance only between the two phases is considered when the phase A and D are excited synchronously. Figures 2.17 and 2.18 are nonlinear self-inductance curves of phase A and D under two-phase excitation mode, respectively. Similarly,

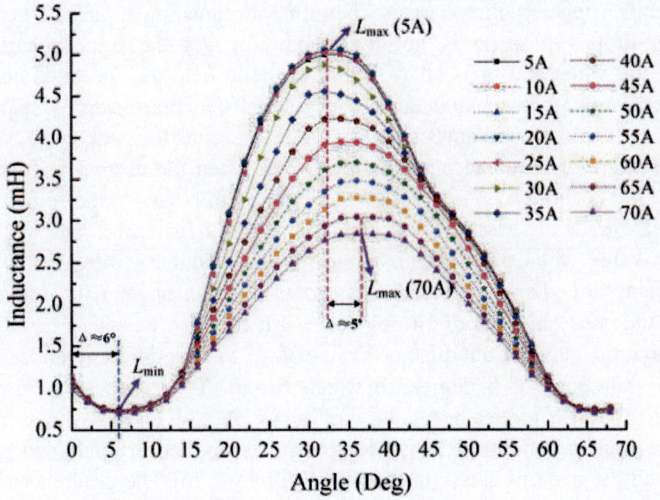

Fig. 2.17 Self-inductance of phase A in two-phase excitation

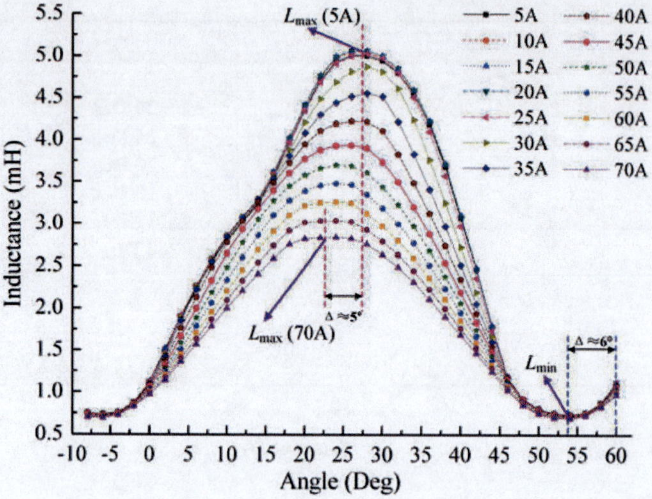

Fig. 2.18 Self-inductance of phase D in two-phase excitation

the mutual inductance curves with respect to various rotor position angles and phase currents are also described in the form of two-dimensional curves as shown in Fig. 2.19.

After analyzing the self-inductance curves of phase A and D shown in Figs. 2.17 and 2.18, the followings can be observed. (1) Due to the differences in relative position relationship and magnetic field distributions between phase A and D when the two phases are excited, the self-induction curves of phase A and D are obviously different from those of single-phase excitation. (2) There is a peak of self-inductance appearing in phase A when the rotor position angle is 0, because the reluctance in this position is not the maximum. As the rotor position angle increases, the valley exists as soon as the magnetic field has the maximum reluctance. Afterwards, the self-inductance can gradually increase step by step as rotor position varies from 0 to about 35 degrees until it reaches the peak value, where the maximum value of inductance is about 5.0 mH when the current is 5 A and the position angle is about 37.5 degrees. (3) Compared to phase A, the self-inductance of phase D gradually rises as rotor position angle increases, and reaches the maximum value when the position angle reaches about 27.5 degrees. Then, the self-inductance of phase D decreases as rotor position angle rises, and a valley occurs in the position angle of 54 degrees. (4) It can also be seen from Figs. 2.17 and 2.18 that the maximum inductance values L_{max} vary at different angles, with the maximum difference of 5 degrees between 5 and 70 A because of the mutual inductance. (5) The inductance values of phase A and D with the same phase current are equal when the SRM is operated under aligned or unaligned positions, even though the exciting phase currents are different. (6) The differences between the self-inductance curves of phase A and D is mainly caused by the rotating

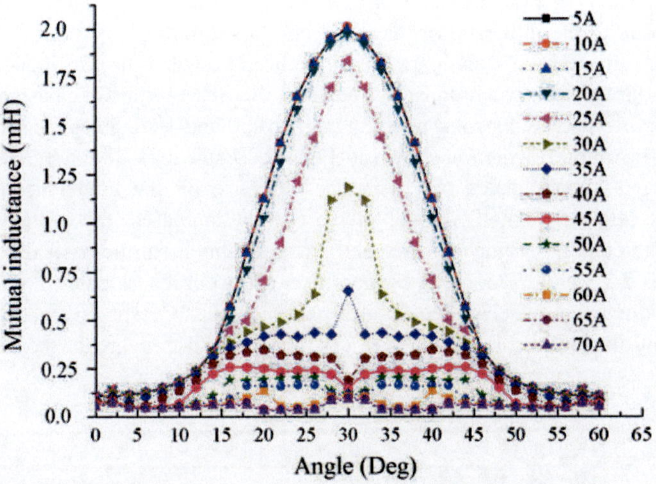

Fig. 2.19 Mutual inductance between phase A and D in two-phase excitation

direction of the motor. Therefore, the self-inductance curves of phase A and D can be interconverted as long as the rotating direction changes. (7) The effects of phase current on self-induction of each phase are the same as those of single-phase excitation, i.e., the higher the phase current is, the smaller the self-inductance can be, when the SRM has the same rotor position angle.

In two-phase excitation mode, the characteristics of mutual inductance between two excited phases are quite different from those of single-phase excitation, which is of great significance for analyzing the inductance model of SRM and the developing the control strategy. Figure 2.19 is the characteristic curves of the mutual inductance between phase A and D, which are drawn with respect to rotor position angles and phase currents. The following features can also be observed from Fig. 2.19. (1) In two-phase excitation, the mutual inductance between the two excited phases is very high in some specific area. The maximum ratio of mutual inductance to self-inductance can exceed 40%, while the minimum ratio is close to 2%. As a result, the mutual inductance under this condition can't be ignored due to the high proportion. (2) Similar to single-phase excitation, the mutual inductance is high enough and can be almost the same in the same position angle within the current range from 0 to 20 A. Namely, the mutual inductance is only related to the rotor position angle when the current is small enough, and has little to do with the phase current. As the phase current gradually increases, the mutual inductance under the same rotor position angle decreases. Meanwhile, the mutual inductance becomes low enough when the phase current is high enough, which is also almost the same in the same position angle even though the phase current changes. (3) Differing from the self-inductance, mutual inductance curves increase first and then decrease as the rotor position angle rises, with the maximum value in 30 degrees. (4) When the rotor position angle varies from 15 to 45 degrees and the phase current is greater than 30 A, the mutual inductance values are basically constant due to the influence of magnetic circuit saturation.

In order to intuitively compare the differences between the self-inductance and mutual inductance between phase A and D of the SRM with two-phase excitation, the inductance data under rotor position angle of 30 and 18 degrees are selected and drawn in histogram, which is shown in Figs. 2.20 and 2.21. It can be clearly seen that the mutual inductance has a greater influence on the inductance when the current is relatively small. As the phase current increases, the influence of the mutual inductance becomes smaller and smaller, which can be basically ignored.

Figures 2.22 and 2.23 are the quantified ratio curves calculated by dividing mutual inductance by self-inductance of phase A and D, respectively, where eight rotor position angles, 30, 26, 22, 18, 14, 10, 6, and 2 degrees, are selected to

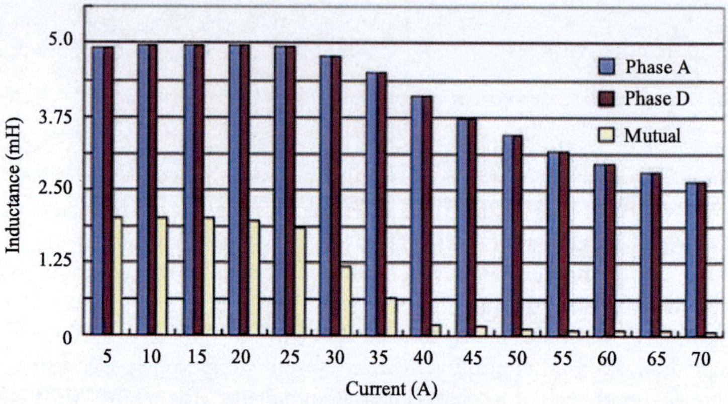

Fig. 2.20 Self-inductance and mutual inductance in 30 degrees

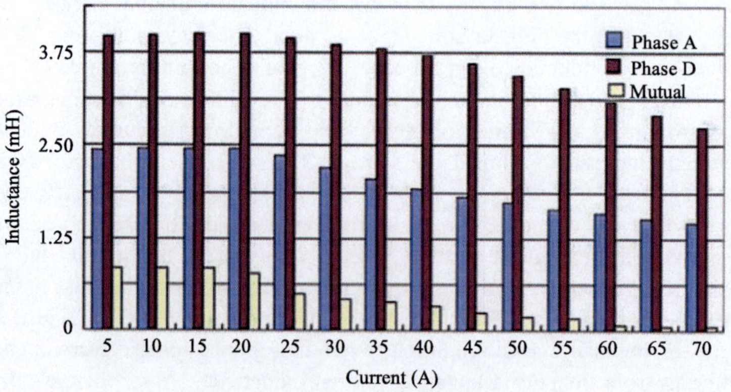

Fig. 2.21 Self-inductance and mutual inductance in 18 degrees

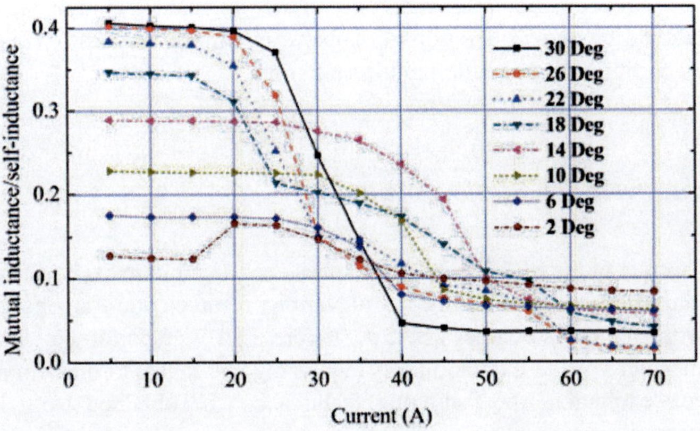

Fig. 2.22 Mutual inductance/self-inductance of phase A in two-phase excitation

comprehensively reflect the ratio situations. The following conclusions can be drawn by the comparative analysis from Figs. 2.22 and 2.23. (1) When the phase current is low, especially smaller than 35 degrees, the ratio of mutual inductance to self-inductance is very large. The maximum relative ratio can even exceed 40% when the rotor position angle is in 30 degrees or nearby. The ratio for phase A is high enough under low phase current, which can exceed 10%, even though the rotor position angle decreases to a low level. As a result, the influence of mutual inductance is quite large and must be considered when the phase current is small. (2) When the phase current increase step by step, the ratio of mutual inductance to the self-inductance for phase A and D gradually decreases and finally maintains at about 5%. (3) As the rotor position angle gradually decreases, the relative proportion of mutual inductance generally becomes smaller and smaller. (4) The

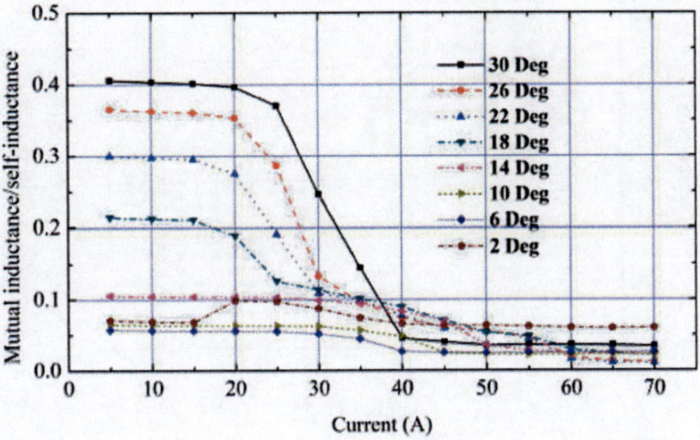

Fig. 2.23 Mutual inductance/self-inductance of phase D in two-phase excitation

influence of magnetic circuit saturation on mutual inductance is high, such that the ratio is greatly high under condition of magnetic circuit unsaturation, while it is low under the condition of magnetic circuit saturation.

2.3.4 Analysis for Static Torque

The static torque of the SRMs with single-phase and two-phase excitation mode can also be calculated by FEA by means of various rotor position angles and phase currents with steps of 2 degrees and 5 A. Figure 2.24 is static torque curves of the SRM with single-phase excitation. By analyzing the static torque curves in the single-phase excitation, the following features can be obtained from Fig. 2.24. (1) The static electromagnetic torque is a nonlinear function of the rotor position angle and phase current, which gradually increases and reaches the maximum value under each phase current, as the increase of the rotor position angle. In this process, the magnetic circuit of the motor gradually changes from the condition of unsaturation to saturation. In addition, it can be obviously seen that there is a flat area for the static torque from 10 to 25 degrees, which could improve the stability of the output dynamic torque when the motor works. (2) With the increase of the rotor position angle, the overlap surface of the stator pole and the rotor pole becomes larger and larger, and the static torque gradually decreases until it reaches the minimum reluctance position.

In order to verify the conclusion that the torque of the SRM with two-phase excitation increases compared to that with single-phase excitation, the static torque in the two-phase excitation is also be calculated and analyzed by FEA. As shown in

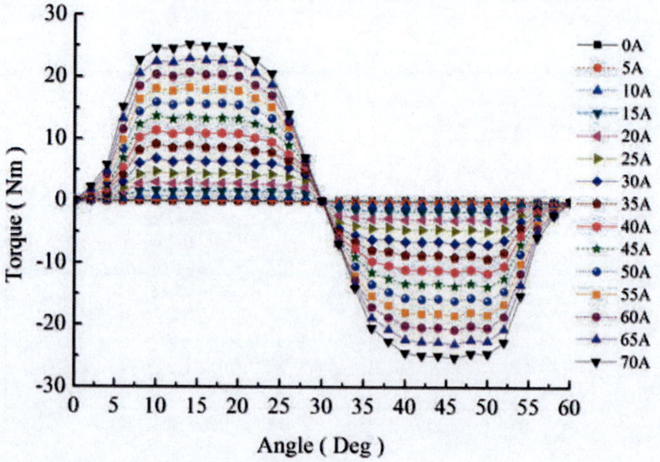

Fig. 2.24 Torque curves in single-phase excitation

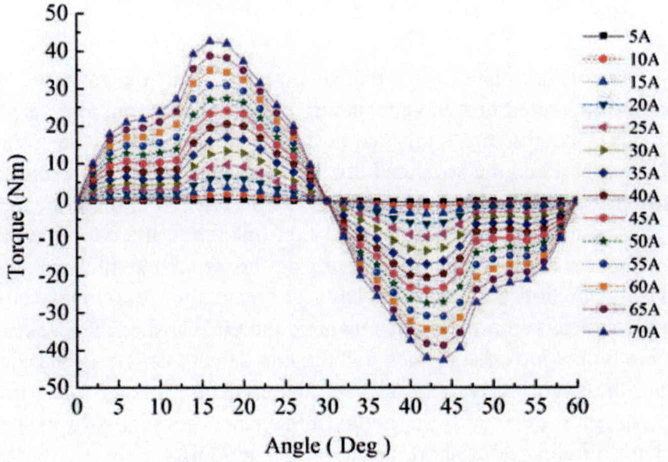

Fig. 2.25 Torque curves in two-phase excitation

Fig. 2.25, the torque values under two-phase excitation are computed with respect to rotor position angles and phase currents. After analyzing the torque curves in Fig. 2.25, the following features can be observed. (1) Compared to the single-phase excitation, the static torque under two-phase excitation is also a non-linear function of phase current and rotor position angle. The static torques in the unsteady position 0 degree and the steady position 30 degrees are both 0, and the maximum torque value appears at about 16 degrees. (2) Differing from the single-phase excitation, the static torque curves don't have a flat area and keeps increasing from rotor position angle of 5–14 degrees. Afterward, it can gradually decrease as the rotor position angle goes up. Because the position relationship between the rotor and two excited stators is complex and completely different from that of the single-phase excitation. (3) Compared to the single-phase excitation, the maximum torque value in the two-phase excitation can exceed 40 N m, which is much larger than that in the single-phase excitation.

2.4 Comparative Analysis Between Two Excitation Modes

Generally, the operation principle and electromagnetic performance of the SRM with single-phase excitation can be accurately described or estimated. However, the electromagnetic fields of the SRM with two-phase excitation are coupled with each other when two phases are energized at the same time. The flux linkage, inductance, static torque as well as dynamic performance under two-phase excitation can change greatly such that it should be compared with that under single-phase excitation, which is greatly significant for modeling, analyzing, and controlling the SRM with two-phase excitation.

2.4.1 Comparative Analysis of Flux Linkage

In order to verify the relationships among the total flux linkage in the two-phase excitation, the separated flux linkage in single-phase excitation, and the sum of the separated flux linkages, the total flux linkage generated when the rotor position angle is 28 degrees, and the separated flux linkages in single-phase excitation at this rotor position are both computed and illustrated in Fig. 2.26. It can be clearly seen from the curves in Fig. 2.26 that the total flux linkage in the two-phase excitation mode is almost overlapped with the sum of the separated flux linkages in the single-phase excitation mode. In this case, the magnetic flux circuits of the SRM with two-phase excitation are short magnetic circuit, in which the yokes of stator and rotor are both wide enough such that the flux density isn't easy to be saturation. In addition, the slight differences between the total flux linkage and sum of separated flux linkages are due to the mutual inductance between two excited phases and flux path differences of short and long magnetic flux.

In order to reflect the relationship between the total flux linkage and single flux linkage in the two different excitation modes, the flux linkage ratio coefficient K_ψ is defined as follows

$$K_\psi = \psi_D(i,\theta)/\psi_S(i,\theta) \tag{2.3}$$

where $\psi_D(i,\theta)$ is the total flux linkage in two-phase excitation mode; $\psi_S(i,\theta)$ represents the single flux linkage in single-phase excitation mode.

Selecting five rotor position angles, 0, 30, 24, 16, and 10 degrees, the flux linkage ratio coefficients in these position angles are calculated according to Eq. (2.3) and illustrated as shown in Fig. 2.27. Combined with the relationships among the total flux linkage in the two-phase excitation, the separated flux linkage in single-phase excitation, and the sum of the separated flux linkages, which are

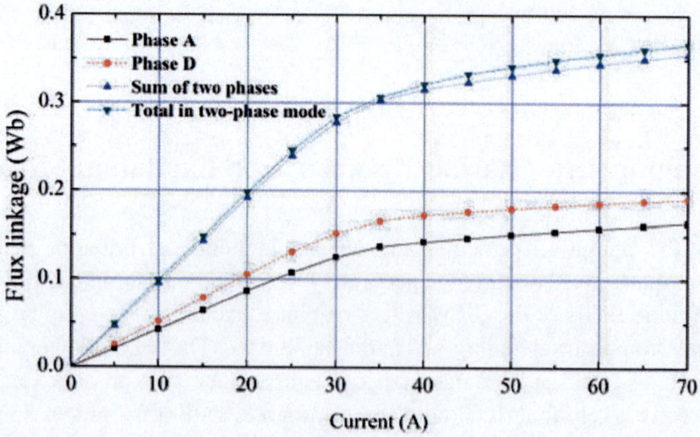

Fig. 2.26 Flux linkage comparison in different excitation mode

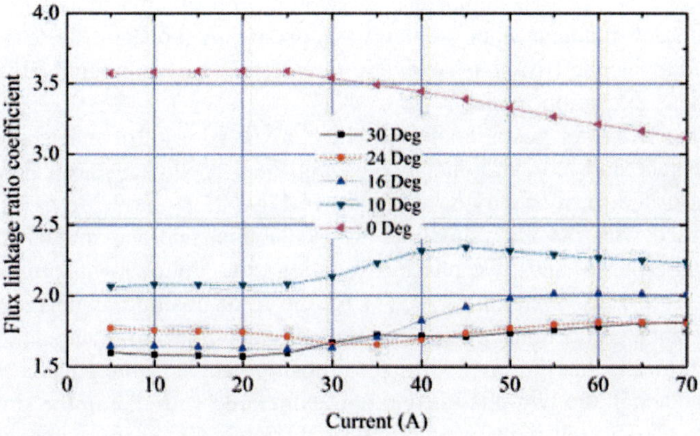

Fig. 2.27 Flux linkage ratio coefficient in different angles

shown in Fig. 2.26, the following features can be obviously observed from Fig. 2.27. (1) Under the same phase current condition, the smaller the rotor position angle is, the larger the obtained flux linkage ratio coefficient is. Namely, the total flux linkage in the two-phase excitation mode is relatively large in small rotor position angle condition, compared to single flux linkage. (2) The flux linkage ratio coefficient changes little, and basically remains within a certain range under the same position angle condition. Only when the angle is small, the flux linkage ratio coefficient changes greatly as the increase of the phase current. (3) From the numerical point of view, the maximum value of the flux linkage ratio coefficient is about 3.6 and the minimum value is about 1.6. Moreover, it varies as the position angle. This means that the variations of total flux linkage in the two-phase excitation are not the sum of variations from two single flux linkages in the single-phase excitation, because of the effects of mutual inductance, position relationship between rotors and stators, and phase current.

2.4.2 Comparative Analysis of Inductance

Similar to the flux linkage, the self-inductance proportional coefficient K_L and mutual inductance proportional coefficient K_M are defined as Eqs. (2.4) and (2.5) to directly reflect the relationship between inductance in two-phase excitation and that in single-phase excitation.

$$K_L = L_D(i,\theta)/L_S(i,\theta) \tag{2.4}$$

$$K_M = M_D(i,\theta)/M_S(i,\theta) \tag{2.5}$$

where $L_D(i,\theta)$ is the self-inductance in two-phase excitation mode; $L_S(i,\theta)$ represents the self-inductance in single-phase excitation mode; $M_D(i,\theta)$ means the mutual inductance in two-phase excitation mode; $M_S(i,\theta)$ is the mutual inductance in two-phase excitation mode.

Selecting five rotor position angles, which are 0, 30, 24, 16, and 10 degrees, the self-inductance and mutual inductance proportional coefficients are calculated and illustrated in the form of curves, as shown in Figs. 2.28 and 2.29.

Combined with the analysis about the self-inductance and mutual inductance under single-phase and two-phase excitation, the following features can be observed from Figs. 2.28 and 2.29. (1) As the rotor position angle increases, the self-inductance proportional coefficient gradually decreases from about 3.5 to about 1.5, which means that the smaller the rotor position angle is, the greater the ratio of self-inductance in the two-phase excitation, compared with that in the single-phase excitation. (2) The self-inductance proportional coefficient changes unevenly under the condition of small phase current, which is greatly affected by the rotor position angle. However, under the condition of large current, the influence of rotor position angle on self-inductance proportional coefficient is small such that the self-inductance ratio coefficient is a monotonous decreasing function of rotor position angle. (3) The mutual inductance proportional coefficient is high in the whole range, which reflect that the mutual inductance in the two-phase excitation is extremely high such that it can not be ignored in general. (4) When the phase current and the rotor position angle are both large, the mutual inductance ratio coefficient can reach 70 or more. This is because the mutual inductance generated in single-phase excitation under this condition is very small and close to 0, such that the calculated coefficient value becomes greatly high. (5) Combining with Figs. 2.28 and 2.29, it can be seen that the mutual inductance proportional coefficient is much greater than the self-inductance proportional coefficient, although both them are higher than 1 indicating that the mutual and self-inductance in

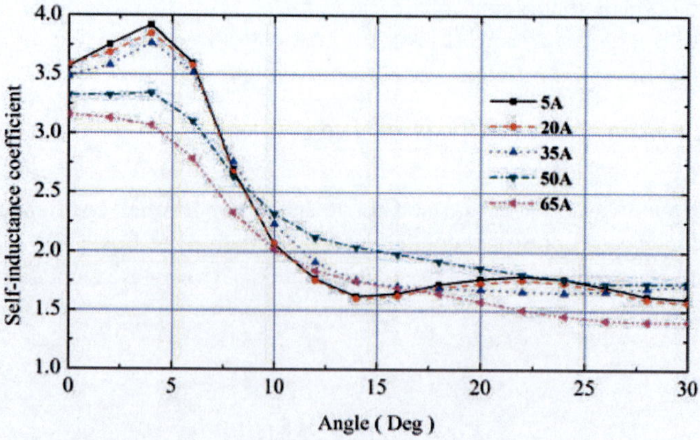

Fig. 2.28 Self-inductance proportional coefficient in different currents

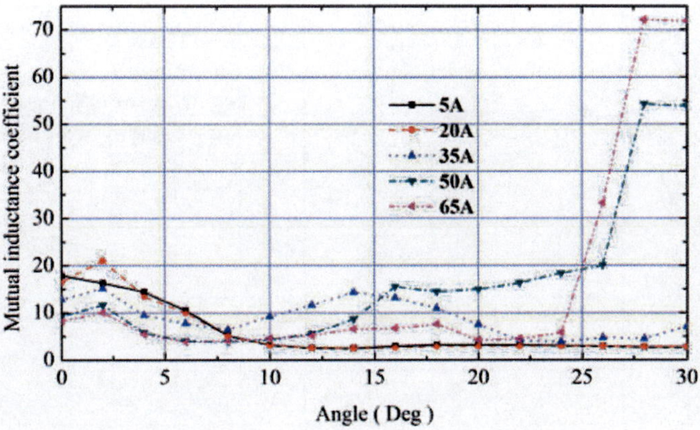

Fig. 2.29 Mutual inductance proportional coefficient in different currents

two-phase excitation are both greater than those in single-phase excitation. That means that the influence of mutual inductance in two-phase excitation is more serious when the rotor position angle and phase current are both the same.

2.4.3 Comparative Analysis of Static Torque

In order to study the torque characteristics of SRM in different excitation modes and the quantitative relationship of the static torques between the two-phase excitation and single-phase excitation, the total torque generated in two-phase excitation, the single torque obtained from single-phase excitation, and the sum of the single torques are computed and shown in Fig. 2.30, where the exciting phase current for each condition is 40 A.

It can be observed from Fig. 2.30 that the changing trends of total torque in two-phase excitation and the sum of single torques are very consistent, although there are certain differences in numerical values. This is caused by the difference of relative position between stators and rotors in two-phase and single-phase excitation, and the coupling of magnetic field. On the other hand, it can prove the superiority of the SRM with short magnetic circuit excitation.

In order to reflect the relationship between the torque values in the two different excitation modes, the expression of the torque proportional coefficient K_T is defined as

$$K_T = T_D(i, \theta)/T_S(i, \theta) \tag{2.6}$$

where $T_D(i, \theta)$ is the torque in two-phase excitation mode, and $T_S(i, \theta)$ represents the torque in single-phase excitation mode.

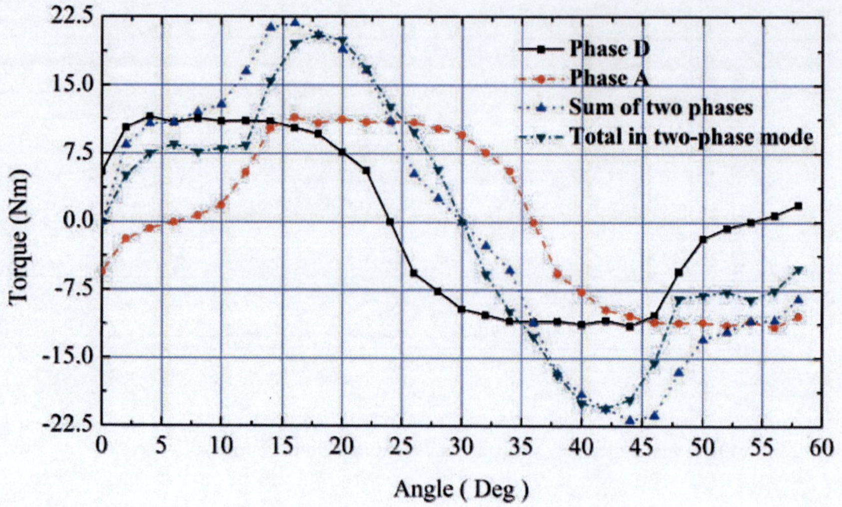

Fig. 2.30 Torque characteristic curves in 40 A

According to Eq. (2.6), five sets of phase current, 5, 20, 35, 50, and 65 A are selected as exciting current to calculate the torque proportional coefficient such that the relationship between the torques in the two excitation modes can be conducted in detail. The curves for torque proportional coefficient in different phase currents are shown in Fig. 2.31, and the following features can be observed after analyzing the values of torque proportional coefficient. (1) Considering different phase currents, the changing trends of the torque proportional coefficient are almost the same, except in the case of small and large rotor position angle. Therefore, the influences of phase currents on torque proportional coefficient are little enough. (2) The changes of the torque proportional coefficient caused by the rotor position angle are relatively large. When the rotor position angle is small, the torque proportional coefficient is very large. Otherwise, the torque proportional coefficient decreases first and then increases with the increase of the rotor position angle. It can be concluded that the rotor position angle has great influence on the torque proportional coefficient. (3) On the other hand, the greater the variation of the torque proportional coefficient is, the higher the torque fluctuation can be, which could directly affect the dynamic characteristics of SRM. Therefore, it can be more difficult to efficiently control the SRM with two-phase excitation than that with single-phase excitation to balance performance between the average torque and torque ripple.

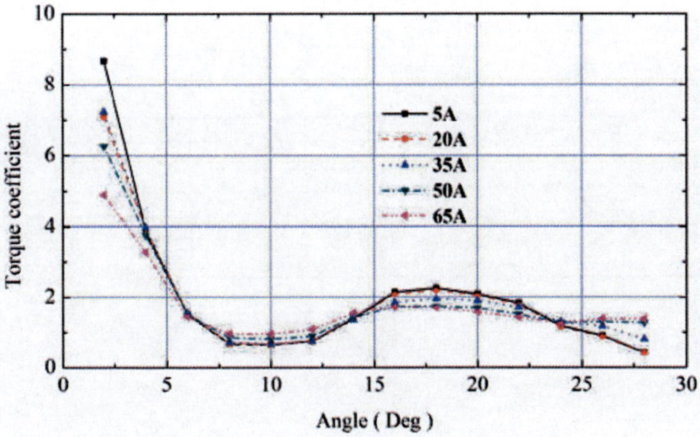

Fig. 2.31 Torque proportional coefficient in different currents

2.4.4 *Comparative Analysis of Dynamic Torque Performance*

To more accurately simulate the dynamic torque characteristics of SRM during operation, the motor electromagnetic models are established in Maxwell by the 2D FEA, the power circuits are executed in Simplorer, and control strategies are established in Matlab, which are shown in Fig. 2.32. The dynamic performance of the SRMs with single-phase and two-phase excitation can be obtained by joint simulation. As shown in Fig. 2.32a, b, the asymmetric half-bridge circuit is used for the SRM with single-phase excitation, while H-bridge circuit is used for the SRM with two-phase excitation. The key design parameters of the SRMs studied in this section are shown in Table 2.2, where both structure parameters are identical except the winding distribution.

Figure 2.33 shows the dynamic phase current and torque profile in a stable operation state of 500 rpm during an electric cycle, when the turn-on and turn-off angles are respectively set as 0 and 15 degrees and the current limit is 50 A. It should be noted that the HESRM in Fig. 2.33b means the SRM with the traditional single-phase winding distribution works under two-phase excitation, compared to the SRM with single-phase excitation (SESRM) and the SRM with two-phase excitation (DESRM). It can be seen from Fig. 2.33 that the average electromagnetic torque of DESRM and HESRM are both more than twice that of SESRM. Meanwhile, torque pulsation of DESRM and HESRM is reduced by more than 150% compared to SESRM, which shows an obvious improvement. It can be seen from Fig. 2.33b that the toque peaks of excitation phase CB are higher than those of AB, AD and CD. This is because the flux circuits generated by excitation phase AB, AD and CD are long paths, while the CB torque peaks belong to a short flux path.

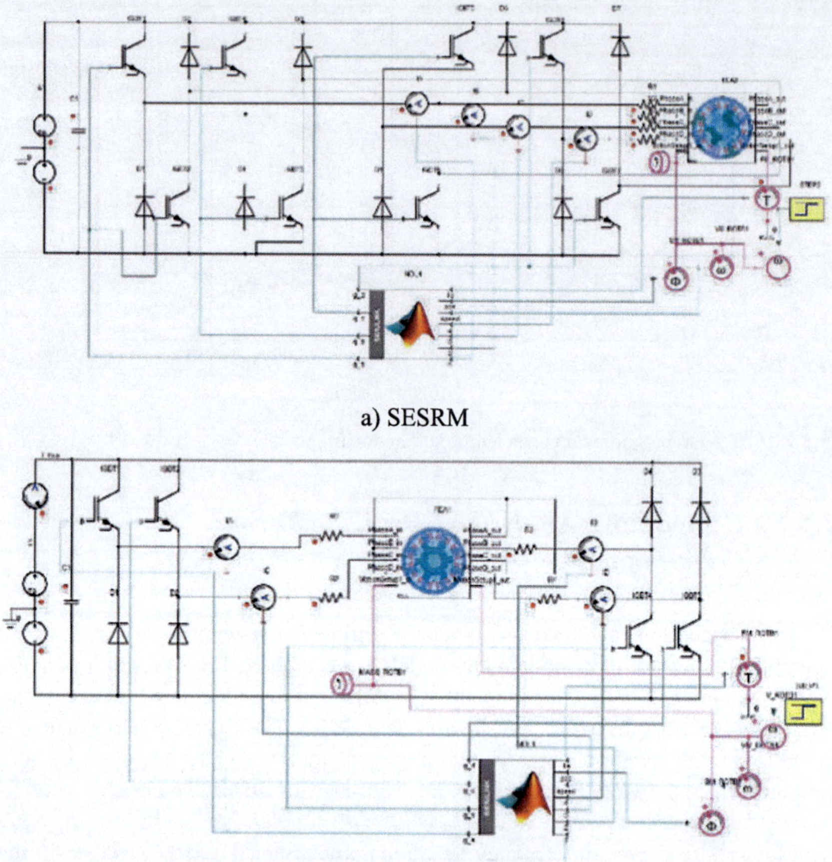

a) SESRM

b) HESRM and DESRM

Fig. 2.32 The co-simulation model of the SRM driving system

Table 2.2 Key specifications of the SRMs

Parameters	Value	Parameters	Value
Rated voltage	72 V	Stator outer diameter	210 mm
Rated power	4 kW	Stator inner diameter	127 mm
Rated torque	25.5 N m	Rotor outer diameter	126.3 mm
Max phase current density	22 A/mm^2	Rotor inner diameter	48 mm
Stator pole arc	18 deg.	Rotor pole arc	19.3 deg.
Max chopping current	180 A	Air gap	0.35 mm
Slot package factors	0.454	Stack length	120

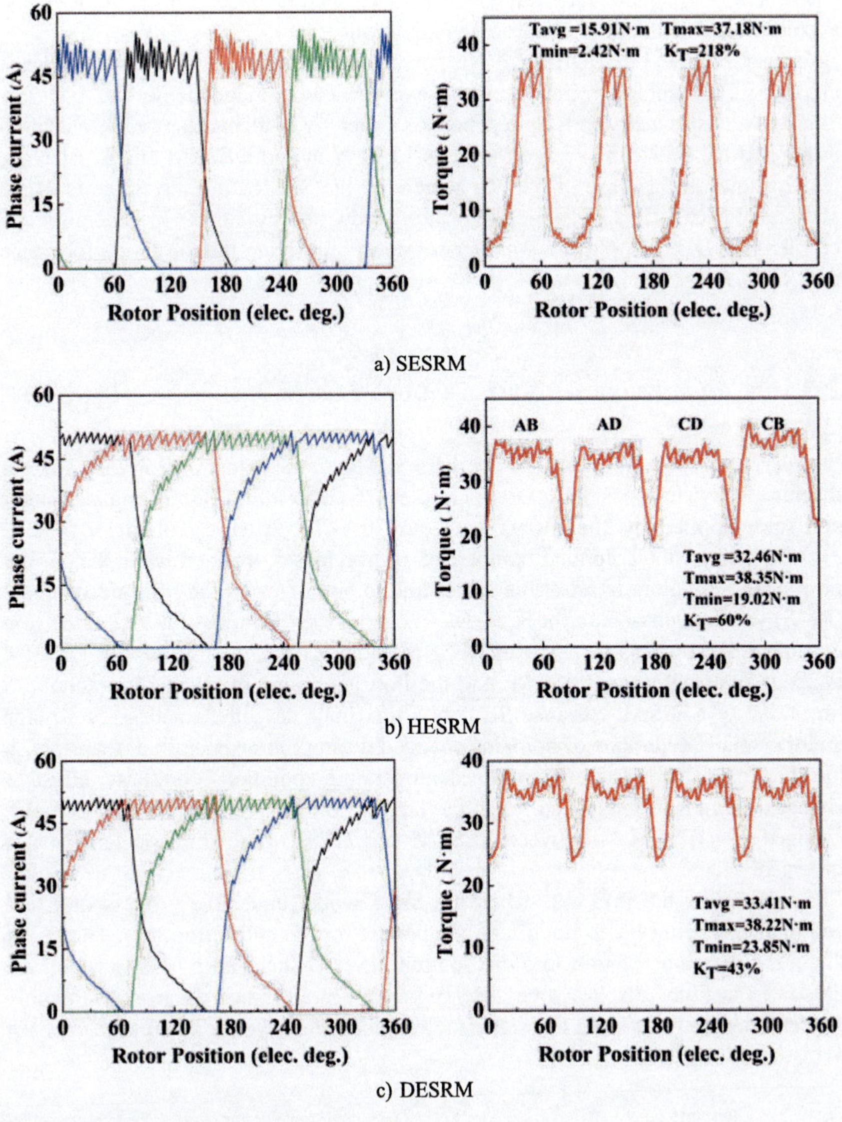

Fig. 2.33 Dynamic torque and phase current waveforms

In order to compare torque pulsation in three motors more directly, torque pulsation coefficient (k_T) is introduced, which is expressed as

$$k_T = \frac{T_{\max} - T_{\min}}{T_{avg}} \times 100\% \tag{2.7}$$

where T_{max}, T_{min} and T_{avg} respectively represent the maximum, minimum and average value of the total electromagnetic torque.

It must be noted that DESRM will guarantee the existence of short flux path at all time. This not only increases the average electromagnetic torque but also removes an asymmetry in torque generation, thereby reducing the torque pulsation in the SRM. SESRM's average torque is 15.91 N m and DESRM's is 33.41 N m, which shows an increase of 52.4%. Moreover, the SESRM torque ripple is 218% and the DESRM's is 43%, which indicates a reduction by 80.3%. In general, it can be seen that DESRM plays a significant role in improving the torque performance of SRM.

2.4.5 Comparative Analysis of Core Losses

In order to observe the impact of the three SRMs on the extent of iron losses in the machine, the values of core loss are calculated in Maxwell. The materials of stator and rotor applied by the SRMs are both 50W470. After calculating by FEA, Fig. 2.34 shows the columnar comparison of iron losses obtained when the current limit of three motors is set at the same time to be 50 A and the root mean square (RMS) value of phase current is 23 A. As can be seen from the Fig. 2.34, the iron loss of SESRM is less than that of DESRM and HESRM, the iron loss of HESRM is about 2 times that of SESRM, and the iron loss value of DESRM is about 1.5 times that of SESRM. Because SESRM's flux lines have the smallest core range and the smallest number of domains changed during motor operation, therefore, it has the smallest iron loss under the same operating conditions. Similarly, HESRM changes the most domains, so it has the largest iron loss. The number of domains changed by DESRM is between SESRM and HESRM, so the iron loss is also intermediate.

Furthermore, the iron loss when the SRM works under the process of phase commutation could be an important component for the entire iron loss. Therefore, the cloud diagrams of iron loss distribution are also analyzed by FEA, which are illustrated in Fig. 2.35. It can be observed from Fig. 2.35 that the iron loss mainly concentrates on rotor tooth tip, stator tooth tip and stator yoke. The reason why the

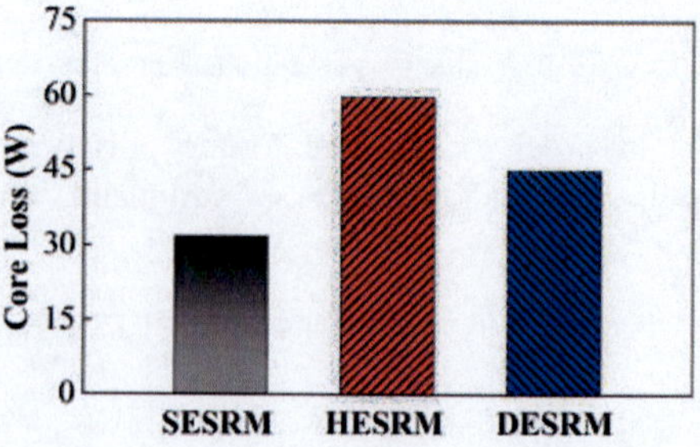

Fig. 2.34 Comparison of iron loss under the same phase RMS current

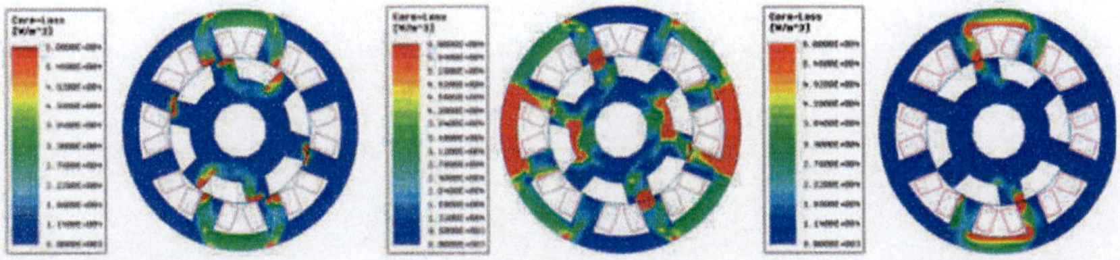

a) SESRM b) HESRM c) DESRM

Fig. 2.35 Distribution of iron loss during commutation

iron loss at the tooth tip of the three motors is large is that the magnetic flux density at the tooth tip will change greatly in the process of motor commutation and alter the direction of the magnetic domain in the tooth tip iron core. As for SESRM shown in Fig. 2.35a, since the flux lines at the yoke are reversed as shown in Fig. 2.36a when SESRM works under phase commutation, SESRM obtains a large amount of iron losses at the yoke during the reversal. It can also be found from Fig. 2.35b that the iron losses in the yoke of HESRM are extremely high. Because, combined with long flux path, there is flux overlap in the partial yoke when commutation occurs, which can be found in Fig. 2.36b. The DESRM avoids the reverse and overlap of the flux lines in yoke, and keeps the motor's flux path in short during the operation. Compared to HESRM, DESRM shortens the flux lines and has relatively small domain changes inside the core, so the iron loss of DESRM is smaller than that of HESRM.

2.4.6 Comparative Analysis of Flux and Force Density

In order to reveal the torque generation mechanism of the SRM under different excitation modes, it is very important to analyse the flux density of air gap. A circle supposed to exit in middle of the air gap is divided into 1000 equidistant points. Through the circle, the values of B_n, B_t, f_n and f_t are calculated by FEA at each point. Because of the symmetric structure of the SRM and the convenience of observation and analysis, the results are only selected from 500 equidistant points, which is half of the circle. The 500 points are determined on the premise that the centre line of the exciting phase or phases should be in the middle of the half circle. The results are shown in Figs. 2.37, 2.38, 2.39 and 2.40, where 0, 15, and 30 degrees represent unaligned position, midway position, and aligned position, respectively. It should be noted that the magnetic field of DESRM is not completely the same with that of HESRM under short path after further analysis, although both are short magnetic path in nature, resulting in different flux and force densities between DESRM and HESRM.

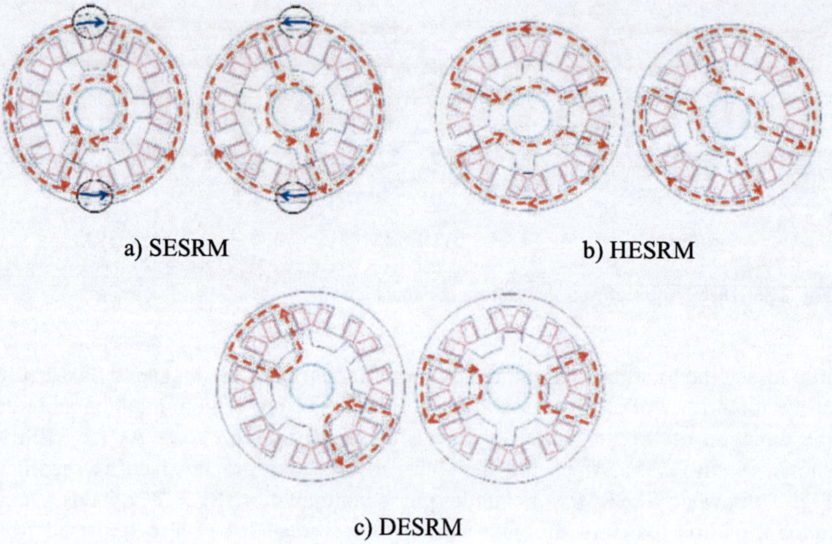

Fig. 2.36 Changes in the direction of flux lines during commutation

Since the initial position of single-phase excitation mode and double-phase excitation mode is different, the aligned position angles (i.e. 15 and 20 degrees) of the two modes are different as well. Figure 2.37a indicates that at the aligned position of SESRM, a substantial, square shaped radial flux density component distributes the entire stator pole arc. As rotor moves to the unaligned position, the radial component of the magnetic field keeps the square wave state but its width decreases continuously. Therefore, the effective width of magnetic flux density radial component can be estimated approximately by the length of overlapped area between stator and rotor. Figure 2.37b, d show the radial flux density of DESRM and HESRM under short flux paths, respectively. Since the excitation phase of DESRM and HESRM are two phases, two wave peaks appear. Their peak values are opposite indicating that the direction of the flux lines is different. However, the wave peaks of HESRM under long flux path mode are on the same side as shown in Fig. 2.37c, indicating the same direction flux lines.

As it can be seen from Fig. 2.38a, the maximum tangential component appears at the two corners of rotor and stator arcs that form the overlapped area. Obviously, the effects of local saturation at these tips are the main reason for the two local maxima of tangential flux density. Under the aligned and unaligned positions, the tangential components of flux density are distributed symmetrically to the central point of the stator pole arc, indicating that the tangential components point in opposite directions on both sides of the central point of the stator pole arc. By comparing the peak values of Fig. 2.38a–d, it can be seen that the tangential magnetic flux density peaks of DESRM and HESRM are both larger than those of the SESRM. Therefore, the conclusion is drawn that the tangential magnetic flux

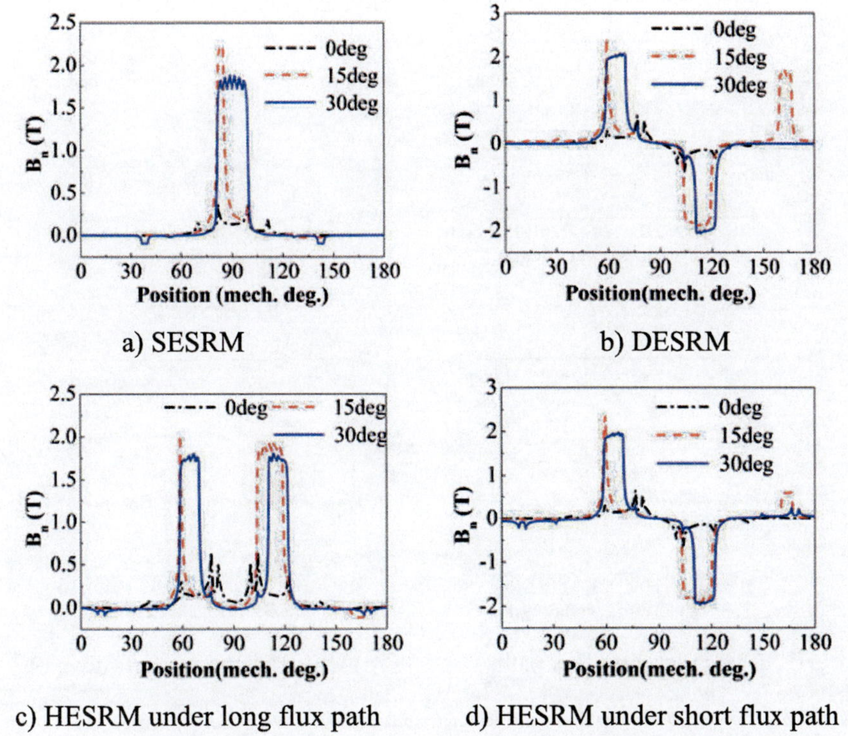

Fig. 2.37 Radial air-gap flux densities under unaligned, midway and aligned positions

density generated by the two-phase excitation mode is greater than that of the single-phase excitation mode. Thus, a greater torque can be generated, and the advantages of DESRM are the most obvious. By comparing Fig. 2.38c, d, the tangential flux density peak value of HESRM short path is about 1.5 times more than that of the long path, showing that short path has more advantages than long path when tangential magnetic flux density is generated.

The radial and tangential force density around the contour is significant for analysing motor performance of torque and torque ripple, which can be calculated based on the air-gap flux density. Using the analysis results above, radial and tangential force density are computed as shown in Figs. 2.39 and 2.40, respectively. From Fig. 2.39a, the largest radial force density appears at the aliened position. As the rotor moves from the midway position to the aligned position, the waveform width and peak value of radial force density increase. It can be seen that the waveform width of radial force density is determined by the area of overlapping part between stator and rotor. And the radial force density is affected by the saturation between rotor and stator. The greater the saturation is, the higher the radial force density peak value will be. This distribution of the radial forces has a unique

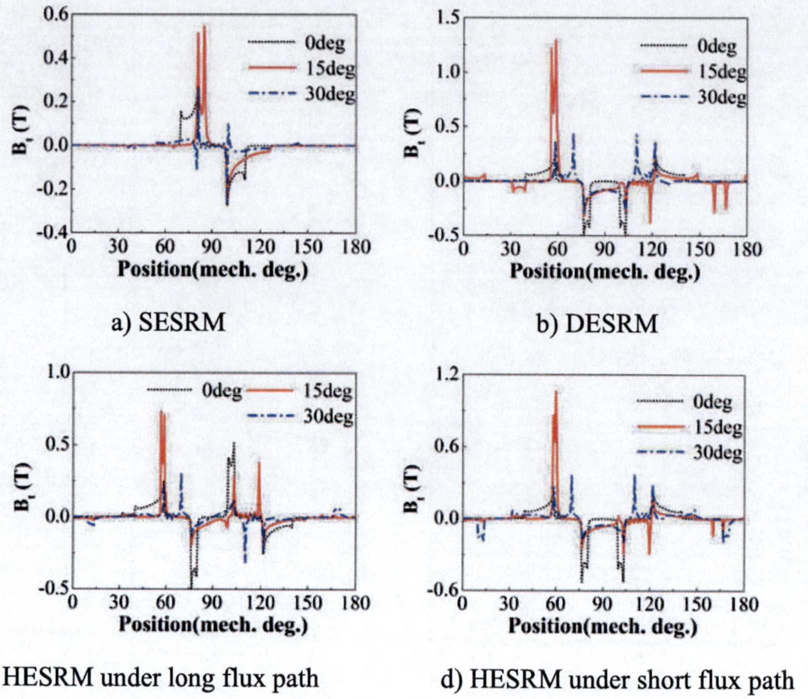

Fig. 2.38 Tangential air-gap flux densities under unaligned, midway and aligned positions

impact on the way radial vibration, which is recognized as a major contributor to acoustic noise in SRM drives. By comparing the four figures in Fig. 2.39, it can be known that the peak value of radial magnetic force density in double-phase excitation mode is more than SESRM due to the increase in the number of excitation poles. For the numerical value of radial force density, the peak value of DESRM is about 2 times of SESRM, 1.5 times of HESRM long path mode, and 1.1 times of HESRM short path mode.

As it can be seen from Fig. 2.40a, under the aligned and unaligned positions of SESRM, the tangential force densities are distributed odd symmetric to the midpoint of stator polar arc. And the quadrature results are zero, indicating that the motor is in equilibrium at this position. It is obvious that there are two local maximum values at the midway position. The first local maximum value is generated from the stator tooth corner of the overlapping area and the second local maximum value is generated from the rotor tooth corner of the overlapping area. Between these maxima, tangential component of the field is concave shape. The main reason of this phenomenon is the numerical decrease of tangential flux density at the midpoint of stator arc. Therefore, the analysis of tangential force density is of great significance to the torque and ripple. Similarly, it can be seen from Fig. 2.40b that two peaks of the tangential magnetic flux density of DESRM were generated

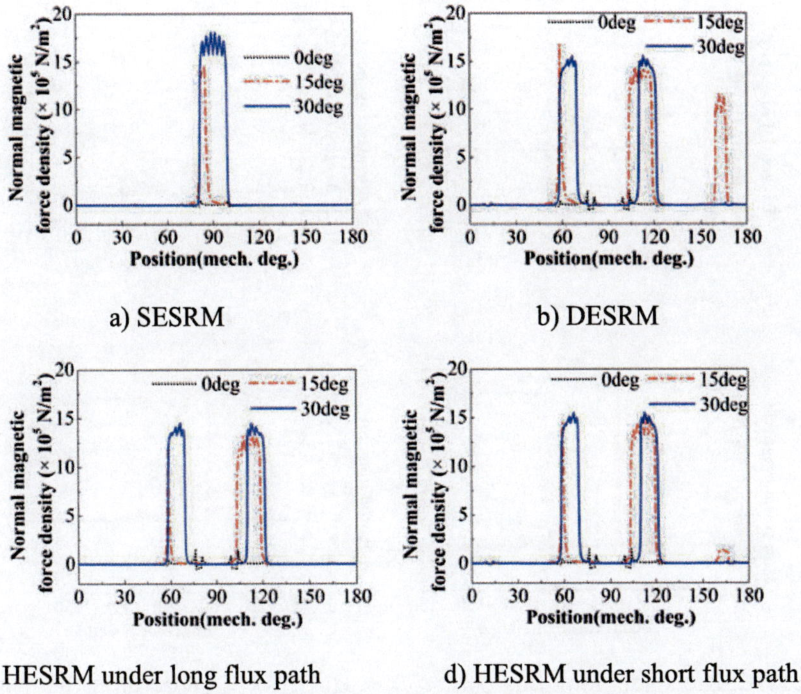

Fig. 2.39 Radial force densities under unaligned, midway and aligned positions

by the parts of a small overlapping area of stator and rotor, and the peak value was about 2.5 times more than that of the SESRM, 2.3 times more than that of the HESRM in the long flux path mode, and 1.4 times more than that of HESRM in the short flux path mode. The tangential force components provided by the regions with large overlap are almost negligible. The tangential force component peak of HESRM short flux path mode is about 1.5 times that of the long flux path mode such that it can be known that the short magnetic circuit mode has advantages in providing tangential flux density.

2.5 Experimental Verification

At present, the most widely used test methods for the flux linkage characteristics of SRM can be divided into indirect measurement and direct measurement [17, 18]. The indirect measurement method obtains the flux linkage characteristic curves indirectly by measuring the static torque or inductance of SRM. Any errors occurring in machining or installation can result in static torque errors such that its accuracy is difficult to guarantee. The method of inductance test needs to use a LCR

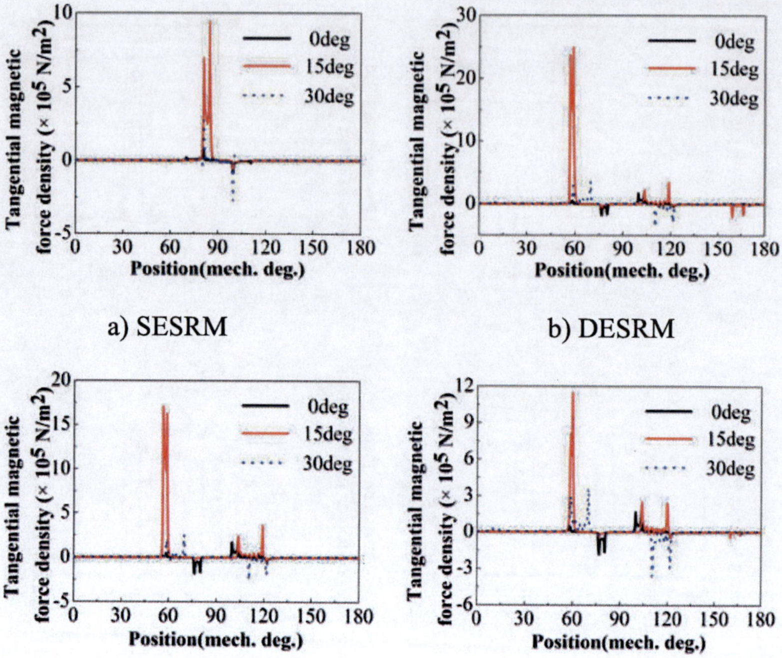

a) SESRM b) DESRM

c) HESRM under long flux path d) HESRM under short flux path

Fig. 2.40 Tangential force densities under unaligned, midway and aligned positions

bridge to test the winding inductance of the motor with different currents to calculate the flux linkage characteristics. However, one of the key disadvantages of this method is that the measurement for inductance is carried out under a very small AC current. As a result, this method is also not available for SRMs, because the inductance of SRM is a function as phase current, i.e., the inductances should be different under various phase currents. The direct measurement method is to apply a small step DC voltage pulse to the winding of the motor and keep the rotor fixed to a special angle during one trial. Then, the data of flux linkage in various angles and currents can be calculated on the basis of the measured data of phase voltage and phase current recorded by an oscillograph. Furthermore, the inductances and torques can be computed by fitting algorithm and numerical differentiation algorithm.

2.5.1 Basic Principle of the Static Test

Figure 2.41 is the experimental schematic diagram and experimental platform of the direct test method designed and adopted in this section, where the DC power supply

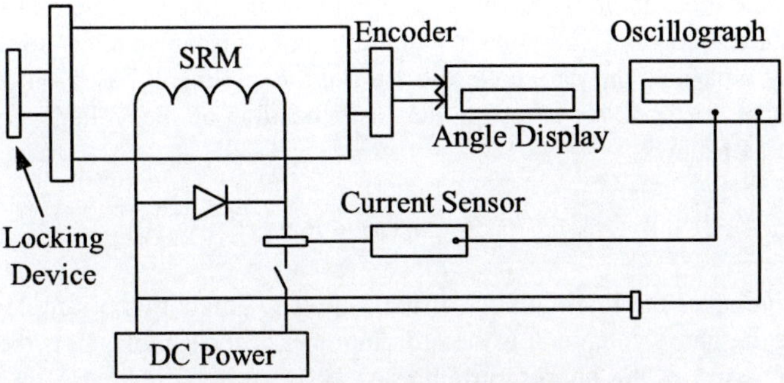

a) Experimental schematic diagram

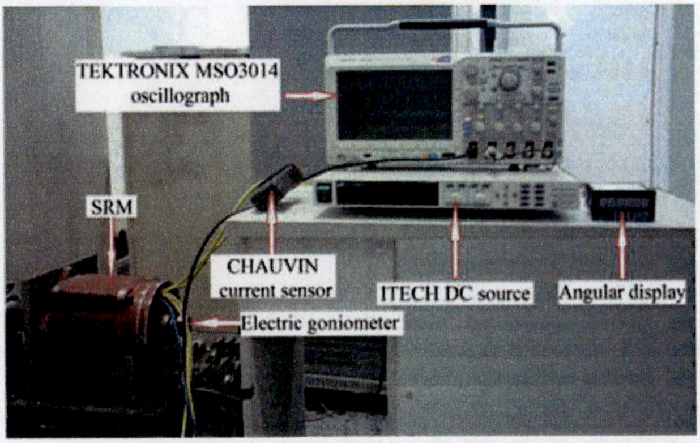

b) Experimental platform

Fig. 2.41 Experiment setup

for applying pulse voltage to the winding is IT6512 from ITECH, and the phase current testing component is a clamp-on current sensor from CHAUVIN, and the oscilloscope employed to record the data of phase voltage and current is MSO 3014 oscilloscope of Tektronix.

The rotor position is fixed at the specific angle with a mechanical locking device and the angle can be displayed by an electronic angle display. Then, a DC voltage is applied to the motor winding from the DC stabilized voltage source, and phase voltage and current can be sampled and recorded by the oscilloscope. According to the instantaneous phase voltage and current, the flux linkage values are obtained by mathematical operation and then the characteristic curves of the flux linkage can be drawn. During the test, special attention should be paid to the influence of temperature on the winding resistance, because the temperature can easily increase in the motor winding due to the high current during the test.

One of the most common methods to compute flux linkage of the motor is based on the voltage equation. The voltage equation can be dispersed to calculate the flux linkage by means of the data collected from the experiment, based on the typical measurement method. As a result, the discretization of the voltage equation is expressed as follows

$$\psi_k(N) = T \times \sum [u_k(n) - R_k i_k(n)] + \psi_k(0) \tag{2.8}$$

where N indicates the total number of the sample, T refers to the sample cycle, u_k represents the phase voltage, n is the order number of the sample, R_k is the winding resistance, i_k means the phases current.

2.5.2 *Static Test Results and Analysis*

When the two phases of the motor are connected in series, the resistance is 0.1 Ω and the rated current is 60 A. In order to prevent the excessive current flowing through the winding from damaging the motor, a pulse voltage of 5.2 V is applied to the two-phase winding. In addition, the flux linkage characteristics are greatly affected by the rotor position angles, thus the angle is subdivided by the angle meter from 0 to 30 degrees with an interval of 1 degree, and the phase voltage and current currents under each angle are measured and recorded. Figure 2.42 is the curves of the phase voltage and current in 2 and 12 degrees collected by MSO 3014 oscilloscope, where the accuracy of the current is 1 A/10 mv. It can be clearly seen that there are great differences in the change rates of phase current under various angle conditions. Obviously, the slope of phase current at 2 degrees is greater than that at 12 degrees, which can also show that the flux linkage characteristics are also very different in various angles.

The rotor position angle is selected from 0 to 30 degrees with a step of 6 degrees. Then, the flux linkage is calculated according to Eq. (2.8) and illustrated in

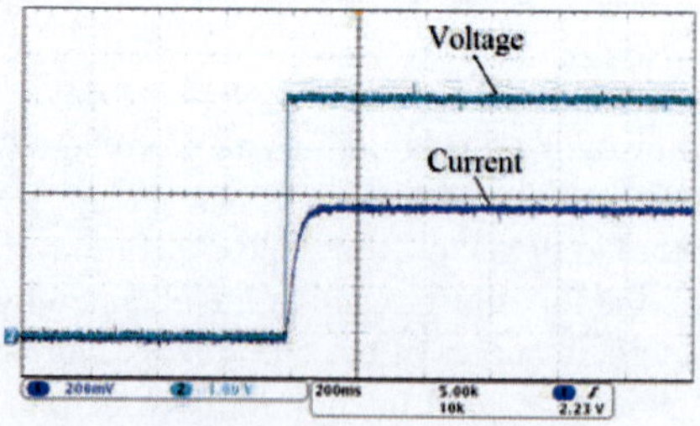

a) Angle is fixed to 2 degrees

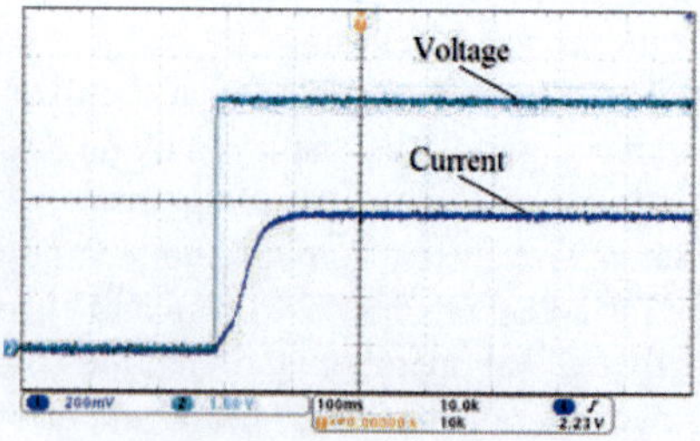

b) Angle is fixed to 12 degrees

Fig. 2.42 Recorded phase voltage and current

Fig. 2.43, where the solid and dotted lines refers to the results from the measurement and FEA, respectively. It can be observed from Fig. 2.43 that the results of electromagnetic characteristic obtained by experiment is basically consistent with those computed by FEA, except that the smoothness of the results from experiment is a little bad because of the calculation errors due to numerical integration, measuring accuracy as well as the defects from manufacture.

On the basis of the flux linkage characteristics shown in Fig. 2.43, the values of magnetic co-energy corresponding to each discrete point from the flux linkages are calculated according to Formula (2.2–2.8). Furthermore, the static torques are computed by means of numerical differential calculation, which are illustrated in Fig. 2.44. It should be noted that the dotted and solid lines also refer to the results obtained by FEA and measurement. It can be analyzed from Fig. 2.44 that the torque characteristics obtained by the test are greatly consistent with those obtained by the FEA. But there are some certain deviations at several points between the two results, because the torque data are obtained indirectly based on two times of numerical calculations, which can lead to certain calculation errors.

2.5.3 Dynamic Performance

In order to reflect the dynamic performance of the SRMs with two-phase and single-phase excitation, the dynamic performance test of the SRMs are carried out and the results are illustrated in Figs. 2.45 and 2.46. As shown in Fig. 2.45, the phase current curves of the SRM with two-phase excitation are obtained under speed of 500 and 750 r/min. It can be seen from Fig. 2.45, the phase current curves

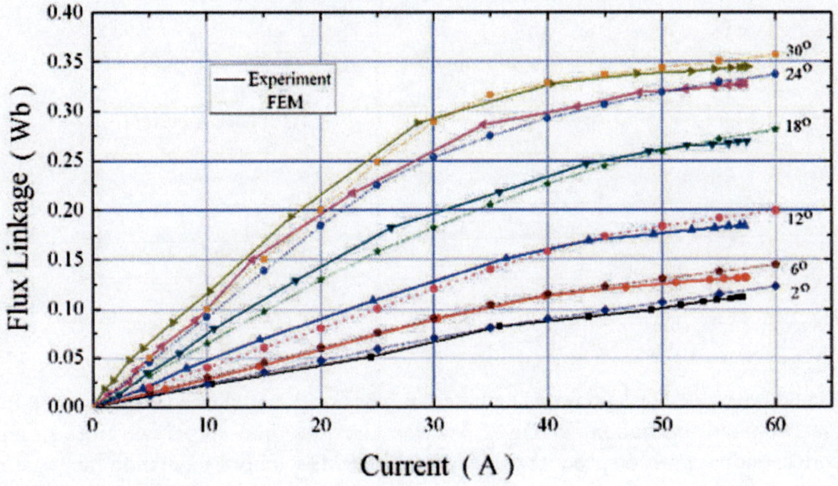

Fig. 2.43 Flux linkage results from experiment and FEA

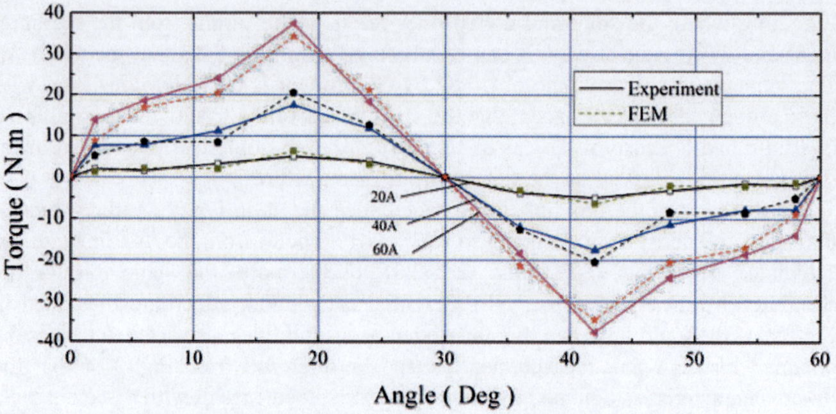

Fig. 2.44 Torque results from experiment and FEA

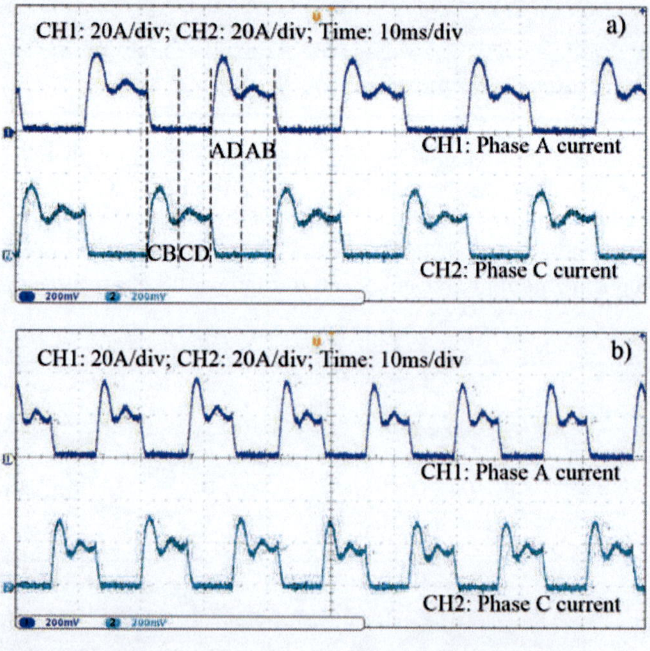

Fig. 2.45 Phase current results

should be divided into phase AD, phase AB, phase CB, and phase CD, according to the two-phase excitation. In Fig. 2.46a, the classic torque–speed waveforms and corresponding power–speed waveforms of the SRMs with two excitation modes are presented respectively, where the constant torque area is 35 N m from 0 to

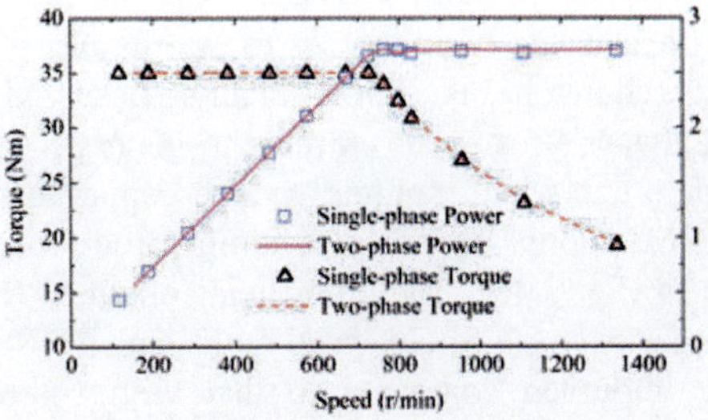

a) Torque–speed and power–speed waveforms

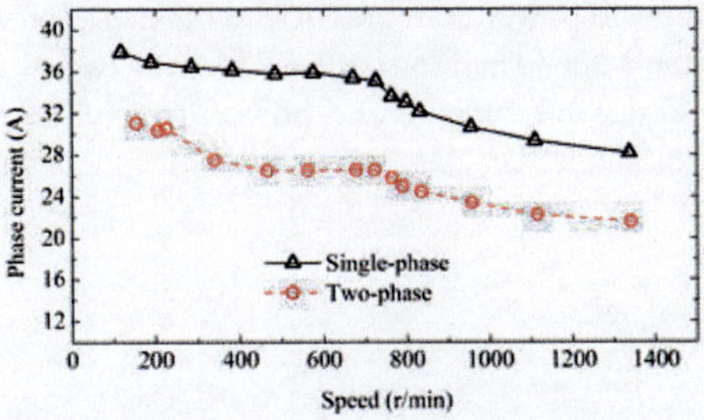

b) Phase currents along the corresponding torque-speed curves

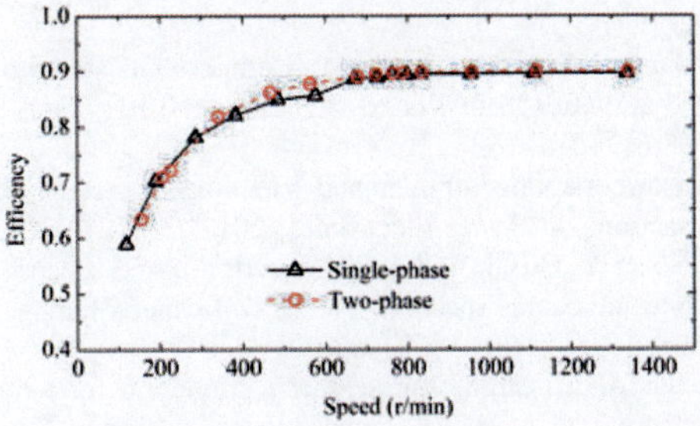

c) Efficiencies along the corresponding torque–speed curves

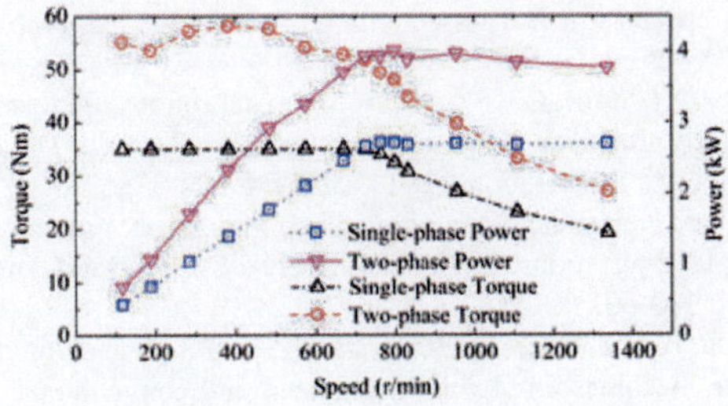

d) Torque–speed and power–speed curves at the same RMS current

Fig. 2.46 Dynamic test results

750 r/min and the constant power area is 2.5 kW from 750 to 1400 r/min. Furthermore, the phase currents with respect to speeds along the corresponding torque-speed curves are shown in Fig. 2.46b. It can be observed that the SRM with two-phase excitation always uses small currents to generate the same torque and power at each working point, which can imply small copper losses. In addition, the efficiencies of the SRMs along the corresponding torque-speed curves are also measured and shown in Fig. 2.46c. The efficiencies obtained from the two SRMs can keep almost the same over the whole speed range, and the SRM with two-phase excitation is slightly improved compared to that with single-phase excitation. Figure 2.46d shows the torque-speed characteristics of two motors at the same phase RMS current. It can be seen that the SRM with two-phase excitation can generate higher torque and power than the SRM with single-phase excitation, when the RMS current is the same. Therefore, the SRM with two-phase excitation has a great advantage on torque per ampere and power density, compared to the SRM with single-phase excitation.

References

1. Hu Y, Song X, Cao W, Ji B. New SR Drive with integrated charging capacity for plug-in hybrid electric vehicles (PHEVs). IEEE Transactions on Industrial Electronics, 2014, 61(10): 5722–5731.
2. Hua W, Hua H, Dai N, Zhao G, Cheng M. Comparative study of switched reluctance machines with half-and full-teeth-wound windings. IEEE Transactions on Industrial Electronics, 2016, 63(3): 1414–1424.
3. Cai J , Deng Z. Sensorless control of switched reluctance motor based on phase inductance vectors. IEEE Transactions on Power Electronics, 2012, 27(7): 3410–3423.
4. Ofori E, Husain T, Sozer Y, Husain I. A pulse-injection-based sensorless position estimation method for a switched reluctance machine over a wide speed range. IEEE Transactions on Industry Applications, 2015, 51(5): 3867–3876.
5. Ye J, Bilgin B, Emadi A. An offline torque sharing function for torque ripple reduction in switched reluctance motor drive. IEEE Transactions on Energy Conversion, 2015, 30(2): 726–735.
6. Hannoun H, Hilairet M, Marchand C. Design of an SRM speed control strategy for a wide range of operating speeds. IEEE Transactions on Industrial Electronics, 2010, 57(9): 2911–2921.
7. Azar Z and Zhu ZQ. Performance analysis of synchronous reluctance machines having nonoverlapping concentrated winding and sinusoidal bipolar with DC bias excitation. IEEE Transactions on Industry Applications, 2014, 50(5): 3346–3356.
8. Ma C and Qu L. Design considerations of switched reluctance motors with bipolar excitation for low torque ripple applications. 2013 IEEE Energy Conversion Congress and Exposition, Denver, CO, 2013: 926–933.
9. Ding W, Hu Y, Fu H and Chen Q. Analysis and evaluation of modular E-shaped stator switched reluctance machines employing segmented and conventional rotor topologies. IET Electric Power Applications, 2016, 10(9): 939–951.
10. Ahn J, Oh S, Moon J, Hwang Y. A three phase switched reluctance motor with two-phase excitation. IEEE Transactions on Industry Applications, 35(5): 1067–1075.

11. Lee C and Krishnan R. New designs of a two-phase E-core switched reluctance machine by optimizing the magnetic structure for a specific application: Concept, design, and analysis. IEEE Transactions on Industry Applications, 2009, 45(5): 1804–1814.
12. Eskandari H and Mirsalim M. An improved 9/12 two-phase E-core switched reluctance machine. IEEE Transactions on Energy Conversion, 2013, 28(4): 951–958.
13. Abdel-Khalik AS, Morsy AS, Ahmed S, and Massoud AM. Effect of stator winding connection on performance of five-phase induction machines. IEEE Transactions on Industrial Electronics, 2014, 61(1): 3–19.
14. Ferreira FJTE and Almeida ATd. Method for in-field evaluation of the stator winding connection of three-phase induction motors to maximize efficiency and power factor. IEEE Transactions on Energy Conversion, 2006, 21(2): 370–379.
15. Deng X, Mecrow B, Martin R, and Gadoue S. Effects of winding connection on performance of a six-phase switched reluctance machine. IEEE Transactions on Energy Conversion, 2018, 33(1): 166–178.
16. Panda D, Ramanarayanan V. Mutual coupling and its effect on steady-state performance and position estimation of even an odd number phase switched reluctance motor drive. IEEE Transactions on Magnetic, 2007, 43(8):3445–3456.
17. Zhang J, Arthur V. Radun. A new method to measure the switched reluctance motor's flux. IEEE Transactions on Industry Applications, 2006, 42(5): 1171–1176.
18. Adrian David Cheok, Wang Z. Flux linkage measurement method for switched reluctance motors and inductor/transformers using a real-time DSP system. International Journal of Electronics, 2010, 89(8): 625–644.

Chapter 3
Optimization Design for SRM in EVs

3.1 Introduction

When the SRM is considered as the drive system for EVs, two kinds of installation position are classical, one is traditional centralized drive and the other is in-wheel drive [1]. Although the two kinds are completely different, both of them are desired to have excellent drive performance, such as high output torque, great torque density, small torque ripple, high efficiency, etc. Because these abilities can meet the special requirements of EVs on power performance, efficiency, and ride comfort [2, 3]. As a result, when the SRMs designed or optimized to use in EVs, these potential objectives should be considered to improve the comprehensive performance of the EV.

Conventionally, Most SRMs in industries are operated in single-phase excitation mode, although the SRM with two-phase excitation mode has better torque performance [4, 5]. However, when two phases of the SRM with two-phase excitation are excited simultaneously, the magnetic field become more complicated resulting in undesirable torque waveforms and extra torque ripple [5]. Meanwhile, as a direct drive motor for EVs, the design principle and optimization for in-wheel SRMs with outer rotor is greatly different from that of the general SRM with inner rotor. Therefore, as for motors used in EVs, not only the SRM with two-phase excitation, but also the in-wheel SRM should be deeply considered to be redesigned or optimized to improve the comprehensive performance of EVs.

In addition, many literatures [6–9] have done great works to improve the dynamic performance of the SRM drive system. However, the dynamic performance improvements should be considered after the structure optimization. On the other hand, the structure optimization and sensitivity analysis for the normal SRM have been conducted in [10–13]. But these researches focused on normal SRM, instead of the in-wheel SRM or the SRM with two-phase excitation mode.

Y. Zhu, *The Key Technologies for Powertrain System of Intelligent Vehicles Based on Switched Reluctance Motors*, Recent Advancements in Connected Autonomous Vehicle Technologies 1, https://doi.org/10.1007/978-981-16-4851-9_3

Therefore, in this chapter, the optimization of SRM with two-phase excitation, the design and optimization for the in-wheel SRM are carried out to improve the comprehensive performance of the EV. During the optimization, focusing on both special SRMs, the effects of structure parameters on objectives are studied and the sensitivity analysis is also performed based on combination of orthogonal experiment design and finite element analysis.

3.2 Requirements of EVs on Driving Motor

Considering the requirements of EVs, the driving electric motors should be with high average torque, large torque density, great efficiency, as well as low torque ripple. Because, these features can effectively meet the tractive performance, economy performance, and ride performance. As a result, these four objectives, which are normal indicators for SRM, should be considered comprehensively when the SRM used in EVs is optimized.

The structure and operating principle of the SRMs with single-phase and two-phase excitation mode have been introduced in Chap. 2. However, after FEA analysis for the SRM with two-phase mode, this kind of SRM could have undesirable static torque and also increase torque ripple, because the magnetic field under two-phase mode is more complicated than that under single-phase mode when two phases are excited at the same time. Also, the conclusion can be easily obtained after compared the static torque curves in Figs. 2.24 and 2.25. Therefore, the structure of the SRM with two-phase excitation mode should be optimized to achieve the best performance of the motor.

In this chapter, based on the comparative analysis of the SRM with single-phase and two-phase mode, four objectives used to optimize the structure of SRM, i.e., static average torque, torque ripple, efficiency, and average torque density, are selected and defined respectively. The performance of the SRM can be evaluated by means of these four objectives. Then, the sensitivity degree of the structure parameters to the objectives are carried out to obtain the relative importance of each parameters of the motor. Moreover, four weight factors are presented and multi-objective optimization for the SRM with two-phase excitation mode is performed based on the sensitivity analysis. Furthermore, taking four-phase 8/6 SRM with two-phase excitation as an example, the motor is optimized through the multi-objective optimization. Finally, the torque performances of the SRM, including static torque and dynamic torque performance, is measured or simulated, and also compared to the initial motor to verify the optimization method.

3.3 Optimization Objectives of the SRM with Two-Phase Mode

3.3.1 *Objectives of the SRM with Two-Phase Mode*

As described in Chap. 2, the SRM with two-phase excitation mode has better performance compared to that of single-phase excitation, because of the torque density and efficiency improvement, as well as low torque ripple. However, in order to meet the requirements of the EV, including static average torque, torque ripple, efficiency, and average torque density, which are described in Sect. 3.1. Four corresponding objectives are selected to achieve the optimization of the SRM with two-phase excitation, i.e., average torque, torque ripple, motor loss, and average torque density. These objectives selected to be considered in the optimization can represent output torque, ride comfort, efficiency, and torque density of the EV, respectively.

Generally, the static torque with respect to position angle and phase current can be calculated by FEA as done in Chap. 1. In order to obtain the output torque as high as possible, the static average torque T_{ave} is selected to represent output torque of the EV, which can be defined as

$$T_{ave} = \frac{\sum_{\theta=\theta_1}^{\theta_n} T_\theta}{n} \tag{3.1}$$

where θ refers to the rotor position angle, T_θ means the static torque of the SRM when the rotor angle is θ, n represents the data number. In order to obtain the significant data of the static torque from the whole, the data locating on the flat region of the torque waveform are considered as effective data. Because, under the common operation condition, this area is the main torque output region. Considering Eq. (3.1), this area is determined by rotor position from θ_1 to θ_n. In addition, it should be noted that the data of static torque can be obtained by FEA or experiment method, although the data are from FEA in this chapter.

As usual, the ride comfort is one of the most important performances for vehicles. Considering EV with SRM driving system, the torque ripple of the SRM should be paid more attention, which can directly affect the ride comfort of EV. From the static torque waveform's point of view, the wider the flat region is, the lower the torque ripple could be. As a result, the torque ripple can be reflected by means of the torque data on the flat region. The torque ripple T_r could be defined by the variance of the torques locating on the flat region of the torque curve, which is shown as

$$T_r = \frac{\sum_{i=0}^{n} \left(T_i - T_{ave}\right)^2}{n} \tag{3.2}$$

where T_i refers to considered static torque value.

The efficiency is also one of the most important performances of EVs because of the requirements of the long driving range. As for SRM, the key influence factor for efficiency is the stray loss, which includes core loss and copper loss. Both losses can be computed by FEA and considered to reflect the efficiency. Therefore, the stray loss is represented by the sum of core and copper losses, and based on which the relative efficiency η is calculated as

$$\eta = \frac{P_R}{P_R + P_{core} + P_{cu}} \tag{3.3}$$

where P_{core}, P_{cu}, and P_R refers to iron loss power, copper loss power, and rated power. In general, both P_{core}, and P_{cu} can be influenced by the structure parameters of SRM. As a result, the iron loss can be computed by the method combined FEA with harmonic analysis, when the structure parameters of the motor determined. It should be illustrated that the material of the rotor and stator lamination is DW 350-35, and the iron loss power is calculated under the condition of rated speed. Moreover, the copper loss power is easier to calculated, because it can be obtained by phase current and winding resistance. The resistance of the motor could change according to the structural variations of stator pole height and arc angle. Therefore, the copper loss power should be recalculated as long as the structure is optimized. On the other hand, there is normally a proportional relationship between copper and iron loss power, thus each one of them could be calculated by empirical formula method when the other one has been obtained.

The torque density of the driving motor is also a significant performance for EV, which can be determined by average torque and motor total mass. Meanwhile, the motor mass and torque waveform are both directly determined by means of the structure parameters of the motor. Therefore, the torque density Q, computed by the static average torque and the motor mass M, should also be considered when the SRM is optimized, which can be expressed as

$$Q = \frac{T_{ave}}{M} \tag{3.4}$$

In Eq. (3.4), the motor mass is computed by summing the masses of stator, rotor, and winding by using the varying volume from each component.

3.3.2 *Optimization Parameters and Constrains*

According to the structure of the SRM with two-phase excitation mode, ten structure parameters can be considered to be optimized to improve the torque performance of the motor, i.e., the air gap and second airgap (g and g_i), the pole arc angle of stator and rotor (β_s and β_r), the yoke thickness of stator and rotor (h_{cs} and h_{cr}), outer diameter of rotor and stator (D_a and D_s), rotor inner diameter (D_i), and

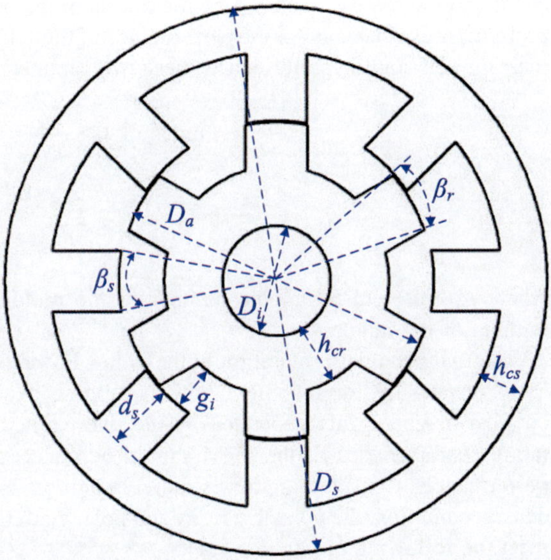

Fig. 3.1 Structure parameters of the 8/6 SRM

stator pole length (d_s). As shown in Fig. 3.1, the structure parameters of the 8/6 SRM with two excitation mode described above are illustrated.

In order to determine the optimization parameters from the ten structure parameters of the motor shown in Fig. 3.1, the parameters should be analyzed and compared in detail. As for the air gap g, which is considered as small as possible during the manufacture process. Meanwhile, the smaller the air gap is, the lower loss the motor generates, such that the airgap is not a proper optimization parameter. Furthermore, it is expected that the stator pole length d_s can be determined as long as outer diameter of rotor and stator, the yoke thickness of stator, and the air gap are ascertained, because the stator pole length can be calculated by means of these parameters. As a result, the stator pole length is also not a proper parameter selected to be optimized. In addition, after further analysis, the second air gap g_i is also not suitable for selecting to be one of the optimization parameters, because it can also be computed by means of other parameters. Moreover, the rotor inner diameter D_i also needn't to be selected as one of the optimization variables, because it is generally limited by the shaft size. Based on the illustrations mentioned above, the rest of the structure parameters, i.e., the pole arc angle of stator and rotor (β_s and β_r), the yoke thickness of stator and rotor (h_{cs} and h_{cr}), and outer diameter of rotor and stator (D_a and D_s), can be selected as the potential optimization parameters to perform the optimization process. It should be noted that these six structure variables, which could be significantly meaningful on improving the performance of torque, torque ripple, torque density, and efficiency, are determined to be optimized in this section, in order to improve the motor performance.

On the other hand, the optimization process should be subjected to some specific constraints because of the basic principle of the SRM. The pole arc angles of the stator and the rotor (β_s and β_r) are the important parameters for effecting the motor performance, especially for the performance of the self-starting. Thus, considering to meet the self-starting performance, these two parameters should be expressed as [14]

$$\min(\beta_s, \beta_r) \geq \frac{2\pi}{qN_r} \tag{3.5}$$

$$\beta_s + \beta_r \leq \frac{2\pi}{N_r} \tag{3.6}$$

where N_r refers to rotor pole number of the motor, and q represents the phase number of the motor.

The magnetic field saturation in the yokes is one of the obvious features for the SRM, especially for the SRM with two-phase excitation mode because of the coupled magnetic flux generated by the two excited phases. Therefore, the saturation characteristics of the SRM should be concerned to ensure the high motor performance. Considering the extreme saturation in the yokes under high flux density condition, the yoke thickness and pole width of the rotor (h_{cr} and b_p) should meet the following constraints [14].

$$h_{cr} \geq b_{pr}/2 \tag{3.7}$$

Compared to the yoke thickness of the rotor, the yoke thickness of the stator is also a key parameter during the design. In addition, both the two parameters can directly influence the iron mass and core loss of the motor such that they can change the torque density and efficiency. As a result, both should be confined by means of pole width of the stator and the rotor (b_{ps} and b_{pr}) [14].

$$h_{cr} = (1.2 \sim 1.4)\frac{b_{pr}}{2} \tag{3.8}$$

$$h_{cs} = (1.2 \sim 1.4)\frac{b_{ps}}{2} \tag{3.9}$$

3.3.3 Goal Function of Optimization

Comprehensively considering the four objectives described in 3.3.1 and the six structure parameters selected in 3.3.2, a comprehensive function is established based on four weight factors, which is computed as

$$\begin{aligned} F(\vec{x}) &= \omega_1 \frac{T_{ave}}{(T_{ave})_{\max}} + \omega_2 \frac{T_r'}{(T_r')_{\max}} + \omega_3 \frac{\eta}{\eta_{\max}} + \omega_4 \frac{Q}{Q_{\max}} \\ \vec{x} &= [\beta_s, \beta_r, h_{cs}, h_{cr}, D_s, D_a] \end{aligned} \tag{3.10}$$

where ω_1, ω_2, ω_3, and ω_4 refer to weight factors, which meet $\omega_1 + \omega_2 + \omega_3 + {}_4 = 1$; T_r' represents the torque ripple, which is the reciprocal of the torque ripple; and $\vec{x}$ means the six structure parameters.

According to the comprehensive function described in Eq. (3.10), the goal function can be easily obtained, which is shown as

$$F(\vec{x}_{opt}) = \max\left\{\omega_1 \frac{T_{ave}}{(T_{ave})_{\max}} + \omega_2 \frac{T_r'}{(T_r')_{\max}} + \omega_3 \frac{\eta}{(\eta)_{\max}} + \omega_4 \frac{Q}{(Q)_{\max}}\right\} \tag{3.11}$$

3.3.4 Sensitive Analysis of the Structure Parameters

In order to comprehensively perform the experiments or simulations, the design of experiment (DOE) method is always employed to achieve the uniform combinations among various design parameters. However, with the increase of the design parameter number, the experiment number rises exponentially based on the parameter number. As a result, the normal DOE could become more and more impractical as parameter number increases [10]. As a significant DOE method, the orthogonal experiment design (OED) method mitigate the rising experiment number, because it can effectively arrange the experiment combinations among various design parameters to reduce the experiment numbers. Meanwhile, OED can involve range analysis for the potential design parameters, which can be used to analyse the sensitivity degree of objectives to the design parameters. After range analysis and sensitive analysis, the effects of each parameter on objectives can be obtained such that the insensitive parameters could be deleted out of the considered design parameters. Then, the experiment arrangements are simplified because of reducing of the parameter number.

To execute the sensitive analysis for each design parameter, the rate of contribution of each parameter to objectives should be researched and defined, because it can reflect the relative influence of each parameter on the objectives. However, the rate of contribution can't be calculated during rang analysis of OED. Therefore, in this chapter, a sensitivity ratio is defined by means of the results from OED, which is shown as

$$\sigma_{m,j} = \frac{\Delta \mathrm{K}_{m,j}}{\sum_{j=1}^{N} \Delta \mathrm{K}_{m,j}} \times 100\% \tag{3.12}$$

where m refers to the number of objectives, N represents the total number of design parameters, $\Delta K_{m,j}$ is the range value.

Combined the results from OED and described Eq. (3.12), the sensitivity ratio of each objectives to various design parameters can be calculated. It should be noted that four objectives are considered to optimize the SRM with two-phase excitation mode, which are T_{ave}, T_r, η, and Q, respectively. In addition, the design parameters should be revaluated after sensitive analysis, because some parameters could be neglected due to the low sensitivity ratio.

According to the constraints described in Eqs. (3.5) to (3.9), the limits of each design parameter can be computed and carried out. As a result, the range of each structure parameter is obtained and discretized, which is shown in Table 3.1. As mentioned above, the OED method is employed to arrange the combinations of structure parameters and carry out the rang analysis. Furthermore, the sensitivity analysis can also be performed by means of Eq. (3.12). It should be noted that an orthogonal table L64(8^7) is used for experiment arrangement in this section. Compared to the normal DOE method having 2097152 combinations, only 64 combinations have to be carried out through OED based on this orthogonal table.

According to range analysis of OED and Eq. (3.1), the contribution rates of each structure parameter on average torque ranges are computed and compared after the average torque data are obtained by FEA under uniform parameter combinations of OED, which is shown in Fig. 3.2a where the AT ranges mean the average torque ranges during range analysis of OED. Analysing the AT ranges in Fig. 3.2a, it can be seen that obvious differences of AT ranges appear among various structure parameters. The maximum and minimum AT ranges are 4.1 N m and 0.2 N m, respectively, which are caused by pole arc angle of the stator β_s, and rotor yoke thickness h_{cr}. The other values are 0.6, 0.4, 0.6, and 2.2 N m, corresponding to the structure of rotor pole arc angle β_r, stator yoke thickness h_{cs}, stator outer diameter D_s, and rotor outer diameter D_a. On the other hand, it can be concluded that the parameter of β_s has the most significant effect on the objective of average torque, while the h_{cr} has the least effect. Moreover, the effect of the D_a on average torque is also relatively crucial, compared to the rest three structure parameters. Thus, compared to other parameters, both β_s and D_a need to be main optimization parameters when the average torque is considered to be improved.

After further reason analysis on effects of each parameter on the average torque, it can be seen that the pole arc angles of the stator and the rotor can directly influence the overlap area between the stator and the rotor when the motor works in or close to the aligned position, such that both parameters can influence the electromagnetic torque. Thus, this is the reason why the AT ranges under the β_s and β_r are relatively high. Considering the rotor outer diameter D_a, it can directly cause changes of the electromagnetic torque, because the arm of the generated

Table 3.1 Discretized values of the structure parameters

Parameters	Minimum value	Maximum value	Discretized step
β_s	20 deg.	27 deg.	1 deg.
β_r	23 deg.	30 deg.	1 deg.
h_{cr}	23 mm	30 mm	1 mm
h_{cs}	18 mm	25 mm	1 mm
D_s	183 mm	204 mm	3 mm
D_a	104 mm	111 mm	1 mm

Fig. 3.2 Effects of each parameter on objectives

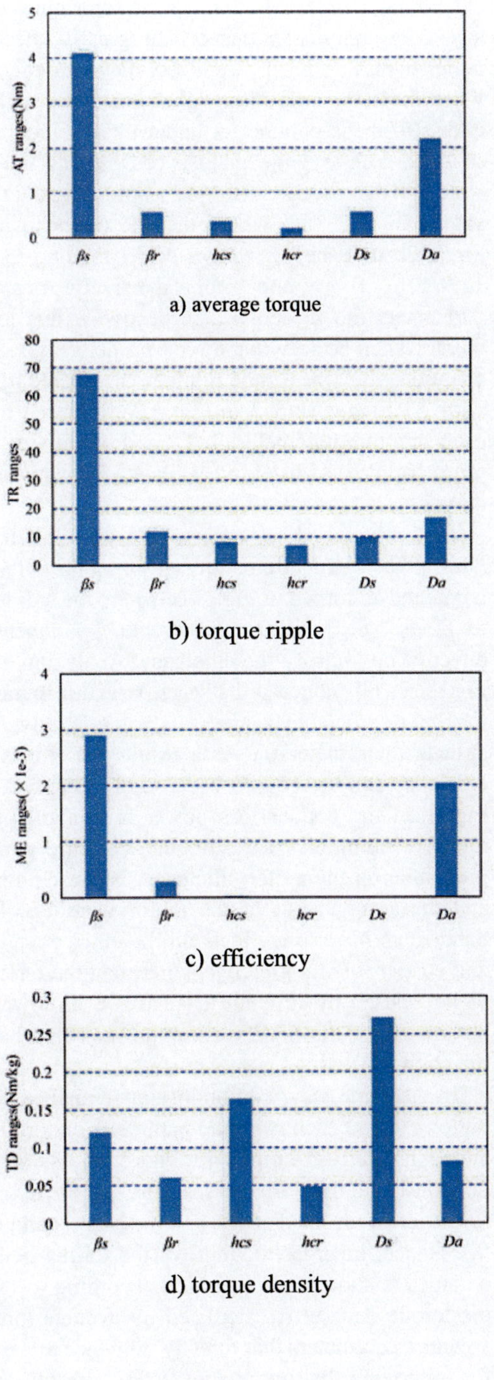

electromagnetic force varies as the rotor outer diameter changes. As a result, D_a is also a key parameter that can be greatly effective on AT ranges. As for the rest parameters, i.e., h_{cs}, h_{cr}, and D_s, they can effectively influence the average torque when the stator and rotor yokes are under magnetic saturation condition. But the effects of these parameters under unsaturation condition are limited.

According to range analysis of OED and Eq. (3.2), the contribution rates of each structure parameter on torque ripple ranges are also computed and compared as shown in Fig. 3.2b where the TR refers to torque ripple. It can be seen from Fig. 3.2b that the TR ranges corresponding to each structure parameter are 67.6, 11.9, 8.1, 7.0, 9.8, and 16.6, respectively. As a result, the most and least significant parameters are β_s and h_{cs}, respectively, due to the values of TR ranges. On the whole, it can be seen that three parameters, i.e., β_s, D_a, and β_r, need to be paid more attention when the torque ripple is considered to be improved. Compared TR ranges with AT ranges shown in Fig. 3.2a, b, it can be observed that the effect trends of each structure parameter on TR and AT ranges are basically the same. Because the torque ripple data used to calculate TR ranges are partially selected from the torque curves.

Similarly, combining range analysis of OED with Eq. (3.3), the contribution rates of each structure parameter on motor efficiency ranges are also computed and compared as shown in Fig. 3.2c where the ME refers to motor efficiency. Compared to β_r, h_{cs}, h_{cr}, and D_s, the structure parameters of β_s and D_a are both greatly effective on improve the efficiency. Analyzing the motor efficiency ranges shown in Fig. 3.2c, all values under each structure parameter are low enough. As for the maximum efficiency ranges, the value is only 2.9×10^{-3}, which is caused by the structure parameter β_s. As a result, the effects of structure parameters on motor efficiency can be ignored because of the low range values. For further analysis on Fig. 3.2c, the copper loss power is primarily generated by the resistance of the motor winding. As a result, the structure parameters of β_s and D_a are greatly influential on the motor efficiency, because both them can directly change the size and the arrangement of the motor windings. However, the rest of the structure parameters have low effects on the efficiency. The loss power, including iron loss and copper loss, is greatly smaller than the rated power of the motor. And the loss power caused by structure change is even lower. Thus, the influences of structure parameters of the SRM on loss power could be ignored during the structure optimization process.

In the same way, combining range analysis of OED with Eq. (3.4), the contribution rates of each structure parameter on torque density ranges are also computed and compared as shown in Fig. 3.2d where the TD refers to torque density. Obviously, almost all the structure parameters have great influences on the performance of torque density, and the highest and lowest values are caused by D_s and h_{cr}. Hence, in order to improve the torque density of the SRM, all the structure parameters should be optimized according to their contribution rates. In addition, the torque density is calculated by average torque and motor mass such that the structure parameters that directly influence the motor mass can be greatly significant for improving the torque density. As a result, the structure parameters from stator parts, which basically locates outside of the motor, is more effective for mass, compared to the parameters from rotor parts locating inside. Therefore, the effects of parameters from stator parts are more crucial than that from rotor parts.

As mentioned above, Fig. 3.2 reflects the effects of each structure parameter on various objectives. But the contribution rate of each structure parameter should also be studied, because it is more obvious to obtain the relative effects for the four objectives. According to results from OED and Eq. (3.12), the sensitivity ratios of the four objectives and comprehensive function established in Eq. (3.10) to each structure parameter are also calculated and shown in Fig. 3.3, where a and b represent sensitivity ratio of the four objectives and comprehensive function, respectively.

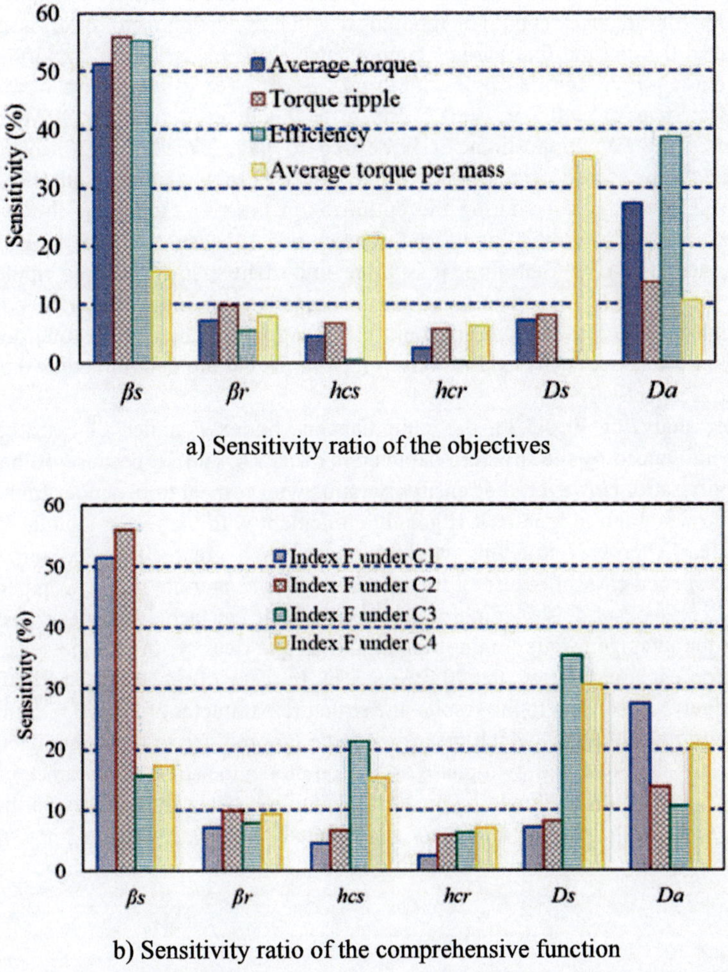

Fig. 3.3 Sensitivity ratios with various structure parameters

As shown in Fig. 3.3a, all the four objectives can be highly influenced by the structure parameters of β_s and D_a because of the high sensitivity ratio. and β_r has basically uniform effects even though the sensitivity ratios are close to 10%. On the contrary, the structure parameter h_{cr} has slight effects on all the objectives. The results from Fig. 3.3a are basically coincident with the effect analysis shown in Fig. 3.2. Therefore, considering the SRM with two-phase excitation mode, the parameters of β_s, β_r, h_{cs}, D_s, and D_a should be optimized to improve the comprehensive performance of the SRM. It should be noted, as an insensitive parameter, h_{cr} is determined according to the constraints expressed in Eq. (3.8).

Similarly, the sensitivity ratios of the comprehensive function to structure parameters are calculated and shown in Fig. 3.3b, where four combinations of the weigh factors are described as C1, C2, C3, and C4. It should be noted that C1, C2, and C3 are employed for single objective optimization, while C4 is for multi objective optimization. According to the analysis results about the efficiency, the objective for efficiency should be ignored such that the weigh factor of ω_3 is always defined to 0 in all the four cases. Comprehensively considering the importance of the average torque, torque ripple, and torque density, the weigh factors in case C4 are selected as $\omega_1 = 0.5$, $\omega_2 = 0.2$, and $\omega_4 = 0.3$. Considering the SRM in this section is for EV applications, it is desired to have excellent performance on average torque, torque ripple, and torque density, such that they all should be improved synchronously during the optimization process. Generally, the average torque is one of the most significant characteristics because of the drive power requirements of EV. Therefore, it is more important than the torque ripple and torque density. On the other hand, as an objective for improving ride comfort performance, the torque ripple is slightly less significant than the torque density. Based on the above analysis, the weight factors for C4 are determined as 0.5, 0.2, and 0.3, respectively.

After analyzing Fig. 3.3b, the comprehensive index F under all cases can be highly influenced by the structure parameters of β_s, D_s, and D_a because of the high sensitivity ratio. However, the sensitivity ratios with respect to h_{cr} under four cases are all low enough to ignore. It is greatly coincident with the results obtained from Fig. 3.3a. Other two structure parameters can have great effects on the index F under special cases, because of the relative higher sensitivity ratios. Compared to C1, C2, and C3, C4 is a compromise case, which can achieve a better balance among the average torque, torque ripple, and torque density. In C4, the sensitivity ratios for each parameter are 17.3%, 9.3%, 15.3%, 7.0%, 30.5%, and 20.6%, respectively. According to the results, the structure parameter of h_{cr} has the smallest sensitivity ratio of 7.0%, which means it can be ignored due to its slight effects. As a result, for the SRM with two-phase excitation mode, the structure parameters, i.e., β_s, β_r, h_{cs}, D_s, and D_a, should be paid more attention when the SRM is optimized. In this section, these five parameters are selected to be optimized to improve the comprehensive performance of the SRM.

3.3.5 *Optimization of the SRM*

According to the objective function expressed in Eq. (3.11), the selected structure parameters are optimized under various cases by means of joint operation between FEA and OED, which are shown in Table 3.2. Obviously, the initial structure parameters and optimized ones under various optimization cases are completely different from each other. It can also be proved that the sensitivity analysis mentioned above is effective. In order to show the differences between the initial and optimized results, the magnetic flux densities of the SRMs are computed by FEA, and the diagrams obtained when the phase current and position angle are 30 A and 26 degrees are illustrated in Fig. 3.4.

Based on the optimized structure parameters shown in Table 3.2, the objectives on average torque, torque ripple, torque density, and comprehensive function are calculated by FEA under various cases and shown in Table 3.3. Distinctly, the results from the initial case are worse than those from other optimized cases, which could directly prove the effectiveness of the optimization. Compared to the initial case, the results of the average torque T_{ave}, torque density Q, and comprehensive function $F(\vec{x})$ under C4 increase by 18.09%, 12.08%, and 6.57%, respectively. Meanwhile, the optimized SRM under C4 is manufactured and the test platform for measuring the static electromagnetic parameters is also established as shown in Fig. 2.37 of Chap. 2. The static torque can be measured according to the experiment principles described in Sect. 2.4.1 and compared with the results from FEA as shown in Fig. 3.5a. Furthermore, the measured results under C4 (C4-M) are illustrated in Table 3.3, which can be used to compare with the computed results by means of FEA. It can be obviously seen from Table 3.3 that the results from FEA and experiment are basically coincident with each other, which can also prove the effectiveness of the results from FEA.

As shown in Fig. 3.5a, the static torque waveforms of the SRMs with initial and optimized structure parameters are compared to the results from experiment under case C4, which are all with respect to rotor position angle from 0 to 60 degrees. Obviously, the waveforms of the torque changes to be flatter in the middle area, compared to the waveforms of the initial design. After further analysis, it can be observed that this variation of the static torque can directly improve the performance on the average torque, torque ripple, and torque density. Furthermore, in order to obtain the detailed torque waveforms of C4 under various phase current,

Table 3.2 Initial and optimal structure parameters

Cases	β_s (deg.)	β_r (deg.)	h_{cs} (mm)	D_s (mm)	D_a (mm)
Initial	22	23	15	180	104
C1	27	27	19	183	111
C2	20	30	17	195	104
C3	26	23	14	183	111
C4	27	27	14	183	111

Fig. 3.4 Magnetic flux density results from FEA

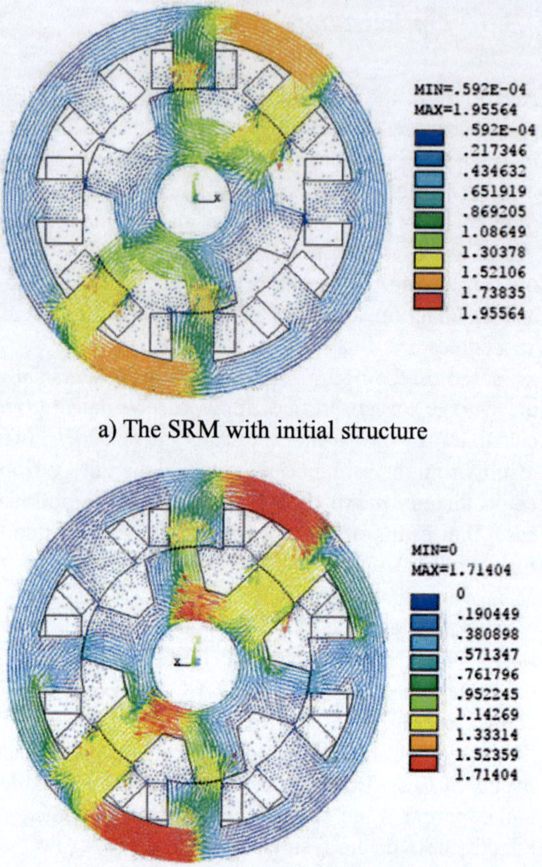

a) The SRM with initial structure

b) The SRM with optimized structure

Table 3.3 Results of objectives under various cases

Cases	T_{ave} (N m)	T_r' ($\times 10^{-3}$)	Q (N m/kg)	F (C4)
Initial	27.8	16	1.71	0.84
C1	33.6	7	1.78	0.88
C2	25.8	24	1.38	0.81
C3	31.5	9	1.93	0.87
C4	32.8	8	1.91	0.89
C4-M	35.1	8	2.04	0.95

the torques with respect to rotor position angles are calculated and shown in Fig. 3.5b. The torque waveforms obtained by the optimized SRM under C4 are all with larger flat region, which can also directly prove the effectiveness of the optimization.

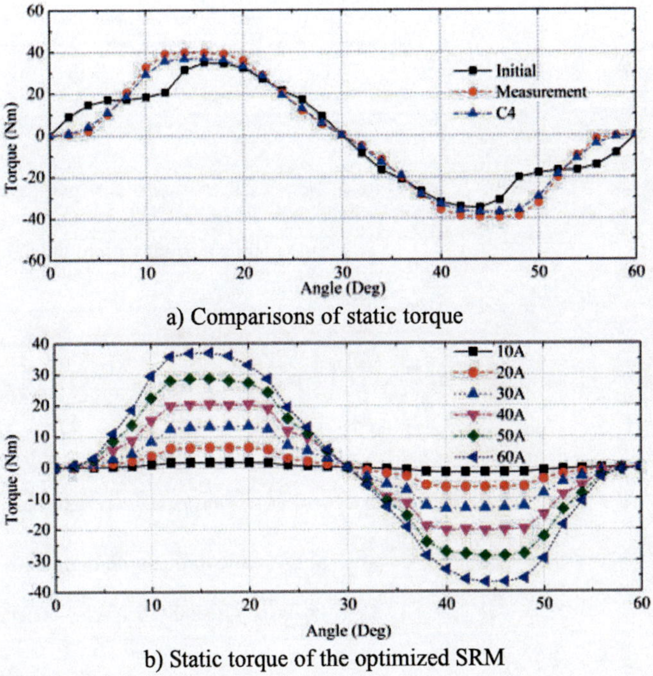

Fig. 3.5 Results of the static torque

In order to show the dynamic performance of the optimized SRM, the dynamic simulation model for the SRM drive system is established. Choosing two normal conditions, i.e., high load torque and low load torque, the phase currents of each phase and total torque waveforms are simulated and obtained as shown in Fig. 3.6, where (a) and (c) are phase current responses, and (b) and (d) are total torque characteristics. The following can be observed that the average torque of the optimized SRM improve by 17.4% in Fig. 3.6b, and 24.6% in d. Meanwhile, the torque ripple under two conditions reduce by 95.8% and 41.3%, respectively, compared to the SRM with initial structure.

3.4 Design of In-Wheel SRM

As we know, EVs and HEVs, having advantages of little environment pollution, low noise, and high efficiency, are two of the best choices for vehicle industry to solve the serious problem of resources and environment. Combined with the advantages of SRM, in-wheel SRM has been paid more and more attention currently, because of simple structure, low cost, and direct drivetrain system for the

Fig. 3.6 Dynamic results

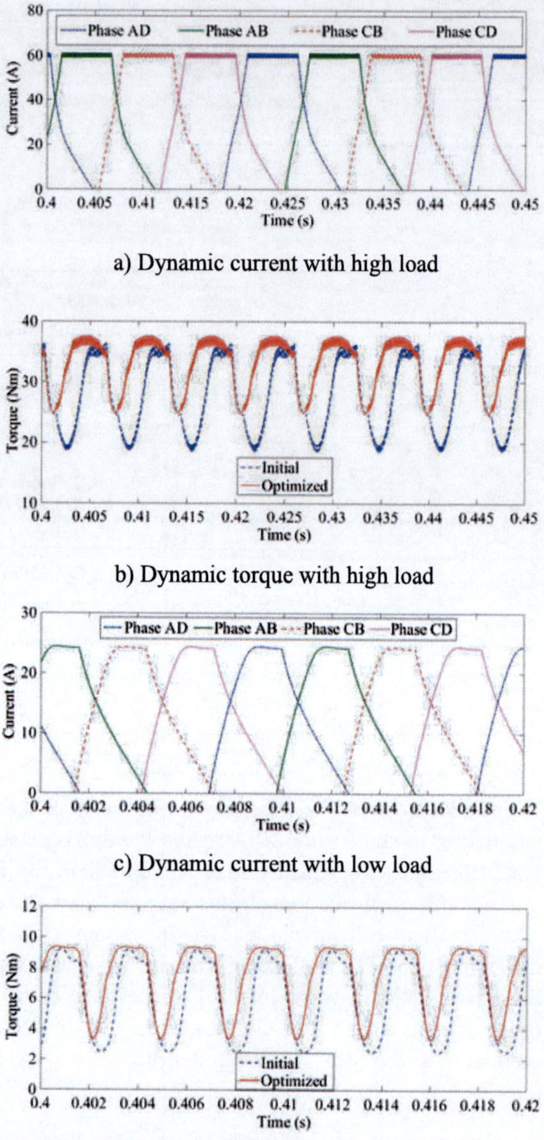

a) Dynamic current with high load

b) Dynamic torque with high load

c) Dynamic current with low load

d) Dynamic torque with low load

in-wheel motor. However, the design for in-wheel SRM is lacking according to the current research status. Therefore, a design example focusing on in-wheel SRM is developed in this section, which is used for EVs.

Generally, there are three key indexes for the power performance of vehicle, i.e., the maximum vehicle velocity, time for acceleration, and grade ability. According to these three indexes, the basic indexes for the in-wheel motor driving the vehicle can be calculated. As usual, the maximum velocity of the vehicle may be frequently employed when the vehicle works, while the power used for accelerating and climbing is always applied for short term work. Therefore, the power for maximum velocity can be used to calculate the rated power of in-wheel motor, while the power for other two indexes is normally for maximum power. In this section, two in-wheel SRMs are considered to drive the EV, and the expected design parameters of the EV are listed in Table 3.4.

The power for the maximum vehicle velocity can be computed by rolling resistance force and air resistance force, which can be used to determine the rated power of the motor. It should be noted that, as for in-wheel motor, the number of the motors can directly influence the parameter determination of each one. Therefore, the total rated power of in-wheel motor can be calculated as

$$P = \frac{V_{\max}}{3600\eta_T}\left(mgf + \frac{C_D A V_{\max}^2}{21.15}\right) \tag{3.13}$$

where, the meanings and corresponding values of each parameter can be found in Table 3.4.

As mentioned above, based on the grade ability of the vehicle, the required power of the in-wheel motor can be computed to achieve the vehicle performance of climbing the maximum grade with a specific driving speed. Furthermore, considering the acceleration ability of the vehicle, another power can also be computed to meet the requirement of acceleration. In order to achieve both requirements of the vehicle, the bigger one between the two powers is finally required, because it can meet both requirements. The separated power and final power can be expressed as

$$P_i = \frac{V_i}{3600\eta_T}\left(mgf + \frac{C_D A V_i^2}{21.15} + mgi\right) \tag{3.14}$$

$$P_a = \frac{V_a}{3600\eta_T}\left(mgf + \frac{C_D A V_a^2}{21.15} + \delta m\frac{dv}{dt}\right) \tag{3.15}$$

$$P_{\max} = \max\{P_i, P_a\} \tag{3.16}$$

Table 3.4 Expected design parameters of the EV

Vehicle parameters	Values
Transmission system efficiency (η_T)	97%
Mass of the EV (m)	400 kg
Rolling friction coefficient (f)	0.01
Drag coefficient of (C_D)	0.23
Windward area A	1.0 m^2
Maximum vehicle velocity $V_{\max}$	34 m/s
Type of the tire	195/65 R15

where P_i, P_a, and $P_{\max}$ refer to the required powers meeting grade ability and acceleration, and maximum power for the motor, V_i, i, V_a, and δ represent the climbing speed, gradient, the accelerating speed, and rotating mass coefficient, respectively.

According to Eqs. (3.13) to (3.16) and vehicle expected design parameters listed in Table 3.4, the rated and maximum power of the in-wheel SRM can be obtained. Meanwhile, the rated speed of the motor can also be determined by means of the maximum vehicle velocity and the tire type. Combined design experience of SRM, the outer diameter of the rotor is obtained according to the tire type. In addition, the voltage level can be easily determined by means of the expected battery voltage. Therefore, the basic design parameters for the in-wheel SRM are obtained. i.e., rated voltage U_N, rated power P_N, rated speed n, rated efficiency η, and outer diameter of the rotor D_a are 96 V, 4 kW, 1000 r/min, 82%, and 360 mm, respectively. In addition, the phase number q, stator pole number N_s, and rotor pole number N_r are also determined empirically, which are 4, 16, and 20, respectively.

Firstly, the pole arc angles of stator and rotor should be determined. Generally, there is an overlapping area when the rotor and stator are in or close to aligned position to achieve the bidirectional self-starting ability of the SRM. Thus, the stator pole arc angle β_s and the rotor pole arc angle β_r should meet the following constrains. According to Eqs. (3.17) and (3.18), the two pole arc angles are determined as β_s = 8 degrees, and β_r = 9 degrees.

$$\min(\beta_s, \beta_r) \geq \frac{2\pi}{qN_r} \tag{3.17}$$

$$\beta_s + \beta_r \leq \frac{2\pi}{N_r} \tag{3.18}$$

Thanks to the battery, the power source is DC for the motor, such that the terminal voltage of the phase winding U is expressed as follows after considering the boosting effect of the capacitive filter.

$$U = 1.1 \times 96\ \text{V} = 105.6\ \text{V} \tag{3.19}$$

Considering the losses of the SRM, the electromagnetic power of the motor can be estimated by means of expected design efficiency η and rated power P_N, which is expressed as

$$P_{em} = P_N \frac{1+\eta}{2\eta} = 4.0 \times \frac{1+0.82}{2 \times 0.82}\ \text{kW} = 4.44\ \text{kW} \tag{3.20}$$

In general, the structural ratio λ can be calculated by the stator outer diameter D_a and iron stack length l_a as $\lambda = l_a/D_a$. The structural ratio of the motor is an important factor, which can directly influence the geometric structure, such as end structure, winding distribution, and even cooling system. Considering that the expected design motor is an outer rotor SRM, the iron stack length l_a is selected as 125 mm and the stator outer diameter D_s is determined as 286 mm.

As one of the most important parameters of SRM, air gap g can influence the motor maximum inductance and the electromagnetic torque. As a result, the air gap is expected as narrow as possible generally. However, on the other hand, the value of the air gap can also be determined by several limitations of motor manufacture and assembly, such that the value can not be designed too small. As for normal industrial motor, the air gap is around 0.5 mm. Considering the in-wheel SRM designed in this section is just in the time of design and validation, the air gap g is currently determined as 0.7 mm.

Then, the stator pole width b_{ps} and the rotor pole width b_{pr} can be calculated as

$$b_{ps} = D_s \sin\frac{\beta_s}{2} = 286 \times \sin\frac{8^\circ}{2}\ \text{mm} = 19.94\ \text{mm} \tag{3.21}$$

$$b_{pr} = (D_s + 2g)\sin\frac{\beta_r}{2} = (286 + 2 \times 0.7)\sin\frac{9^\circ}{2}\ \text{mm} = 22.54\ \text{mm} \tag{3.22}$$

The second air gap g_i, which is the distance from stator pole surface to rotor slot bottom, is also an important factor influencing the inductance. Differing from the air gap, it is mainly relative to the minimum inductance instead of the maximum one. In order to obtain smaller minimum inductance that could improve the output power, the second air gap should be wide enough, approximately 20 to 30 times of the air gap. On the other hand, it cannot be designed too wide because it also directly influences the rotor yoke thickness and shaft diameter. Therefore, it should meet the constrain shown as Eq. (3.23). Considering the requirements of rotor yoke thickness and shaft diameter, the second air gap g_i is selected as 17 mm.

$$g_i \geq \frac{b_{ps}}{2} \tag{3.23}$$

Generally, the rotor yoke thickness h_{cr} is determined based on that the yoke thickness can avoid excessive saturation when the SRM work under the condition with high magnetic flux density. Therefore, the value should meet Eq. (3.24). On the other hand, the saturation condition can be more serious when two or more phase are excited synchronously. As a result, the rotor yoke thickness should also meet the requirement shown as Eq. (3.25). According to the requirements from Eqs. (3.24) and (3.25), the rotor yoke thickness is determined as 20 mm.

$$h_{cr} \geq b_{pr}/2 \tag{3.24}$$

$$h_{cr} = (1.2 \sim 1.4)\frac{b_{pr}}{2} \tag{3.25}$$

Similarly, the stator yoke thickness h_{cr} should also be determined according to the stator pole width b_p. However, it is normally expected that the SRM with higher stator yoke thickness is helpful for increasing the vibration and noise of the motor. As a result, the value of the stator yoke thickness is determined as 35 mm after comprehensive consideration.

The shaft diameter D_i can influence the mechanical strength of the shaft, such that the value should not too small resulting in extra rotor vibration and high dynamic eccentricity. Meanwhile, it cannot also be designed too high, because of the size requirements of second air gap g_i and rotor yoke thickness h_{cr}. Consequently, the shaft diameter D_i is determined as 160 mm.

The motor winding needs wide space such that the thick cable can be used, which can effectively decrease the copper loss of the motor. Thus, the stator slot depth d_s should be as high as possible in the reasonable range. After determination of stator outer diameter, stator yoke height, and shaft diameter, the stator slot depth d_s can be calculated as

$$d_s = \frac{1}{2}(D_s - 2h_{cs} - D_i) = \frac{1}{2}(286 - 2 \times 20 - 160)\ \text{mm} = 28\ \text{mm} \tag{3.26}$$

Then, the effective iron length l_{Fe} can be computed by

$$l_{Fe} = 0.93 l_s = 0.93 \times 125\ \text{mm} = 116\ \text{mm} \tag{3.27}$$

The stator pole pitch is also calculated by

$$\tau_r = \frac{\pi D_a}{N_r} = \frac{3.14 \times 360}{20}\ \text{mm} = 56.52\ \text{mm} \tag{3.28}$$

Currently, the main structure parameters have been determined as mentioned above. Then, the winding parameters need to be calculated. Generally, the in-wheel SRM need to work continuously because of the EV requirements, based on which the current density J should be around 4–5.5 A/mm^2. Therefore, the preliminary value for current density J' is selected as 5 A/mm^2. In addition, the preliminary number of winding turns N_{ph} is also determined as 80. Thus, the sectional area of the cable can be computed as

$$S'_a = \frac{I}{J'} = \frac{41.7}{5}\ \text{mm}^2 = 8.37\ \text{mm}^2 \tag{3.29}$$

Then, the total cable sectional area from one phase in each slot S_{Cu} is determined as

$$S_{Cu} = \frac{1}{4} N_{ph} S_a = \frac{1}{4} \times 80 \times 8.37\ \text{mm}^2 = 167.4\ \text{mm}^2 \tag{3.30}$$

The window area between two adjacent stator poles S_W can be calculated as

$$
\begin{aligned}
S_W &= \frac{1}{2}\cdot\frac{\pi}{N_s}\left[\left(\frac{D_s}{2}\right)^2-\left(\frac{D_s}{2}-d_s\right)^2\right]-\frac{1}{2}b_{ps}d_s \\
&= \left\{\frac{1}{2}\cdot\frac{\pi}{16}\left[\left(\frac{286}{2}\right)^2-\left(\frac{286}{2}-28\right)^2\right]-\frac{1}{2}\times 19.94\times 28\right\}\ \mathrm{mm}^2 \\
&= 429.7\ \mathrm{mm}^2
\end{aligned}
\tag{3.31}
$$

After determining the basic geometrical parameters of the motor, the slot filling factor k_s can be obtained as 78% be means of S_{Cu} and S_W.

$$
k_s = \frac{S_{Cu}}{S_W} = \frac{167.4}{429.7} = 0.39 \tag{3.32}
$$

In order to obtain the detailed geometrical variables, the structure is shown in Fig. 3.7, where the potential parameters are all marked. And the variables are calculated as follows.

$$
b_w = \frac{D_s}{2}\tan\frac{\pi}{N_s}-\frac{1}{2}b_{ps} = \left(\frac{286}{2}\tan\frac{\pi}{16}-\frac{1}{2}\times 19.94\right)\ \mathrm{mm} = 18.5\ \mathrm{mm} \tag{3.33}
$$

$$
a = b_{ps}+b_w = (19.94+18.5)\ \mathrm{mm} = 38.4\ \mathrm{mm} \tag{3.34}
$$

$$
b = l_s+2R+b_w = (125+2\times 4+18.5)\ \mathrm{mm} = 151.5\ \mathrm{mm} \tag{3.35}
$$

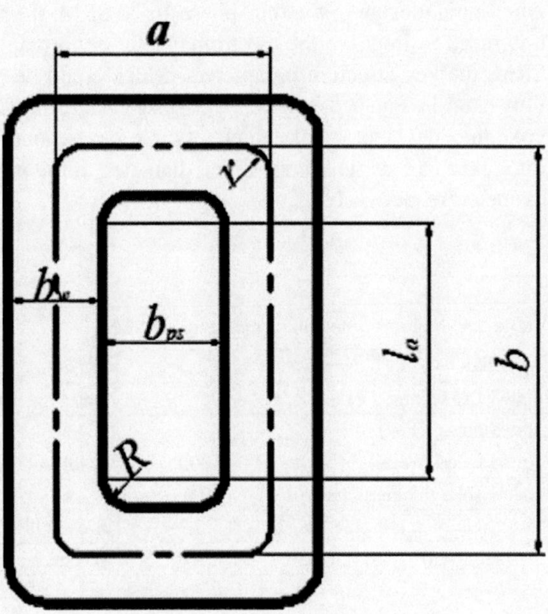

Fig. 3.7 Structure and variables of the winding

$$r = R + \frac{1}{2}b_w = \left(4 + \frac{1}{2} \times 18.5\right) \text{ mm} = 13.2 \text{ mm} \tag{3.36}$$

Then the average length of the winding l_{av} is computed as

$$l_{av} = 2l_s + 2\left(b_{ps} - 2R\right) + 2\pi r = 357 \text{ mm} \tag{3.37}$$

The total winding length of each phase is

$$l = N_{ph} l_{av} = 80 \times 0.357 \text{ m} = 29 \text{ m} \tag{3.38}$$

The winding resistance of each phase is calculated as

$$R_p = \rho \frac{l}{S_a} = 0.0217 \frac{80}{8.37} \Omega = 0.07\Omega \tag{3.39}$$

According to the design process mentioned above, the main parameters for the designed outer rotor in-wheel SRM are listed in Table 3.5, and the lamination material for the motor body is DW 350-35.

3.5 Optimization of In-Wheel SRM

3.5.1 Objectives and Optimization Variables

After initial design of the four-phase 16/20 SRM, the structure parameters should be optimized to improve the electromagnetic performance of the motor. As for 16/20 SRM, the key structure parameters, which could be potential to be optimized, are illustrated in Fig. 3.8, where, h_{cs}, h_{cr}, g, g_i, β_s, β_r, D_s, D_a, and D_i, refer to stator yoke thickness, rotor yoke thickness, air gap, second air gap, stator pole arc angle, rotor pole arc angle, stator outer diameter, rotor outer diameter, and rotor inner diameter, respectively.

Table 3.5 Main parameters for the designed SRM

Variables	Values	Variables	Values
Rated DC voltage (V)	96	Outer diameter of stator (mm)	290
Rated power (kW)	4	Stator inner diameter (mm)	145
Rated speed (r/min)	1000	Rotor outer diameter (mm)	360
Stator yoke thickness (mm)	29	Rotor yoke thickness (mm)	18
Arc angle of stator pole (deg.)	5.5	Arc angle of rotor pole (deg.)	8.5
Air gap (mm)	0.35	Second air gap (mm)	17

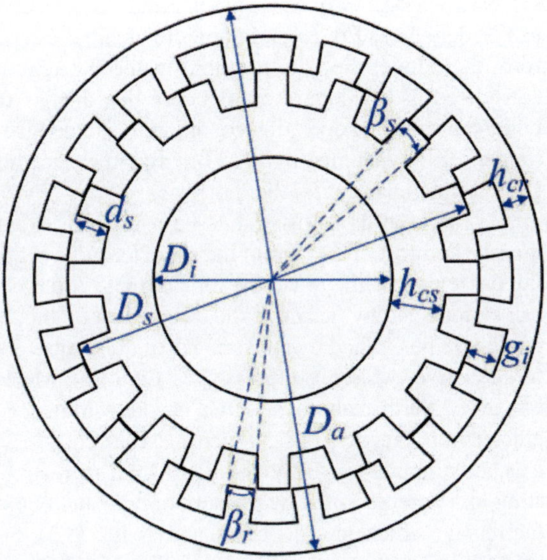

Fig. 3.8 Potential optimized parameters of typical in-wheel SRM

Theoretically, all the structure parameters shown in Fig. 3.8 could be potentially considered as optimized variables. However, considering the special structure of the in-wheel SRM and constrains for the vehicle rims, it is expected that rotor outer diameter D_a should meet the size requirements of EV rims, such that it is not a proper optimization parameter. Furthermore, it is also expected that the stator outer diameter D_s can be determined as long as the second airgap g_i and the rotor outer diameter D_a are ascertained. As a result, the stator outer diameter D_s is also not suitable for selecting to be one of the optimization parameters, because it can also be computed by means of other parameters. Based on the illustrations mentioned above, the rest of the structure parameters, i.e., the pole arc angle of stator and rotor (β_s and β_r), the yoke thickness of stator and rotor (h_{cs} and h_{cr}), outer diameter of rotor and stator (D_a and D_s), and rotor inner diameter (D_i), can be selected as the potential optimization parameters to perform the optimization process. It should be noted that these seven structure variables, which could be significantly meaningful on improving the performance of torque, torque ripple, torque density, and efficiency, are determined to be optimized in this section, in order to improve the motor performance.

Being similar with selected objectives considering for the SRM with two-phase excitation mode, in order to meet the requirements of the EV, including static average torque, torque ripple, efficiency, and average torque density, which are described in Sect. 3.1. Four corresponding objectives are selected to achieve the optimization of the four-phase 16/20 in-wheel SRM, i.e., average torque, torque ripple, motor loss, and average torque density. These objectives selected to be considered in the optimization can represent output torque, ride comfort, efficiency, and torque density of the EV, respectively. Therefore, focusing on in-wheel SRM, the static average torque T_{ave} is selected to represent output torque of the EV, which can be defined as Eq. (3.1). In order to obtain the significant data of the static torque from the whole, the data locating on the flat region of the torque waveform are considered as effective data. Because this area is the main torque output region under the common operation condition. it should also be noted that the data of static torque can be obtained by FEA or experiment method, although the data are from FEA in this chapter. Furthermore, considering EV with SRM driving system, the torque ripple of the SRM should be paid more attention, which can directly affect the ride comfort of EV. From the static torque waveform's point of view, the wider the flat region is, the lower the torque ripple could be. As a result, the torque ripple can be reflected by means of the torque data on the flat region. The torque ripple T_r could also be defined by the variance of the torques locating on the flat region of the torque curve, which is expressed as Eq. (3.2). Moreover, in order to improve the efficiency performances of EVs, the key influence factor of in-wheel SRM for efficiency is the stray loss, which includes core loss and copper loss. Both losses can be computed by FEA and considered to reflect the efficiency. Therefore, the stray loss is represented by the sum of core and copper losses, and based on which the relative efficiency η is calculated as Eq. (3.3). It should be noted that the iron loss can be computed by the method combined FEA with harmonic analysis, when the structure parameters of the motor determined. And the copper loss power is easier to calculated, because it can be obtained by phase current and winding resistance. The resistance of the motor could change according to the structural variations of stator pole height and arc angle. Therefore, the copper loss power should be recalculated as long as the structure is optimized. In addition, the torque density of the driving motor is also a significant performance for EV, which can be determined by average torque and motor total mass. Meanwhile, the motor mass and torque waveform are both directly determined by means of the structure parameters of the motor. Therefore, the torque density Q, computed by the static average torque and the motor mass M, should also be considered when the SRM is optimized, which can be expressed as Eq. (3.4). It should be noted that the motor mass is computed by summing the masses of stator, rotor, and winding by using the varying volume from each component.

As for the constrains during the optimization process, the pole arc angles of the stator and the rotor (β_s and β_r) are the important parameters for effecting the motor performance, especially for the performance of the self-starting. Thus, considering to meet the self-starting performance, these two parameters should be expressed as Eqs. (3.5) and (3.6). Furthermore, the magnetic field saturation in the yokes is one of the obvious features for the SRM, especially for the in-wheel SRM because of the compact structural size. Therefore, the saturation characteristics of the SRM should be concerned to ensure the high motor performance. Considering the extreme saturation in the yokes under high flux density condition, the yoke thickness and pole width of the rotor (h_{cr} and b_p) should meet the constraints shown as Eq. (3.7). Compared to the yoke thickness of the rotor, the yoke thickness of the stator is also a key parameter during the design. In addition, both the two parameters can directly influence the iron mass and core loss of the motor such that they can change the torque density and efficiency. As a result, both should be confined by means of pole width of the stator and the rotor (b_{ps} and b_{pr}), as expressed in Eqs. (3.8) and (3.9).

3.5.2 Goal Function of Optimization for In-Wheel SRM

Comprehensively considering the four objectives described above and the seven structure parameters selected in 3.5.1, a comprehensive function is established based on four weight factors, which is computed as

$$F(\vec{x}) = \omega_1 \frac{T_{ave}}{(T_{ave})_{\max}} + \omega_2 \frac{T'_r}{(T'_r)_{\max}} + \omega_3 \frac{\eta}{(\eta)_{\max}} + \omega_4 \frac{Q}{(Q)_{\max}} \qquad (3.40)$$
$$\vec{x} = [\beta_s, \beta_r, g, g_i, h_{cs}, h_{cr}, D_i]$$

where $\omega_1, \omega_2, \omega_3$, and ω_4 refer to weight factors, which meet $\omega_1 + \omega_2 + \omega_3 + \omega_4 = 1$, and $\vec{x}$ means the seven structure parameters.

According to the comprehensive function described in Eq. (3.40), the goal function can be defined as Eq. (3.11).

3.5.3 Sensitivity Analysis of the Structure Parameters

As described in Sect. 3.3, the OED method mitigate the rising experiment number, because it can effectively arrange the experiment combinations among various design parameters to reduce the experiment numbers. Meanwhile, OED can involve range analysis for the potential design parameters, which can be used to analyse the sensitivity degree of objectives to the design parameters. After range analysis and sensitive analysis, the effects of each parameter on objectives can be obtained such that the insensitive parameters could be deleted out of the considered design parameters to simplify the experiment arrangements. Therefore, the OED method is also used in this section to achieve the arrangements and effect analysis for the in-wheel SRM. In addition, to execute the sensitive analysis for each design parameter, the rate of contribution of each parameter to objectives should be researched and defined, because it can reflect the relative influence of each parameter on the objectives. However, the rate of contribution can't be calculated during rang analysis of OED. As a result, the sensitivity ratio defined as Eq. (3.12) is also employed to carry out the sensitivity analysis for the in-wheel SRM. It should be noted that the sensitivity ratio of each objectives to various design parameters can be calculated, combined the results from OED and described Eq. (3.12).

Table 3.6 Discretized values of the structure parameters

Parameters	Minimum value	Maximum value	Discretized step
β_s	4.5 deg.	9 deg.	0.5 deg.
β_r	4.5 deg.	9 deg.	0.5 deg.
h_{cr}	16 mm	20 mm	0.5 mm
h_{cs}	14 mm	35 mm	3 mm
D_i	125 mm	160 mm	5 mm
g_i	17 mm	22.5 mm	0.69 mm
g	0.3 mm	0.7 mm	0.05 mm

According to the constraints described in Sect. 3.5.1, the limits of each design parameter can be computed and carried out. As a result, the range of each structure parameter is obtained and discretized, which is shown in Table 3.6. As mentioned above, the OED method is employed to arrange the combinations of structure parameters and carry out the rang analysis. Furthermore, the sensitivity analysis can also be performed by means of Eq. (3.12). It should be noted that an orthogonal table L64(8^7) is also used for experiment arrangement in this section.

According to range analysis of OED and Eq. (3.1), the contribution rates of each structure parameter on average torque ranges are computed and compared after the average torque data are obtained by FEA under uniform parameter combinations of OED, which is shown in Fig. 3.9a. Analysing the AT ranges, it can be seen that obvious differences of AT ranges appear among various structure parameters. The maximum and minimum AT ranges are 20.48 N m and 1.61 N m, respectively, which are caused by airgap g and second airgap g_i, respectively. The other values are 17.24, 10.89, 5.13, 4.66, and 2.41 N m, corresponding to the structure of stator pole arc angle β_s, rotor pole arc angle β_r, stator yoke thickness h_{cs}, rotor yoke thickness h_{cr}, and rotor inner diameter D_i. Obviously, the parameter of g has the most significant effect on the objective of average torque, while the g_i has the least effect. Moreover, the effects of the β_s, β_r, h_{cs}, and h_{cr} on average torque are also relatively crucial, compared to the D_i. Therefore, g, β_s, β_r, h_{cs}, and h_{cr} need to be paid more attentions when the average torque is considered to be improved, compared to the structure parameters g_i and D_i. After further reason analysis on effects of each parameter on the average torque, it can be seen that, as one of the most important parameters of SRM, air gap g can influence the motor maximum inductance and the electromagnetic torque, such that the effect from air gap is highest. In addition, the pole arc angles of the stator and the rotor can directly influence the overlap area between the stator and the rotor when the motor works in or close to the aligned position, such that both parameters can influence the electromagnetic torque. Thus, this is the reason why the AT ranges under the β_s and β_r are also relatively high. As for the rest parameters, i.e., h_{cs}, h_{cr}, D_s, and g_i, they can effectively influence the average torque when the stator and rotor yokes are under magnetic saturation condition. But the effects of these parameters under unsaturation condition are limited.

Similarly, the contribution rates of each structure parameter on torque ripple ranges are also computed and compared as shown in Fig. 3.9b. It can be seen that the TR ranges corresponding to each structure parameter are 334, 310, 221, 62, 71, 89, and 96, respectively. As a result, the most and least significant parameters are β_s and g_i, respectively, due to the values of TR ranges. On the whole, it can be seen that three parameters, i.e., β_s, β_r, and g, need to be paid more attention when the torque ripple is considered to be improved. Compared TR ranges with AT ranges shown in Fig. 3.9a, b, it can be observed that the effect trends of each structure parameter on TR and AT ranges are basically the same. Because the torque ripple data used to calculate TR ranges are partially selected from the torque curves.

Furthermore, combining range analysis of OED with Eq. (3.3), the contribution rates of each structure parameter on motor efficiency ranges are also computed and compared as shown in Fig. 3.9c. Compared to β_s, g, h_{cs}, and D_i, the structure parameters of β_r, h_{cs}, and g_i are all greatly effective on improve the efficiency. Analyzing the motor efficiency ranges shown in Fig. 3.9c, all values under each structure parameter are low enough. As for the maximum efficiency ranges, the value is only 8×10^{-5}, which is caused by the structure parameter β_r. As a result, the effects of structure parameters on motor efficiency can be ignored because of the low range values.

In the same way, combining range analysis of OED with Eq. (3.4), the contribution rates of each structure parameter on torque density ranges are also computed and compared as shown in Fig. 3.9d. Obviously, the air gap has highest influences on the performance of torque density with the value more than 0.5 N m/kg, while the second air gap has least influence because of the value of around 0.02 N m/kg. As for the β_s, the torque density range is relatively high with the value of 0.19 N m/kg. For the rest parameters, the ranges are all around 0.1 N m/kg, which is high enough when focusing on the torque density value, although they are greatly smaller than that obtained from the air gap. Hence, in order to improve the torque density of the SRM, all the structure parameters except g_i should be optimized according to their contribution rates.

As mentioned above, Fig. 3.9 reflects the effects of each structure parameter on various objectives. But the contribution rate of each structure parameter should also be studied, because it is more obvious to obtain the relative effects for the four objectives. According to results from OED and Eq. (3.12), the sensitivity ratios of the four objectives and comprehensive function established in Eq. (3.40) to each structure parameter are also calculated and shown in Fig. 3.10, where (a) and (b) represent sensitivity ratio of the four objectives and comprehensive function, respectively.

As shown in Fig. 3.10a, almost all the four objectives can be highly influenced by the structure parameters of β_s, β_r, and g, because of the high sensitivity ratio. However, all the structure parameters have significant effects on one or more objectives, although β_s, g, h_{cs}, and D_i have slight effects on the efficiency. Therefore, considering the in-wheel SRM, all the structure parameters of the motor should be optimized to improve the comprehensive performance of the SRM.

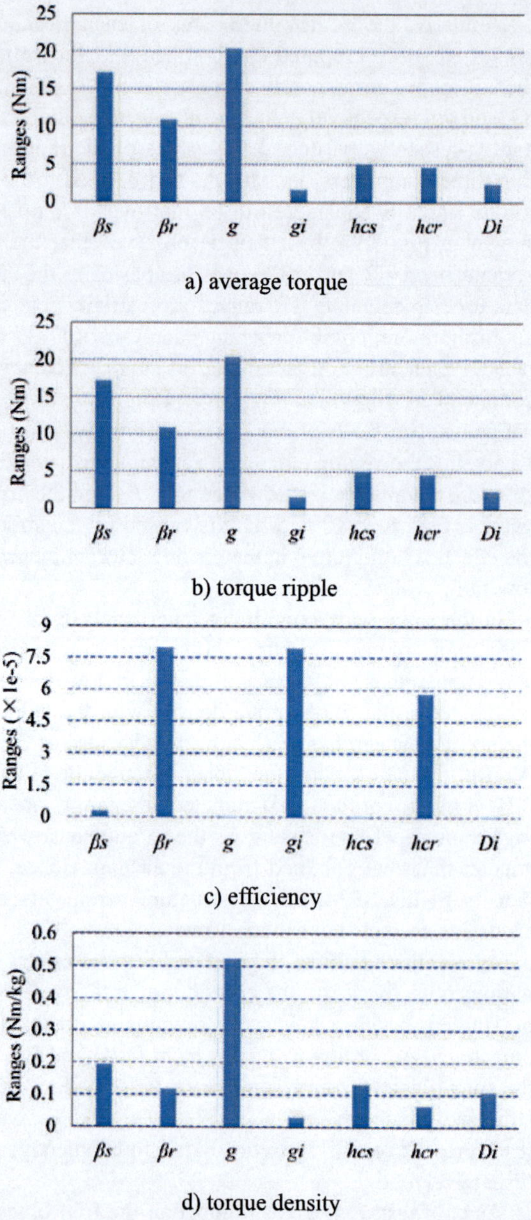

Fig. 3.9 Effects of each parameter on objectives

Similarly, the sensitivity ratios of the comprehensive function to structure parameters are calculated and shown in Fig. 3.10b, where four combinations of the weigh factors are described as C1, C2, C3, and C4. It should be noted that C1, C2, and C3 are employed for single objective optimization, while C4 is for multi

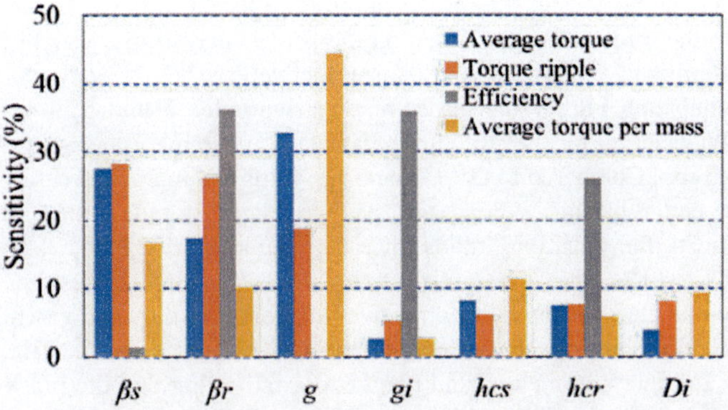

a) Sensitivity ratio of the objectives

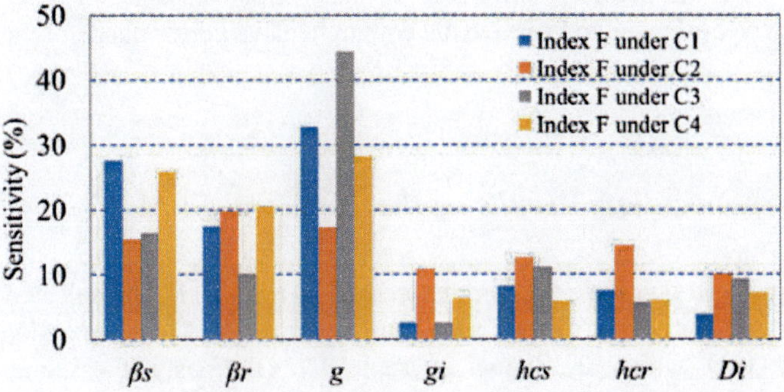

b) Sensitivity ratio of the comprehensive function

Fig. 3.10 Sensitivity ratios with various structure parameters

objective optimization. According to the analysis results about the efficiency, the objective for efficiency should be ignored such that the weigh factor of ω_3 is always defined to 0 in all the four cases. Comprehensively considering the importance of the average torque, torque ripple, and torque density, the weigh factors in case C4 are selected as $\omega_1 = 0.4$, $\omega_2 = 0.3$, and $\omega_4 = 0.3$. Considering the SRM in this section is for EV applications, it is desired to have excellent performance on average torque, torque ripple, and torque density, such that they all should be improved synchronously during the optimization process. Generally, the average torque is one of the most significant characteristics because of the drive power requirements of EV. Therefore, it is more important than the torque ripple and torque density. Differing from the selection for the SRM with two-phase excitation, the torque ripple performance of the in-wheel SRM is much important for the suspension and in-wheel motor system, because the in-wheel motor is installed under the vehicle sprung. Therefore, in this section, as an objective for improving ride comfort performance, the torque ripple is expected to have the same significance with the torque density. Based on the above analysis, the weight factors for C4 are determined as 0.4, 0.3, and 0.3, respectively.

After analyzing Fig. 3.10b, the comprehensive index F under all cases can be highly influenced by all the structure parameters in summary, because of the high sensitivity ratio. Compared to C1, C2, and C3, C4 is a compromise case, which can achieve a better balance among the average torque, torque ripple, and torque density. In C4, the sensitivity ratios for each parameter are 25.8%, 20.6%, 28.2%, 6.3%, 5.8%, 6.1%, and 7.1%, respectively. According to the results, the more effective parameters for the multi-objective function are β_s, β_r, and g, while the rest parameters have less effects because of the relatively low values. But, the total values for the four parameters could be 25.4%, indicating that the index F can be highly impacted by the four parameters. As a result, for the 16/20 in-wheel SRM, the structure parameters, i.e., β_s, β_r, g, g_i, h_{cs}, h_{cr}, *and* D_i, should be paid more attention when the SRM is optimized. In this section, these seven parameters are selected to be optimized to improve the comprehensive performance of the in-wheel SRM.

3.5.4 Optimization Results of the In-Wheel SRM

According to the objective function expressed in Eq. (3.11), the selected structure parameters are optimized under various cases by means of joint operation between FEA and OED, which are shown in Table 3.7. Obviously, the initial structure parameters and optimized ones under various optimization cases are completely different from each other. It can also be proved that the sensitivity analysis mentioned above is effective.

Based on the optimized structure parameters shown in Table 3.7, the objectives on average torque, torque ripple, torque density, and comprehensive function are calculated by FEA under various cases and shown in Table 3.8. Distinctly, the results from the initial case are worse than those from other optimized cases, which could directly prove the effectiveness of the optimization. Compared to the initial case, the results of the average torque T_{ave}, torque density Q, and comprehensive function $F(\vec{x})$ under C4 increase by 18.95%, 7.73%, and, 39.89%, respectively.

Table 3.7 Initial and optimal structure parameters of in-wheel SRM

Cases	β_r (deg.)	β_s (deg.)	h_{cs} (mm)	h_{cr} (mm)	D_i (mm)	g_i (mm)	g (mm)
Initial	8.5	5.5	29	18	145	17	0.35
C1	7.5	9	17	18	135	19.8	0.3
C2	7.5	9	35	20	160	22.5	0.7
C3	7	6.5	17	18	160	20.5	0.3
C4	7.5	9	35	20	160	22.5	0.3

Table 3.8 Results of objectives of in-wheel SRM under various cases

Cases	T_{ave} (N m)	T_r' ($\times 10^{-3}$)	Q (N m/kg)	F (C4)
Initial	55.3	5	1.3	0.59
C1	66.9	17	1.4	0.69
C2	41.3	1560	0.9	0.84
C3	63.7	7	1.7	0.72
C4	65.7	1160	1.4	0.99

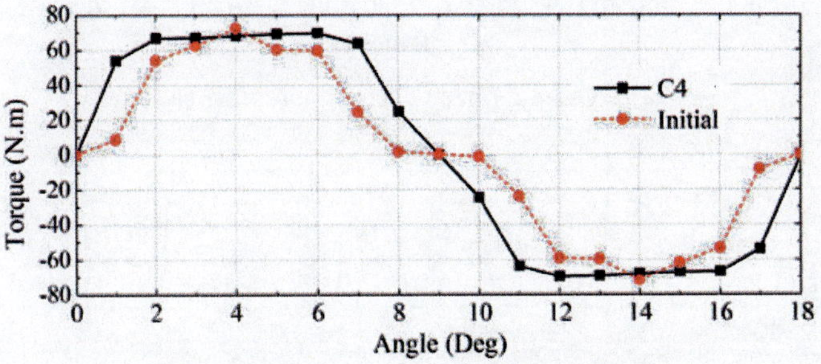

a) Comparisons of static torque under 42 A

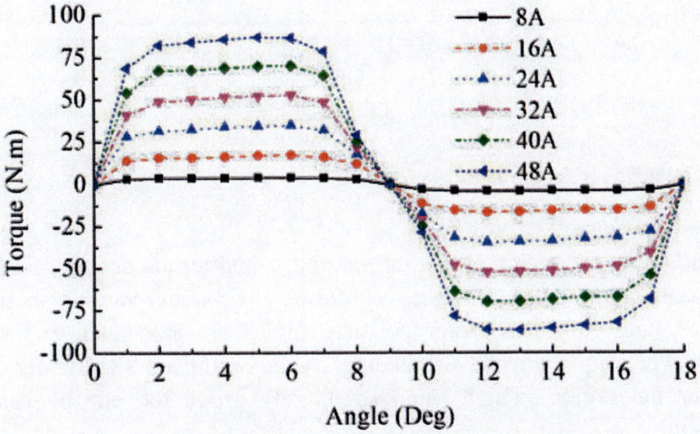

b) Static torque of the optimized SRM under C4

Fig. 3.11 Results of the static torque of the in-wheel SRM

As shown in Fig. 3.11a, the static torque waveforms of the SRMs with initial structure parameters and optimized structure parameters under case C4 are illustrated and compared, which are all with respect to rotor position angle from 0 to 18 degrees. Obviously, the waveforms of the torque change to be flatter in the middle area, compared to the waveforms of the initial design. After further analysis, it can be observed that this variation of the static torque can directly improve the

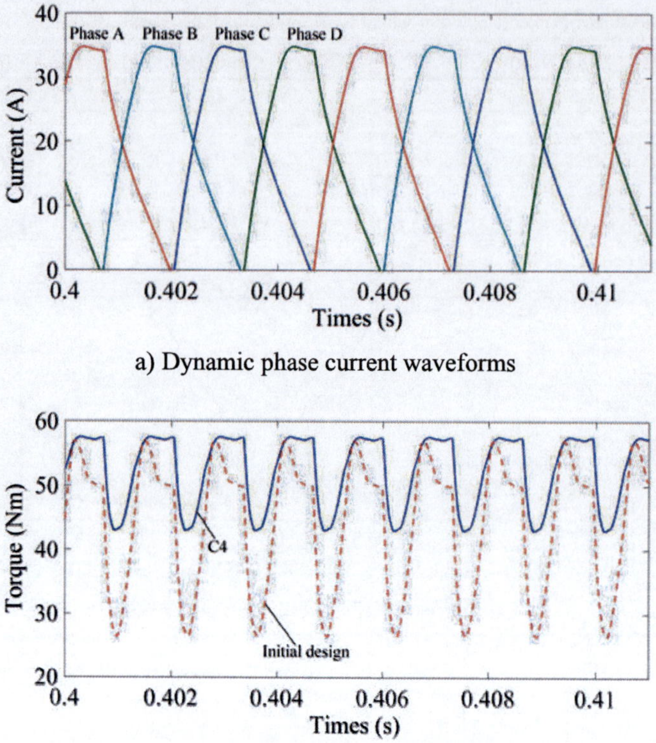

a) Dynamic phase current waveforms

b) Dynamic torque waveforms

Fig. 3.12 Dynamic results of the in-wheel SRM

performance on the average torque, torque ripple, and torque density. Furthermore, in order to obtain the detailed torque waveforms of C4 under various phase current, the torques with respect to rotor position angles are calculated and shown in Fig. 3.11b. The torque waveforms obtained by the optimized SRM under C4 are all with larger flat region, which can also directly prove the effectiveness of the optimization.

In order to show the dynamic performance of the optimized in-wheel SRM, the dynamic simulation model for the SRM drive system is established. Fixing the turn-on and turn-off angle to 1.5 and 6 degrees, the phase currents of each phase and total torque waveforms are simulated and obtained as shown in Fig. 3.12, where (a) is phase current responses, and (b) is total torque characteristics. The following can be observed from Fig. 3.12 that the average torque of the optimized in-wheel SRM under C4 improve by 18.8%, compared to the SRM with initial structure. Meanwhile, the torque ripple of the optimized in-wheel SRM under C4 reduce by 45.8%, compared to the SRM with initial structure.

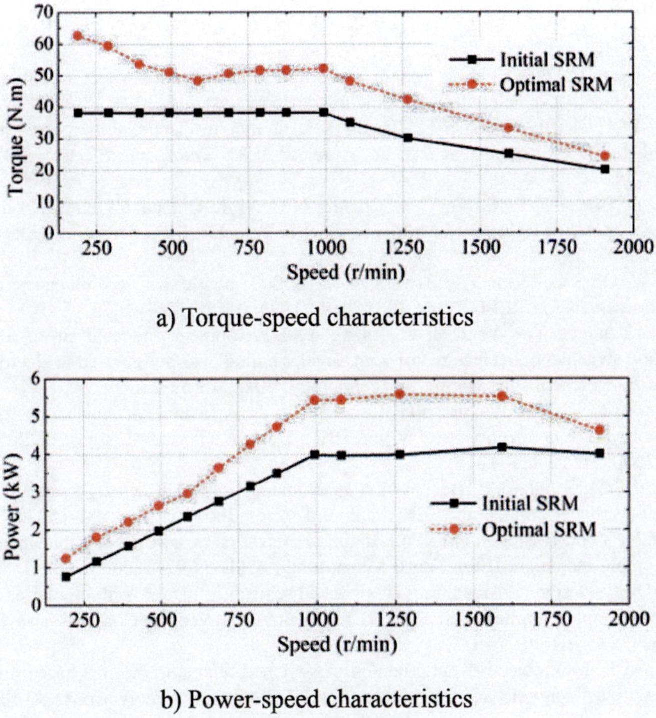

Fig. 3.13 External characteristics of the in-wheel SRMs

Furthermore, the initial and optimized SRMs and their drive systems are established according to the electromagnetic characteristics obtained above to achieve the external characteristics of the SRMs. As shown in Fig. 3.13, the torque-speed and power-speed characteristics are obtained, where the initial SRM is operated under classic condition with a constant torque of 38.2 N m below 1000 r/min and a constant power of 4 kW over 1000 r/min. In addition, the same phase current and switch angles of the initial SRM are employed to the optimized one. Then, the torque-power-speed curves of the optimized in-wheel SRM are generated as shown in Fig. 3.13. Obviously, with the same phase current and switch angles, the optimized motor has better performance on output torque and power than the initial motor, which means the optimized SRM could effectively improve the dynamic performance of EV.

References

1. Zhu Y, Yang C, Yue Y, et al. Design and optimisation of an In-wheel switched reluctance motor for electric vehicles. IET Intelligent Transport Systems, 2019, 13(1): 175–182.
2. Boldea I, Tutelea L N, Parsa L, Dorrell D. Automotive electric propulsion systems with reduced or No-Permanent magnets: an overview. IEEE Trans. Ind. Electron, 2014, 61(10): 5696–5711.
3. Xue X D, Cheng K W E, Ng T W, Cheung N C. Multi-Objective Optimization Design of In-Wheel Switched Reluctance Motors in Electric Vehicles. IEEE Trans. Ind. Electron, 2010, 57(9): 2980–2987.
4. Ahn J W, Oh S G, Moon J W, Hwang Y M. A three-phase switched reluctance motor with two-phase excitation. IEEE Trans. Ind. Appl., 1999, 35(5): 1067–1075.
5. Zhu Y, Yang C, Yue Y, Zhao C, Zhang Y. Development and analysis of a two-phase excitation switched reluctance motor with novel winding distribution used in electric vehicles. Journal of Electrical Engineering & Technology, 2018, 13(6): 2364–2375.
6. Rahman K M, Fahimi B, Suresh G, Rajarathnam A V, Ehsani M. Advantages of switched reluctance motor applications to EV and HEV: Design and control issues. IEEE Trans. Ind. Appl, 2000, 36(1): 111–121.
7. Rahman K M, Schulz S E. Design of high efficiency and high density switched reluctance motor for vehicle propulsion. IEEE Trans. Ind. Appl, 2002, 38(6): 1500–1507.
8. Choi Y K, Yoon H S, Koh C S. Pole-shape optimization of a switched-reluctance motor for torque ripple reduction. IEEE Trans. Magn, 2007, 43(4): 1797–1800.
9. Ma C, Qu L. Design considerations of switched reluctance motors with bipolar excitation for low torque ripple applications. Proc. IEEE Energy Conversion Congress and Exposition, Denver, USA, 2013: 926–933.
10. Ma C, Qu L. Multiobjective optimization of switched reluctance motors based on design of experiment and particle swarm optimization. IEEE Trans. Energy Convers, 2015, 30(3): 1144–1153.
11. Xue X D K, Cheng W E, Lin J K, Zhang Z, Luk K F, Ng T W, Cheung N C. Optimal control method of motoring operation for SRM drives in electric vehicles. IEEE Trans. Veh. Technol, 2010, 59(3): 1191–1204.
12. Shaked N T, Rabinovici R. New procedures for minimizing the torque ripple in switched reluctance motors by optimizing the phase-current profile. IEEE Trans. Magn, 2005, 41(3): 1184–1192.
13. Xue X D K, Cheng W E, Ho S L. Optimization and evaluation of torque-sharing functions for torque ripple minimization in switched reluctance motor drives. IEEE Trans. Power Electron, 2009, 24(9): 2076–2090.
14. Krishnan R. Switched reluctance motor drive: modeling, simulation, analysis, design, and applications. CRC Press, New York, 2001.

Chapter 4
Optimization and Control of SRM Drive System for EV Applications

4.1 Introduction

As mentioned in the previous chapters, switched reluctance motor is becoming one of the first choices for electric vehicles because of its advantages of simple structure, low cost and flexible control, compared to other motors [1–3]. However, the torque ripple of the switched reluctance motor is much harmful to the motor and even the vehicle drive system [4]. In order to obtain the good traction performance of electric vehicle, which depends on the performance of SRM, it is very important to study the dynamic characteristics of motor to reduce the torque ripple of SRM and improve the average torque of SRM.

Because of the special structure of the stator rotor and the saturation nonlinear characteristic of the magnetic circuit, it is difficult to establish the driving system model of switched reluctance motor accurately [5, 6]. On the other hand, the establishment of dynamic model of SRM will directly affect the optimization design of the motor, dynamic performance analysis and evaluation, efficiency evaluation, as well as the development of high-performance controller [7]. Therefore, the establishment of accurate and reliable nonlinear dynamic model for SRM has very important theoretical and engineering significance.

The torque dynamic characteristics of SRM for electric vehicles play an important role in improving the dynamic performance and ride comfort of the whole electric vehicle. However, the excessive torque ripple of SRM is very harmful to the motor itself and the transmission mechanism of the EVs [8–10]. Therefore, in the driving system of EVs, it is of great significance to reduce the torque pulsation of SRM, increase its average torque, and improve the working efficiency of the motor, which is of great significance for obtaining good traction characteristics of EVs. In order to improve the comprehensive driving performance of electric vehicles driven by switched reluctance motors, the goals mentioned above are considered to be optimized in this chapter.

Y. Zhu, *The Key Technologies for Powertrain System of Intelligent Vehicles Based on Switched Reluctance Motors*, Recent Advancements in Connected Autonomous Vehicle Technologies 1, https://doi.org/10.1007/978-981-16-4851-9_4

Based on the new winding excitation mode of the SRM and its flux linkage characteristics, inductance characteristics, as well as torque characteristics, which are described in Chap. 2, the dynamic characteristics of the new SRM drive system are explored. Aiming at the performance requirements of high torque, low ripple, and high efficiency of SRM for EVs, a control strategy that can realize the synchronization optimization of torque, torque ripple and efficiency is proposed in this chapter. In general, it is of great significance to improve the comprehensive performance of EVs driven by SRM under driving or braking conditions.

4.2 Basic Equation of the SRM

The electromagnetic principle of SRM can be clearly expressed based on basic electromagnetic equations, although the nonlinear electromagnetic characteristics are serious for the SRM due to the special structure. Therefore, the basic electromagnetic principle can be illustrated by voltage equation, flux linkage equation, mechanical dynamic equation, and electromagnetic torque equation, which can be found in Chap. 1. In order to clearly illustrate the relationship among the basic electromagnetic equations, a schematic diagram for the electromagnetic coupling of the SRM are shown in Fig. 4.1, when the iron loss of the motor is not considered.

According to the illustration in Fig. 4.1 and basic electric circuit principle, the voltage balance equation can be calculated as Eq. (4.1) by taking the kth phase as an example, where u_k, i_k, R_k, and Ψ_k refer to terminal voltage, phase current, phase resistance, and flux linkage of the kth phase.

$$u_k = R_k i_k + \frac{d\psi_k}{dt} \tag{4.1}$$

In general, the self-inductance of SRM is greatly higher than the mutual inductance, although the flux linkage of SRM is a nonlinear function on phase current, rotor position angle, self-inductance and mutual inductance. Therefore, the

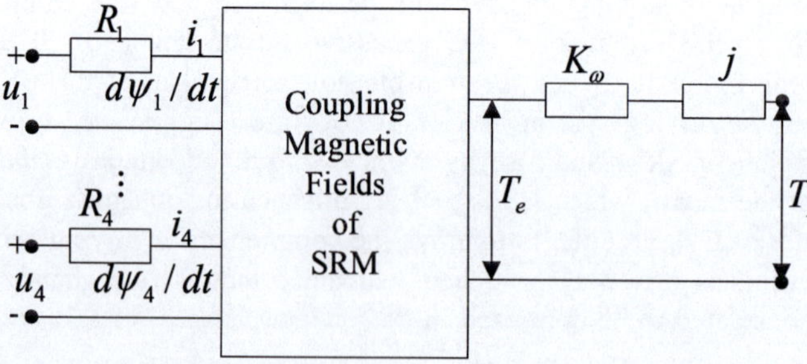

Fig. 4.1 Schematic diagram for the electromagnetic coupling of the SRM

mutual inductance of SRM normally is neglected when the flux linkage is expressed by means of inductance. As a result, the flux linkage can be expressed as Eq. (4.2), where L_k and θ_k refer to the self-inductance of the kth phase and rotor position angle of the motor.

$$\psi_k = L_k(\theta_k, i_k)i_k \tag{4.2}$$

According to Fig. 4.1 and basic mechanics principle, the electromagnetic torque T_e can be computed by means of moment of inertia J, friction coefficient K_Ω, motor speed Ω, and load torque T_L. The expression of the electromagnetic torque is shown in Eq. (4.3), which can be also computed according to Eq. (4.4) by means of co-energy W'_m.

$$T_e = J\frac{d^2\theta}{dt^2} + K_\omega\frac{d\theta}{dt} + T_L \tag{4.3}$$

$$T_e(i, \theta) = \left.\frac{\partial W'(i, \theta)}{\partial \theta}\right|_{i=const} \tag{4.4}$$

4.3 Dynamic Model of SRM Drive System

Because the flux linkage characteristics and torque characteristics of SRM are nonlinear functions of rotor position angle and phase current, it is difficult to establish an accurate SRM model. At present, scholars all over the world have conducted a large number of studies on the modeling methods of switched reluctance motor based on the flux and torque characteristics. These methods can be summarized as the followings.

(1) Analytical method: The method of linearizing or piecewise linearizing switched reluctance motor model is adopted to establish the flux and torque model [11], but this model ignores the nonlinearity of magnetic circuit and the edge effect of magnetic field, which leads to large errors. Using the magnetization curves of a switched reluctance motor at several special positions, an analytical model of SRM is established by modularizing current or flux. The accuracy of these models is improved, but they are far from meeting the high-performance control requirements of the SRM drive system.

(2) Neural network method: Neural network has the ability to learn approximating nonlinear mapping, such that the application of neural network to SRM model is of great help to solve the nonlinear problem of flux linkage characteristics. Therefore, many works have studied the neural network method to construct the nonlinear model of SRM.

(3) The look-up table method: The magnetic flux linkage characteristics and torque characteristics under various rotor position angles and phase current conditions

are calculated by experimental measurement or by using finite element analysis method [12, 13]. The data are stored in the form of table to establish the driving system model, and then the numerical values of magnetic flux linkage and torque are solved in real time by the look-up table method. This method is relatively convenient and direct, and quite accurate.

4.3.1 Block Diagram of the SRM Drive System

Compared to ordinary motors, the SRM drive system can achieve different kinds of control for SRM by adjusting the power converter according to the signals from controller. Figure 4.2 is the block diagram of the control system designed on the basis of the new winding excitation mode proposed, in which the reference speed is the set speed, and the motor speed, angle and current parameters are the feedback signals from the motor or power converter. The controller realizes the operation of the SRM drive system by means of speed regulation and current regulation, which are also shown in Fig. 4.2.

4.3.2 Nonlinear Dynamic Model of the SRM

The speed regulator is the basis for dynamic model analysis of SRM. It adopts proportional integral control (PI) in this chapter, whose input is the difference between the reference speed and the actual speed of SRM, and output is the given value of the phase current of SRM. Its control law can be expressed as Eq. (4.5).

$$I(t) = k_p\left(\Delta w + \frac{1}{T}\int_0^t \Delta w dt\right) \tag{4.5}$$

where $I(t)$ is the phase current output, K_p represents proportional coefficient, Δw means speed error, and T refers to integral constant.

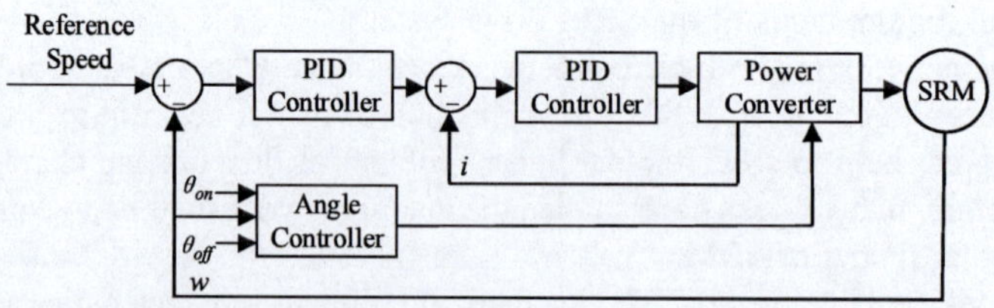

Fig. 4.2 Schematic diagram of the SRM drive system

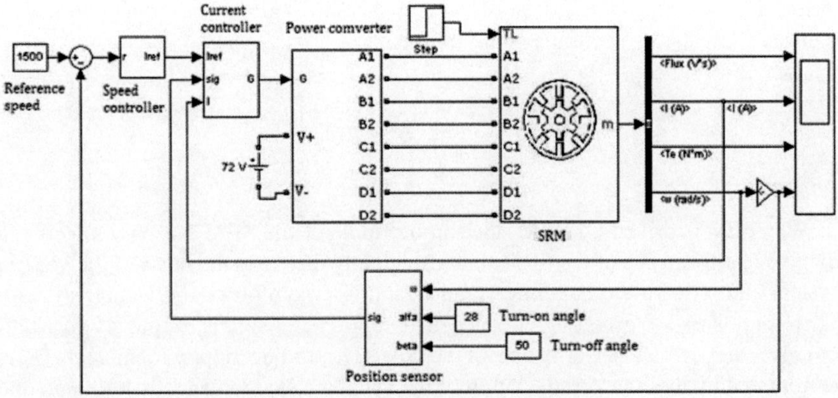

Fig. 4.3 Dynamic model of the SRM drive in MATLAB/Simulink

By means of the flux linkage characteristic and torque characteristic data obtained by numerical calculation for the SRM with the new two-phase excitation mode described in Chap. 2, the two-phase coupling data of the SRM could be considered as on phase to established the SRM model.

It should be noted that the flux linkage model and torque model of the SRM drive system with the new winding excitation mode are established by means of look-up table method. On this basis, the dynamic simulation model of SRM drive system is established in MATLAB/Simulink environment. As shown in Fig. 4.3, it includes five parts, i.e., motor power converter model, position and angle detection model, current control model, speed control model, and motor body model.

4.4 Dynamic Characteristics of the SRM Drive System

The average output torque and torque ripple are the main factors that influence the dynamic performance of SRM. In order to improve the dynamic and steady state characteristics of SRM, these two indexes must be studied in depth in order to design high performance motor controller. On the other hand, the vehicle driving distance is an important factor that restricts the development of EVs. As a drive system used for EVs, the optimization control of motor efficiency is also of great theoretical and engineering significance. Therefore, the average output torque, torque ripple, and driving distance are important indexes for dynamic performance evaluation of SRM for EVs.

The average output torque is very important for the dynamic characteristics of motor. For the four-phase winding SRM, the synthetic average electromagnetic torque can be expressed by the average value within a period. Meanwhile, the average value can be computed by integral form. As a result, the average output torque T_a is calculated as Eq. (4.6), where, θ is the angle of rotor position, and T_t represents synthetic electromagnetic torque.

$$T_a = \frac{3}{\pi}\int_0^{\pi/3} T_t(i,\theta)d\theta \tag{4.6}$$

When EV is driven by SRM, the torque ripple of the SRM not only affects the dynamic performance of the SRM, but also greatly increases the noise of the vehicle system and reduces the handling stability and ride comfort of the vehicle system. Therefore, it is of great significance to suppress the torque ripple of SRM to improve the overall performance of vehicle. The torque ripple coefficient TR is represented by the standard deviation of the synthetic electromagnetic torque in one cycle. Using the mathematical expression, the torque ripple coefficient TR can be expressed as Eq. (4.7), where, N is the number of samples within a period.

$$TR = \left(\sum (T_t - T_a)^2/(N-1)\right)^{1/2} \tag{4.7}$$

In order to improve the dynamic characteristics of SRM, the concept of torque smoothness coefficient of SRM is put forward based on the introduction of torque ripple coefficient TR expressed in Eq. (4.7), which is defined as the reciprocal of torque ripple. As a result, the torque smoothness coefficient TS is computed as Eq. (4.8).

$$TS = \sqrt{\frac{1}{\sum (T_t - T_a)^2/(N-1)}} \tag{4.8}$$

In the drive system of EVs, the output dynamic characteristics of motor have to meet the requirements of the vehicle, including vehicle acceleration time, maximum speed, maximum grade ability, as well as other dynamic performance indexes. At the same time, it is necessary to improve the drive efficiency and increase the driving range of EVs. These indexes are closely related to the working characteristics of the motor. The equivalent power coefficient can not only reflect the real-time power output characteristics of the motor, but also describe the comprehensive performance of the motor drive system, which can be specifically expressed as Eq. (4.9).

$$PF = \frac{T_a\omega}{UI_{rms}} \tag{4.9}$$

where U is power input voltage, ω refers to output speed of SRM, I_{rms} denotes root mean square (RMS) bus current of motor.

For the SRM used for EVs, the higher average torque, torque smoothness coefficient, and equivalent power coefficient can represent the higher the output torque, the lower torque pulsation, and the higher efficiency of SRM. Therefore, in the optimization process of the control parameters of SRM, it is expected that the values of these three indexes can be as large as possible.

4.4.1 Effects of the Load Torque

In literature [14] and [15], they not only studied the influence of motor main parameters on the dynamic characteristics of the motor, but also studied and analyzed the dynamic change relationship between various currents and dynamic characteristics of the torque. However, the change of current cannot reflect the dynamic characteristics of the motor in real time, especially when the motor is running under variable working conditions. Because, the change of vehicle load torque could directly cause the current to change sharply in a short time during these conditions, and the special working environment of EVs requires that the driving switched reluctance motor needs to adapt the various working conditions of the vehicle in real time. Therefore, it is obviously insufficient for using current to evaluate the dynamic performance index of the SRM driving EVs. On the other hand, if the current is used as a reference factor to optimize the turn-on angle and turn-off angle of SRM, the variation range of the turn-on and turn-off angle can obviously become more frequent such that the calculation amount of the motor controller will be increased under the special working conditions of the real-time variable operation of EVs. Meanwhile, it could not be conducive to the reduction of the motor pulsation.

Considering the special operating environment of EVs, the change of load torque can not only reflect the dynamic load of the motor, but also reflect the steady state current under different conditions in real time. Because, a certain load torque normally corresponds to a certain steady state current when the SRM works under the condition of the target speed. Therefore, the influence of load torque on the dynamic characteristics of the drive system have to be studied clearly.

Figure 4.4 shows the numerical curves of average torque versus load torque at different speeds when the SRMs' turn-on and turn-off angles are −6 and 24 degrees, respectively. It can be seen that the average torque value obtained under a specific load torque value condition is basically the same as the load torque value, and the speed has little effect on it. In other words, the value of the average torque mainly depends on the size of the load torque, such that it is not necessary to consider the average torque when optimizing the motor control parameters based on the load torque and speed. It should be noted that the RMS current values under various working conditions are selected to reflect the average torque characteristics under specific load torque in this chapter, in order to investigate the average torque characteristics of the SRM under different load torque and speed conditions.

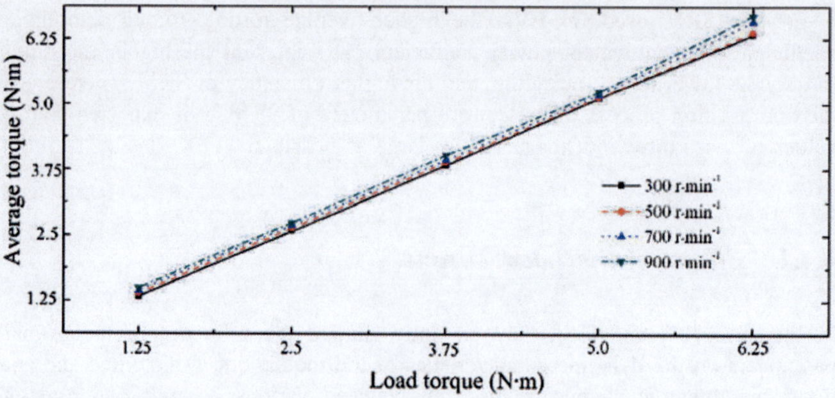

Fig. 4.4 The average torque versus load torque

Therefore, it is expected that the smaller the RMS current can be, the better the average torque is.

Figure 4.5 shows the variation curves of the three target parameters: RMS current value, torque comfort coefficient, and equivalent power coefficient with load torque at different speeds when the turn-on angle and turn-off angle of the SRM are fixed at −6° and 24° respectively, as shown in Fig. 4.5a–c. As can be seen from Fig. 4.5, the dynamic performance of the SRM varying with load torque at various speeds are observed as follows.

(1) The RMS current value increases gradually with the increase of load torque, and the RMS values are also different under various load torque conditions.
(2) The torque smoothness coefficient decreases with the increase of load torque. The variation is relatively smooth at high speed and large at low speed.
(3) In general, the equivalent power coefficient value increases with the increase of load torque. Under the same load torque condition, the higher the rotational speed is, the larger the coefficient value will be.

Given that the motor speed is 500 r/min and the turn-on angle is −4 degrees, the dynamic performance of the motor changing with load torque under different turn-off angles is studied and illustrated as shown in Fig. 4.6, in which Fig. 4.6 a–c are the curves of RMS current, torque smoothness coefficient and equivalent power coefficient, respectively. From the varying curves shown in Fig. 4.6, the followings can be observed.

(1) With the increase of load torque, the RMS current value has a trend of gradual increase. When the turn-off angle is 28 degrees, the RMS current value changes greatly. Under the same load torque condition, the RMS current increases with the increase of the turn-off angle within the given turn-off angle range.

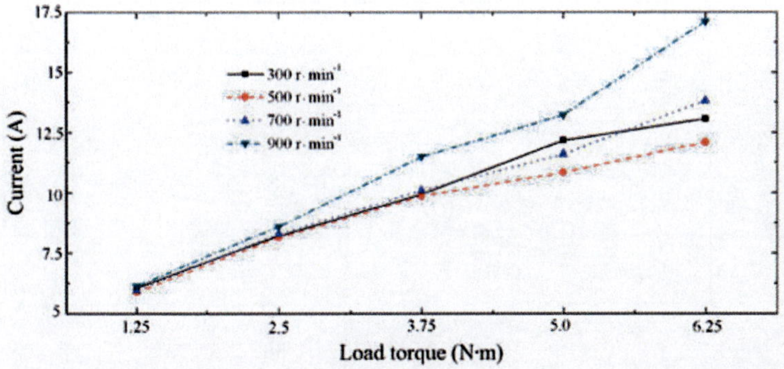

a) RMS current

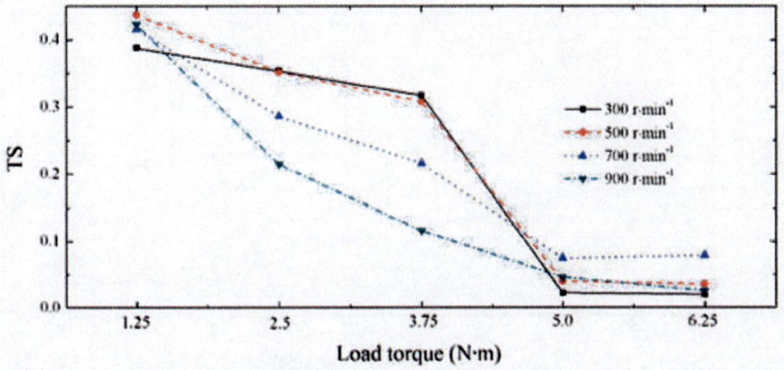

b) Torque smoothness coefficient

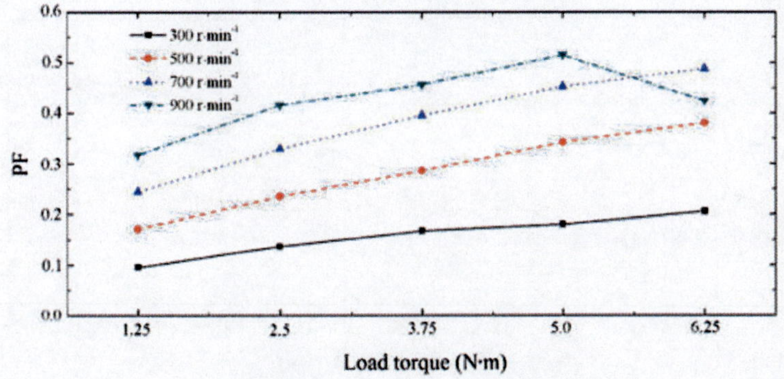

c) Equivalent power coefficient

Fig. 4.5 The effects of the load torque at various speeds

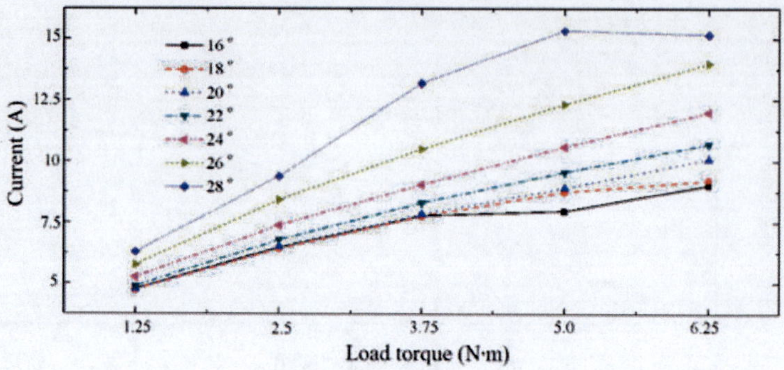

a) RMS current

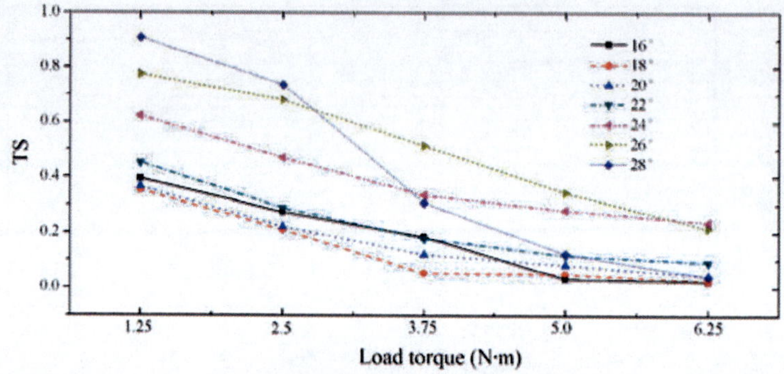

b) Torque smoothness coefficient

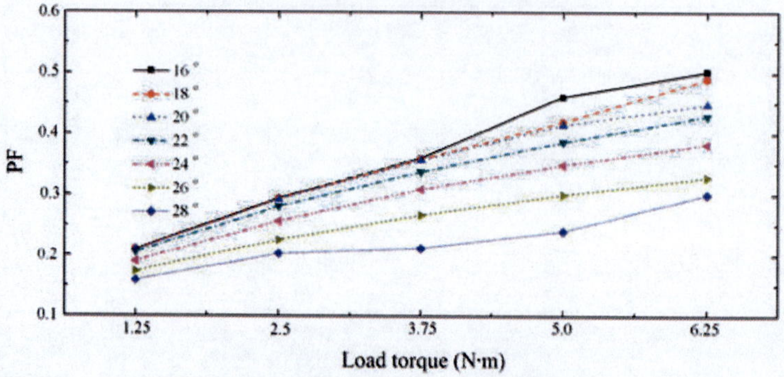

c) Equivalent power coefficient

Fig. 4.6 The effects of the load torque with various turn-off angles

(2) The torque smoothness coefficient decreases gradually with the increase of load torque. Although the degree of decrease varies with different turn-off angles, the trend is the same.
(3) The equivalent power coefficient value increases with the increase of the load torque. Moreover, under the same load torque and within the given range of turn-off angle, the equivalent power coefficient value decreases as the turn-off angle increases.

According to the methods described above, the SRM speed was also set at 500 r/min and the turn-off angle was set at 24 degrees. Then, the variation of main dynamic performance of the motor with load torque under different turn-on angles was studied. As can be seen in Fig. 4.7, Fig. 4.7a–c respectively show the change curves of RMS current, torque smoothness coefficient, and equivalent power coefficient. According to the variation trend of the three performance indicators in Fig. 4.7 with the load torque, the influence law of the load torque on the three performance characteristics under different turn-on angles is analyzed as follows.

(1) The RMS current gradually increases with the increase of load torque, which basically shows a linear relationship. Within the given turn-on angle range, the RMS current decreases as the turn-on angle increases.
(2) The change trend of the torque smoothness coefficient is basically similar to that under different rotation speeds and different turn-off angles. With the increase of the load torque, the torque smoothness coefficient gradually decreases. When the turn-on angle is small enough and the turn-on angle is large enough, the range of change is relatively large, but the overall change trend is basically the same as that under other turn-on angle conditions.
(3) Similar to the change trend of RMS current, the value of equivalent power coefficient gradually increases with the increase of load torque, which also basically shows a linear relationship. Within the given turn-on angle range, the equivalent power coefficient decreases as the turn-on angle increases.

Through the analysis for the influence law of load torque on torque performance indexes under the conditions of different rotation speeds, different turn-off angles, and different turn-on angles, it can be seen that different load torques have obvious effects on the dynamic performance of the SRM. As a result, it is very necessary to select different load torque to optimize the dynamic performance indexes of the SRM, respectively, which is of great significance to improve the dynamic characteristics of the motor, reduce energy consumption, and improve the driving range of EVs.

4.4.2 *Effects of the Turn-On Angle*

The dynamic model of the SRM is established and simulated, and the variation law between turn-on angle and RMS current, torque smoothness coefficient, and

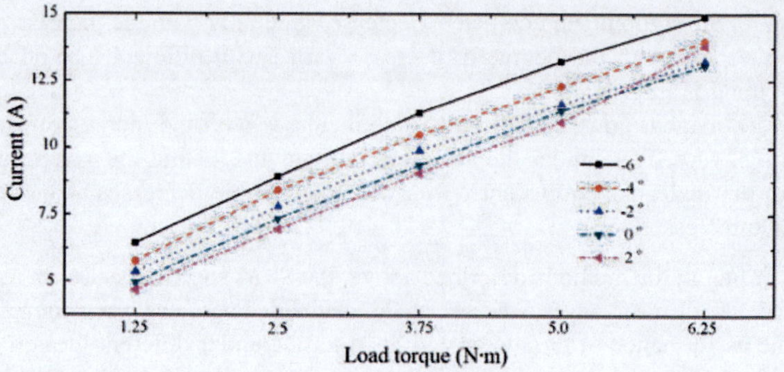

a) RMS current

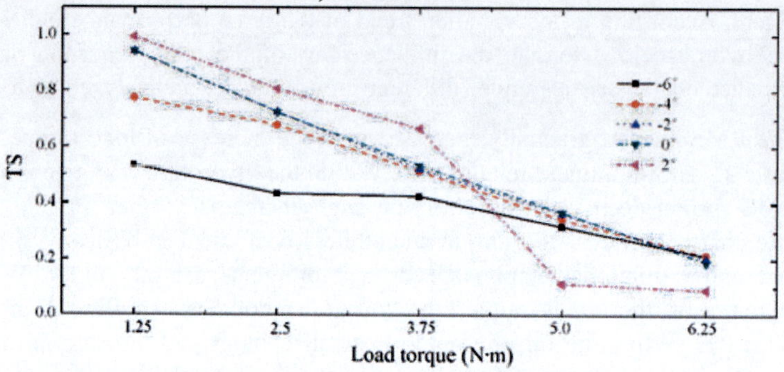

b) Torque smoothness coefficient

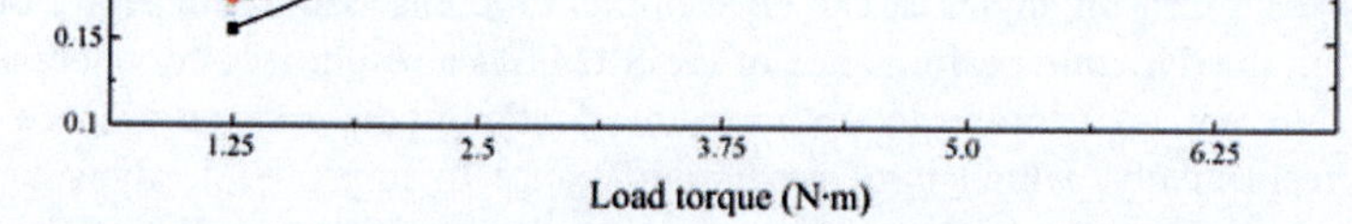

c) Equivalent power coefficient

Fig. 4.7 The effects of the load torque with various turn-on angles

equivalent power coefficient are studied under different speed conditions. Figure 4.8 is the change curves of the motor torque performance with the change of the turn-on angle under different speed conditions, where the values of turn-off angle and load torque are 26 degrees and 1.25 N m, respectively. Figure 4.8a–c

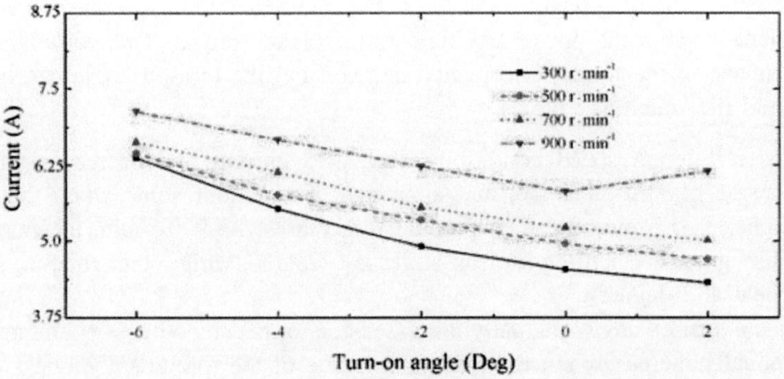

a) RMS current

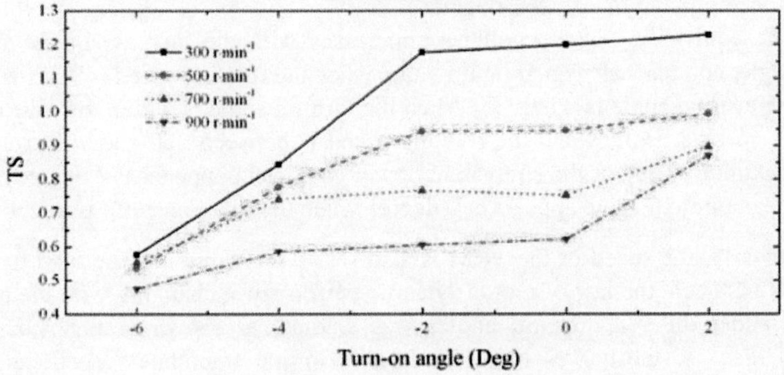

b) Torque smoothness coefficient

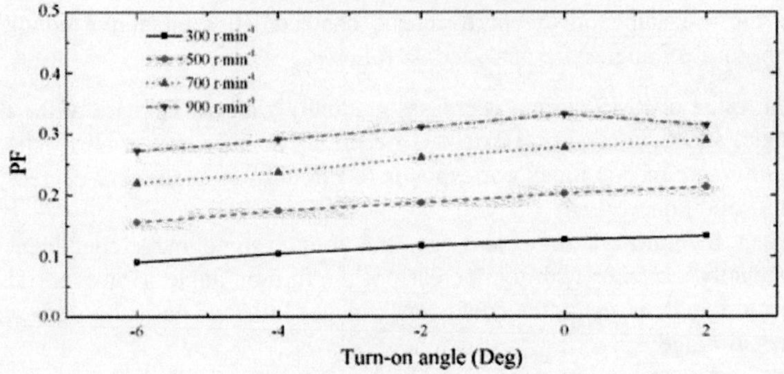

c) Equivalent power coefficient

Fig. 4.8 The effects of the turn-on angle at various speeds respectively show the relation curves between RMS current, torque smoothness coefficient, equivalent power coefficient, and load torque. The variation trend between one of the three performance indexes and the turn-on angle can be analyzed and illustrated as follows.

(1) Under different speed conditions, the RMS current first decreases with the increase of turn-on angle, and appears the minimum value when the speed reaches 900 r/min, and then gradually increases. At 900 r/min, the minimum value appears when the turn-on angle is 0 degree, while other rotating speeds appear at 2 degrees.
(2) As the turn-on angle gradually increases, the torque smoothness coefficient also gradually increases, but the maximum value of the torque smoothness coefficient at different rotational speeds appears at 2 degrees when the turn-on angle is in the range of −6 to 2 degrees.
(3) The equivalent power coefficient increases with the increase of the turn-on angle, and reaches its maximum value when the rotating speed is 900 r/min and the turn-on angle is 0 degree. When the turn-on angle is within the given range of −6 to 2 degrees and the rotating speed is between 300 and 700 r/min, the maximum value of the equivalent power coefficient appears at 2 degrees, while the maximum value appears at 0 degree when the rotating speed is at 900 r/min.

Similarly, the speed of the SRM is also set at 500 r/min and the load torque is 1.25 N m. Then, the law of motor dynamic performance changing with the turn-on angle under different turn-off angles was studied, as shown in Fig. 4.9, where Fig. 4.9a–c are the curves of RMS current, torque smoothness coefficient, and equivalent power coefficient, respectively. According to the variation trend of the three performance indicators with the turn-on angle in Fig. 4.9, the influence trends of the turn-on angle on the performance characteristics under the condition of different turn-off angles are analyzed as follows.

(1) The value of RMS current decreases gradually with the increase of the turn-on angle, and the decreasing extent is becoming smaller and smaller. The maximum value of this index corresponds to the turn-on angle of 2 degrees under this condition.
(2) When the turn-off angle is large, the torque smoothness coefficient value gradually increases with the increase of the turn-on angle. However, when the turn-off angle is small, the coefficient changes little and basically stays within a certain range.
(3) The equivalent power coefficient value increases with the increase of the turn-on angle, and the maximum value of the coefficient appears at 2 degrees at different turn-off angles.

In order to study the law of motor performance parameters changing with the turn-on angle under different load torque conditions, the speed of the SRM is set at 500 r/min and the turn-off angle is 26 degrees, as shown in Fig. 4.10. According to

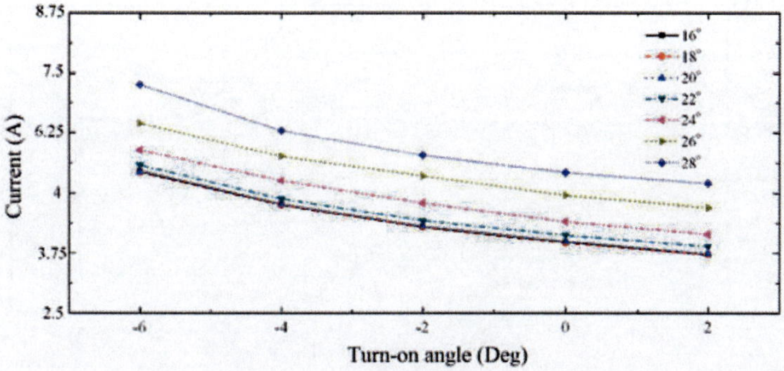

a) RMS current

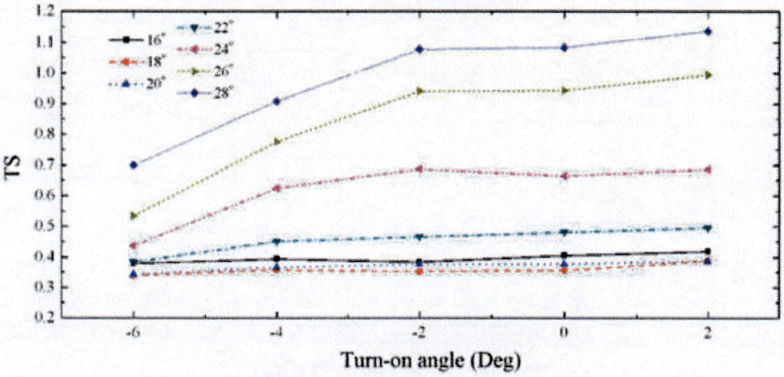

b) Torque smoothness coefficient

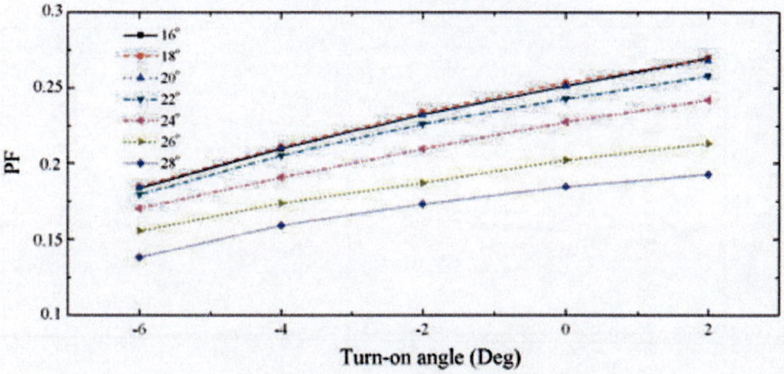

c) Equivalent power coefficient

Fig. 4.9 The effects of the turn-on angle with various turn-off angles

the trend curve of the three performance indexes as the turn-on angle changes in Fig. 4.10, the influence of the turn-on angle on the performance characteristics under different load torque is analyzed as follows.

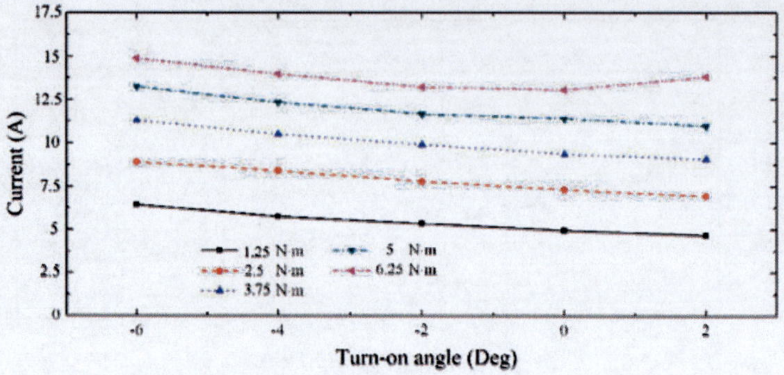

a) RMS current

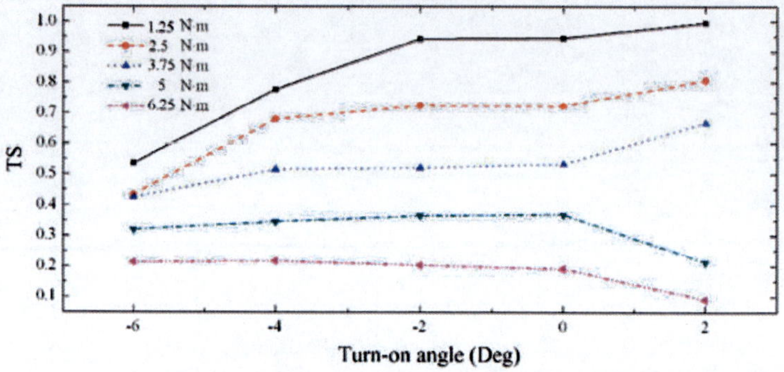

b) Torque smoothness coefficient

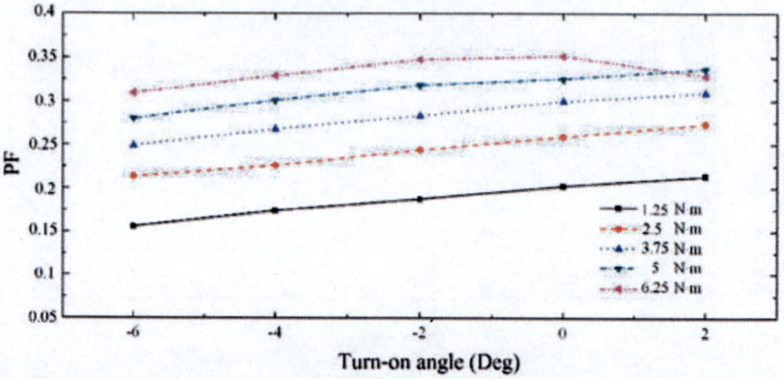

c) Equivalent power coefficient

Fig. 4.10 The effects of the turn-on angle with various load torque

(1) The RMS current value decreases gradually with the increase of the turn-on angle. When the load torque is 6.25 N m, the value of this coefficient tends to decrease first and then increase. The minimum value of the coefficient corresponding to various turn-on angle is also different under different load torque conditions.
(2) The variation rule of the torque smoothness coefficient with the turn-on angle is as follows: the value of the coefficient changes greatly when the load torque is small, while the value of the coefficient changes little and it basically maintains within a certain range when the load torque is large. In addition, the value of the coefficient decreases with the increase of the motor load torque.
(3) The value of the equivalent power coefficient increases with the increase of the turn-on angle, and there is a certain relationship between the value change and the load torque, that is, when the load torque is 6.25 N m, its maximum value appears at 0 degree, and when the load torque is other values, its maximum value appears at 2 degrees.

Through the analysis of the influence law of turn-on angles on torque performance indexes under different rotation speeds, different turn-off angles, and different load torque conditions, it can be seen that different turn-on angles have obvious influence on the dynamic performance of the SRM. In addition, the turn-on angle corresponding to the optimal value of each motor performance index obtained under different speeds and loads is also different. Therefore, it is very necessary to select the appropriate turn-on angle to optimize the dynamic performance parameters of the SRM, which has very important theoretical and engineering significance to improve the dynamic characteristics of the motor, reduce energy consumption, and improve the driving range of EVs.

4.4.3 Effects of the Turn-Off Angle

Just like the turn-on angle, the choice of the turn-off angle has a great influence on the dynamic characteristics of the SRM drive system. In order to study the influence of the turn-off angle on the torque characteristics, the strategy similar to the influence analysis method of the turn-off angle was adopted to study the influence of the turn-off angle on the three performance indexes of the SRM under three conditions of different speeds, different turn-on angles, and different load torques.

The established dynamic model of the SRM is given specific constraints, and then the driving system model is simulated to obtain the calculation data of the selected performance indexes. The obtained data are calculated and processed, and the variation rule is shown in the form of curves. On this basis, the variation rule between the turn-off angle and RMS current, torque smoothness coefficient, and equivalent power coefficient is studied and analyzed under different constraint conditions.

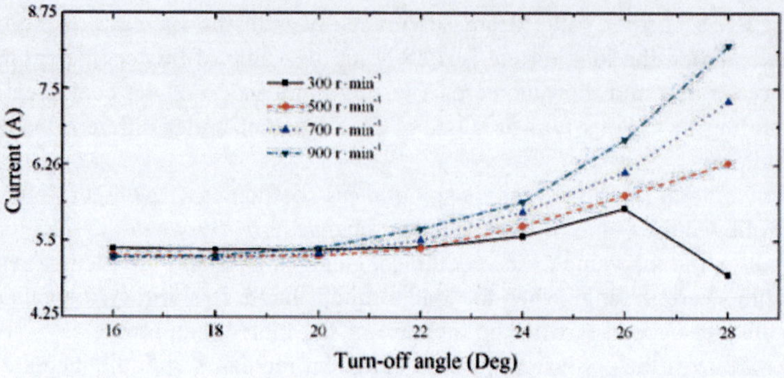

a) RMS current

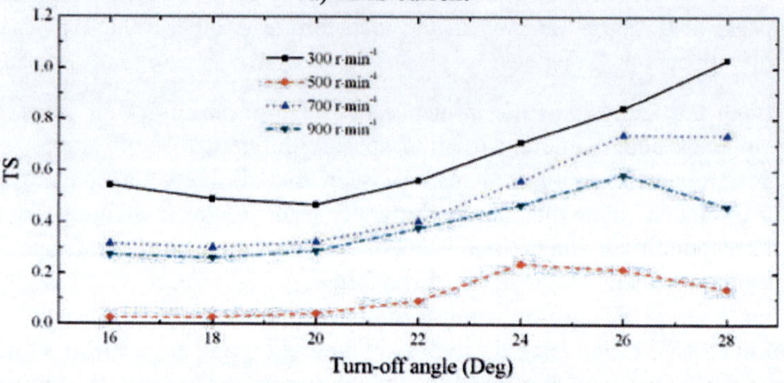

b) Torque smoothness coefficient

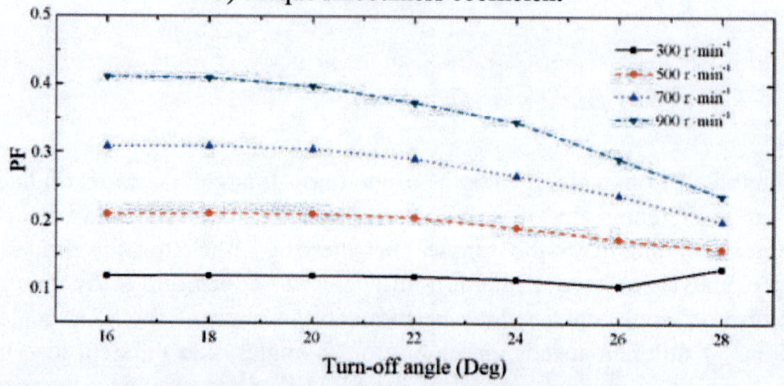

c) Equivalent power coefficient

Fig. 4.11 The effects of the turn-off angle at various speeds

Figure 4.11 shows the variation curves of motor torque performance indexes with turn-off angle at different speeds, in which the values of turn-on angle and load torque are set as –4 degrees and 1.25 N m, respectively. In the same process as the previous analysis, Fig. 4.11a–c respectively show the relationship curves between RMS current, torque smoothness coefficient, and equivalent power coefficient and load torque. The curve variation trend can be used to analyze the variation rule between the three performance indexes and the turn-off angle, which are illustrated as follows.

(1) Within the range of turn-off angle from 16 to 20 degrees, the RMS current value is basically constant at different speeds. When the value is greater than 20 degrees, the value increases with the increase of the turn-off angle, and the change range is larger and larger. Only when the rotation speed is 300 r/min, the RMS current tends to decrease again as long as the turn-on angle is above 26 degrees. Based on the different speeds, the minimum value of the RMS current corresponding to the turn-off angle is not the same.
(2) With the gradual increase of the turn-off angle, the variation trend of the torque smoothness coefficient at different rotating speeds is basically the same, which increases first and then decreases. In addition, based on the different speeds, the corresponding maximum value of this index appears at 24 degrees or 26 degrees.
(3) Except for the condition that the rotation speed is 300 r/min, it can be seen that the equivalent power coefficient has a similar rule with the change of the turn-off angle. When the rotating speed is 300 r/min, the coefficient is basically unchanged when the turn-on angle is below 26 degrees, and starts to increase above 26 degrees.

Similarly, given a speed of 500 r/min and a load torque of 1.25 N m, on the basis of which, the variation rules of the three performance indexes of the motor, RMS current, torque smoothness coefficient, and equivalent power coefficient, with the turn-off angle are studied under the conditions of different turn-on angles, as shown in Fig. 4.12. According to the change curves of the three indexes shown in Fig. 4.12 with the turn-off angle, the influence rule of the turn-off angle on the performance characteristics under different turn-on angle conditions is analyzed as follows.

(1) With the change of the turn-off angle, the change trend of RMS current values is basically the same at different turn-on angles, and the turn-off angle corresponding to the minimum value of this index is basically at 16 degrees.
(2) Under different turn-on angles, the value of the torque smoothness coefficient gradually increases with the increase of the turn-off angle. Since the numerical curve of this index increases monotonically, the turn-off angle corresponding to the maximum value of this index at each turn-on angle is 28 degrees.
(3) The equivalent power coefficient value decreases with the increase of the turn-off angle, and the variation trend of this index is the same under different turn-on angles. In addition, the value of turn-off angle corresponding to the maximum value of this coefficient roughly appears at 16 and 18 degrees with the increase of the turn-on angle.

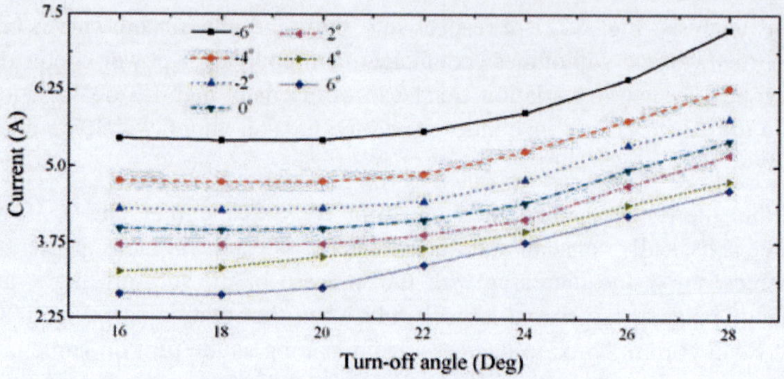

a) RMS current

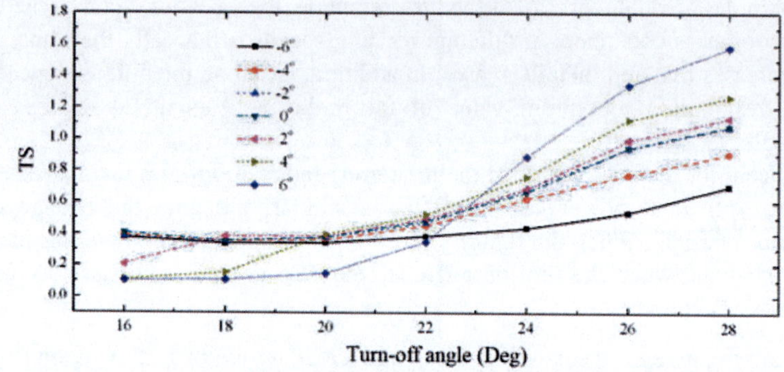

b) Torque smoothness coefficient

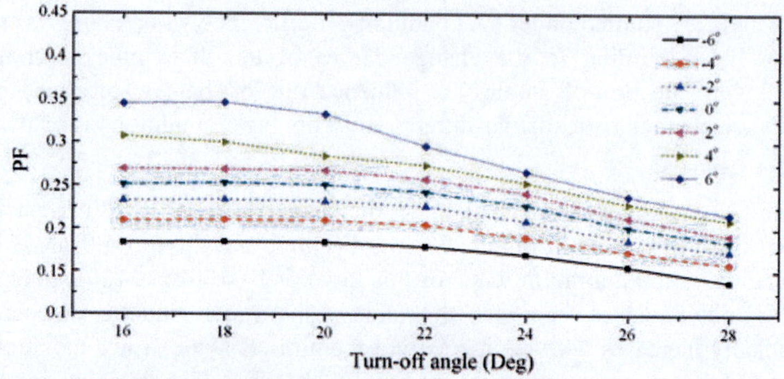

c) Equivalent power coefficient

Fig. 4.12 The effects of the turn-off angle with various turn-on angles

Figure 4.13 shows the curves of motor torque performance indexes changing with the turn-off angle under different load torque conditions, in which the values of turn-on angle and motor speed are set as −4 degrees and 500 r/min, respectively. As

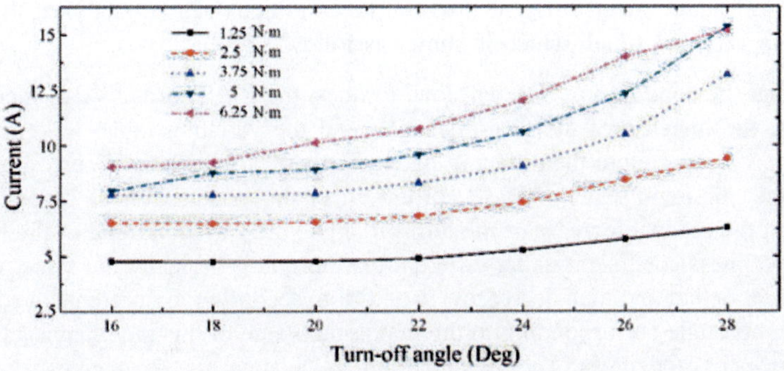

a) RMS current

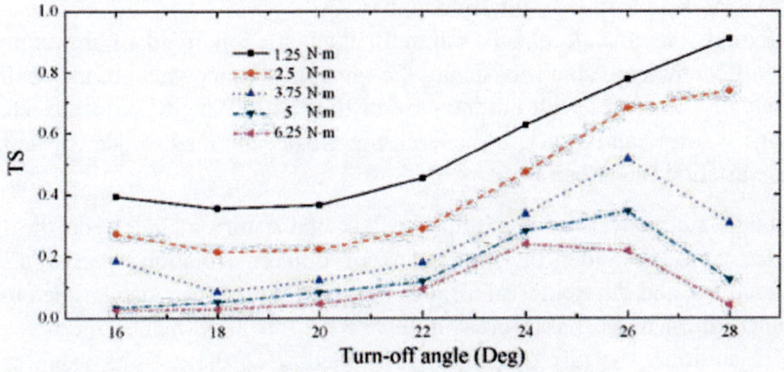

b) Torque smoothness coefficient

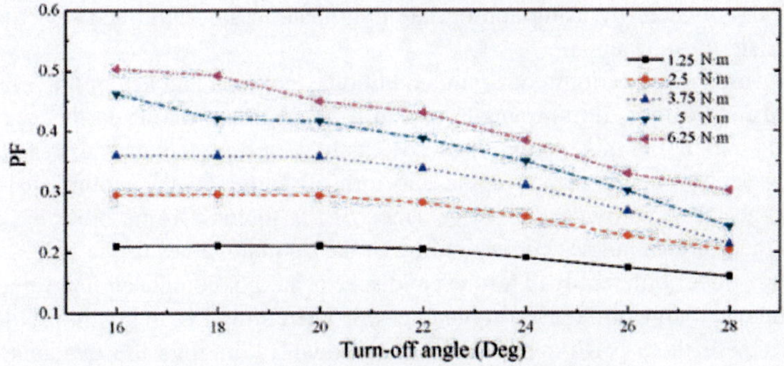

c) Equivalent power coefficient

Fig. 4.13 The effects of the turn-off angle with various load torque

in the previous analysis process, Fig. 4.13a–c respectively show the curve relation diagram of RMS current, torque smoothness coefficient, and equivalent power coefficient with load torque. The variation rule of the three performance indexes and the turn-off angle under different load torque conditions can be analyzed through the curve variation trend, which is shown is follows.

(1) Under the condition of different load torques, the RMS current value increases with the increase of the turn-off angle, and the variation range is larger and larger. According to the different load torques, the turn-off angle corresponding to the minimum value of RMS current under this condition is at 16 degrees.
(2) With the gradual increase of the turn-off angle, the variation trend of the torque smoothness coefficient under different load torque is basically the same, which increases first and then decreases. In addition, according to different speeds, the turn-off angle corresponding to the maximum value of this index appears at 28 degrees or 26 degrees. And the maximum value appears at 26 degrees when the load torque is large enough, while the maximum value appears at 28 degrees when the load torque is small enough.
(3) Although the load torque is different, the variation trend of the equivalent power coefficient value is basically the same, that is, the value of the coefficient gradually decreases with the increase of the turn-off angle, and the decreasing range is larger and larger. In general, the greater the load torque is, the higher the variation range can be.

Through the analysis of the influence law of the turn-off angle on the torque performance index under the conditions of different rotation speeds, different turn-on angles, and different load torques in this section, it can also be seen that the different turn-off angle has a great influence on the performance indexes of the SRM. In addition, the turn-off angle corresponding to the optimal value of each motor performance index obtained under different speeds and load torques is also different. Therefore, it is very necessary to optimize the dynamic performance parameters of SRM by using appropriate optimization algorithm and selecting the appropriate turn-off angle.

In a word, under various constraint conditions, the trend and law of the influence of SRM load torque, turn-on angle, and turn-off angle on motor performance are analyzed in detail in this section. It can be easily seen from the analysis results that proper selection of the urn-on angle and turn-off angles of the motor is of great significance for improving the performance of the motor. On the other hand, the turn-on and turn-off angles corresponding to the optimal values of the performance indicators under different load torques and speeds are not completely the same, and there are even large differences in many cases. Therefore, it is an urgent problem to balance these three performance indexes reasonably, improve the dynamic characteristics of the motor under various working conditions, and optimize the comprehensive performance of the motor. It has very important theoretical and engineering significance to improve the average torque of the motor, reduce the torque ripple, and improve the efficiency of the motor. At the same time, it will be helpful to improve the dynamic performance and economic performance of EVs.

4.5 Parameters Optimization of the SRM Drive System

According to the analysis in Sect. 4.4, it can be seen that load torque, turn-on angle, and turn-off angle have great influences on RMS current, torque smoothness coefficient, and equivalent power coefficient of SRM. Therefore, in order to improve the performance of motor drive system, an optimal design method is proposed, that is, the turn-on and turn-off angles that satisfy the objective function under different load torque and speed conditions are selected. In practical application, the optimized turn-on angle and turn-off angle can be obtained through the combination calculation of load torque and speed, so as to achieve the purpose of optimizing the whole system.

In order to optimize the performance of SRM drive system, the methods of single-index optimization and multi-index synchronous optimization are proposed in the chapter. On the basis of establishing their respective objective functions, the motor's turn-on and turn-off angles are optimized.

4.5.1 Single Objective Function for SRM Dynamic Performance

In order to improve the dynamic characteristics of SRM drive system, maximizing the average torque, minimizing the torque ripple, and maximizing the motor efficiency are the fundamental objectives of motor performance optimization. According to the analysis in Sect. 4.4, the parameters reflecting these three indicators are RMS current, torque smoothness coefficient, and equivalent power coefficient, respectively. Therefore, to improve the dynamic performance, minimizing RMS current, maximizing torque smoothness coefficient, and maximizing equivalent power coefficient can be considered as the optimization objectives.

According to the proposed optimization purpose, the optimization objective functions of RMS current, torque smoothness coefficient, and equivalent power coefficient are respectively expressed as Eqs. (4.10), (4.11), and (4.12).

$$K_I(\theta_{on}, \theta_{off}) = \max\{1/I_{rms}\} \quad (4.10)$$

where K_I is objective function of RMS current optimization, θ_{on} refers to motor turn-on angle, and θ_{off} denotes turn-off angle.

$$K_{TS}(\theta_{on}, \theta_{off}) = \max\{TS\} \quad (4.11)$$

where K_{TS} is the objective function of optimization of torque smoothness coefficient.

$$K_{PF}(\theta_{on}, \theta_{off}) = \max\{PF\} \tag{4.12}$$

where K_{PF} is optimization objective function of equivalent power coefficient.

The establishment of the objective function is the basis of motor performance optimization. Through the calculation and analysis of the established objective function, the optimal turn-on angle and turn-off angle under each working condition can be found, so as to improve the performance of parameter indexes reflected by the objective function.

The objective functions based on the established Eqs. (4.10–4.12) can be described from the three aspects of average torque, torque ripple, and motor efficiency, respectively. Therefore, the performance characteristics of SRM are optimized based on these three objective functions.

4.5.2 Average Torque Optimization and Analysis

The average torque is basically constant when the load torque is fixed, but there is numerical difference in the RMS current. It can be regarded that he smaller the RMS current under the same average torque value is, the average torque could be. Therefore, the RMS current value can be used to reflect the performance index of the average torque.

Using the established dynamic model of the SRM drive system, performance simulation tests under different combinations between turn-on and turn-off angles are carried out within the range of the given rotation speed varying from 300 to 1500 r/min, and the load torque varying from 1.25 to 6.25 N m. The obtained RMS current data are calculated under different constraints. Table 4.1 shows the RMS current values obtained by model simulation and mathematical operation of the SRM drive system at different turn-on and turn-off angles when the motor speed is 1100 r/min and the load torque is 2.5 N m.

The RMS current values obtained under different speeds and load torques are listed in the form of table. Then, according to the objective function Formula (4.10),

Table 4.1 RMS current results when the speed and torque are 1100 r/min and 2.5 N m

RMS current (A)		Turn-on angle				
		−12 deg.	−10 deg.	−8 deg.	−6 deg.	−4 deg.
Turn-off angle	16 deg.	9.161835	8.362878	7.651295	7.116585	6.606943
	18 deg.	9.332355	8.609195	7.86777	7.25284	6.936203
	20 deg.	9.783765	8.89156	8.27942	7.58404	7.290745
	22 deg.	10.68796	9.593965	9.12538	8.269523	7.795995
	24 deg.	12.98036	11.241	10.06537	9.53307	9.717978
	26 deg.	12.73755	12.85982	12.57356	12.71103	12.01092

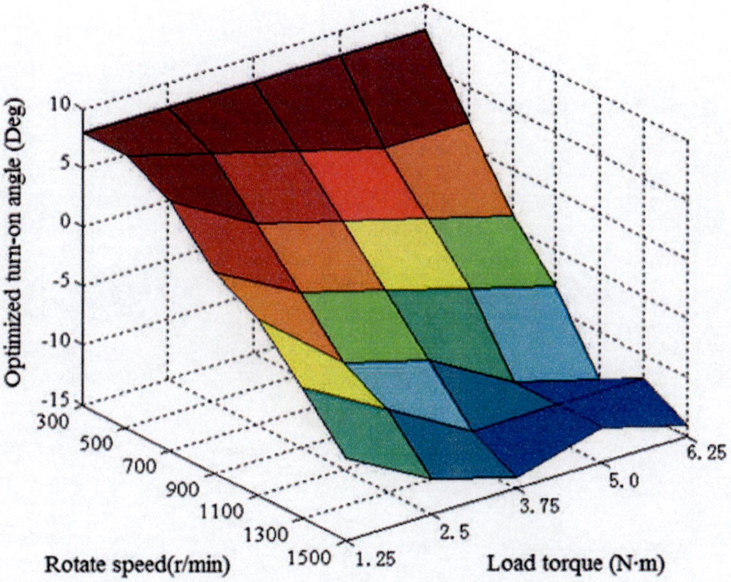

Fig. 4.14 Optimal turn-on angles based on maximization of average torque

the optimal turn-on angle and turn-off angle values under each constraint condition are obtained, so as to complete the optimization of the motor's average torque index.

In order to visually display the optimization results of the system, the optimized turn-on and turn-off angles obtained under various states are expressed in the form of three-dimensional surface based on the motor speed and load torque. Figure 4.14 shows the optimized turn-on angle changing curve with motor speed and load torque. Figure 4.15 shows the three-dimensional curved surface relation diagram of the optimized turn-off angle varying with motor speed and load torque.

It can be seen from Figs. 4.14 and 4.15 that the turn-on angle and the turn-off angle are not constant with the change of speed and load torque when the average torque is optimized, which is corresponding to the research results in Sect. 4.4. At the same time, the change trends of the switch angles are very different. The optimized turn-on angle decreases with the increase of the speed, and the decrease range is large. On the contrary, the optimal turn-off angle has a small variation range and basically changes within a small range.

4.5.3 Torque Ripple Optimization and Analysis

The optimization of torque ripple is also studied by using the dynamic model of the SRM drive system, which is the same as that of average torque optimization.

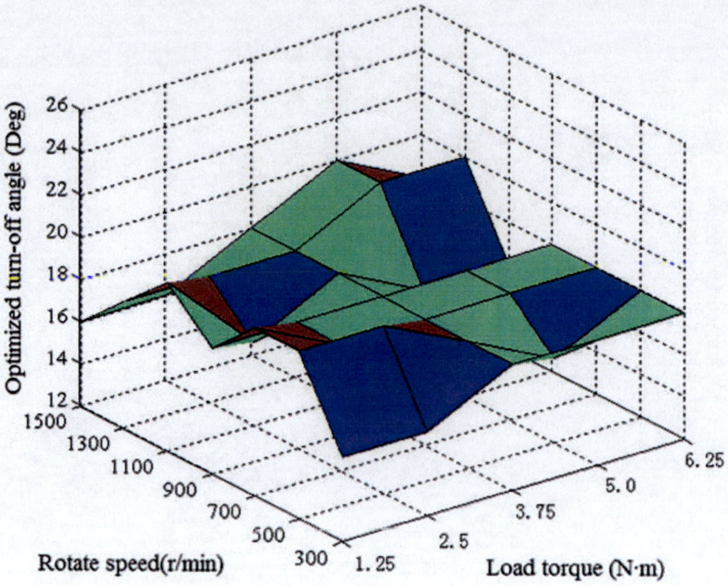

Fig. 4.15 Optimal turn-off angles based on maximization of average torque

In the range of the given rotation speed varying from 300 to 1500 r/min and load torque varying from 1.25 to 6.25 N m, performance simulation tests are carried out on the combinations between different turn-on angles and turn-off angles, and the real-time torque data are calculated mathematically. Therefore, the numerical values of the torque smoothness coefficient are obtained under different constraint conditions. As shown in Table 4.2, the torque smoothness coefficient values are obtained by system model simulation and mathematical operation under different turn-on and turn-off angles when the motor speed is 1100 r/min and the load torque is 2.5 N m.

The torque smoothing coefficient under different speeds and load torques is calculated, and the values are expressed in table form. Then, according to the

Table 4.2 *TS* when the speed and torque are 1100 r/min and 2.5 N m

Torque smoothness coefficient		Turn-on angle (deg.)				
		−12 deg.	−10 deg.	−8 deg.	−6 deg.	−4 deg.
Turn-off Angle	16 deg.	0.094313	0.119436	0.142663	0.149234	0.152173
	18 deg.	0.08579	0.105	0.124113	0.144694	0.164692
	20 deg.	0.084928	0.106085	0.127818	0.151871	0.176232
	22 deg.	0.094129	0.116813	0.139303	0.167099	0.192651
	24 deg.	0.067023	0.109522	0.140177	0.179015	0.218252
	26 deg.	0.068568	0.079232	0.111592	0.130658	0.199514

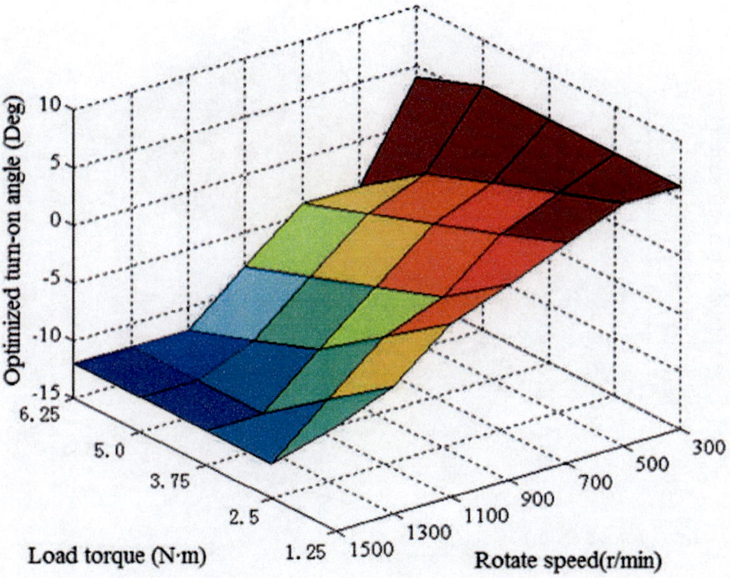

Fig. 4.16 Optimal turn-on angles with maximization of TS

objective function formula (4.12), the optimal turn-on angle and turn-off angle values under each constraint condition are obtained, so as to complete the optimization of the motor torque ripple index.

In order to show the optimization results of torque ripple indexes more intuitively, the optimized turn-on angle and optimized turn-off angle under various states are expressed based on the motor speed and load torque in the form of three-dimensional surface. Figures 4.16 and 4.17 are the three-dimensional curved surfaces of the optimized turn-on and turn-off angles with the motor speed and load torque.

From Figs. 4.16 to 4.17, it can be seen that the optimized turn-on angle and turn-off angle both show a trend of decreasing with the increase of motor speed. The optimized turn-on angle decreases from 6 to −12 degrees, while the optimized turn-off angle decreases from 28 degrees to the minimum value of 16 degrees.

4.5.4 SRM Efficiency Optimization and Analysis

Similar to the optimization process in Sect. 4.5.3, based on the dynamic model of switched reluctance motor drive system, the performance simulation tests are carried out in the given speed range of 300 to 1500 r/min and load torque range of 1.25 to 6.25 N m through the combinations between different turn-on angles and turn-off

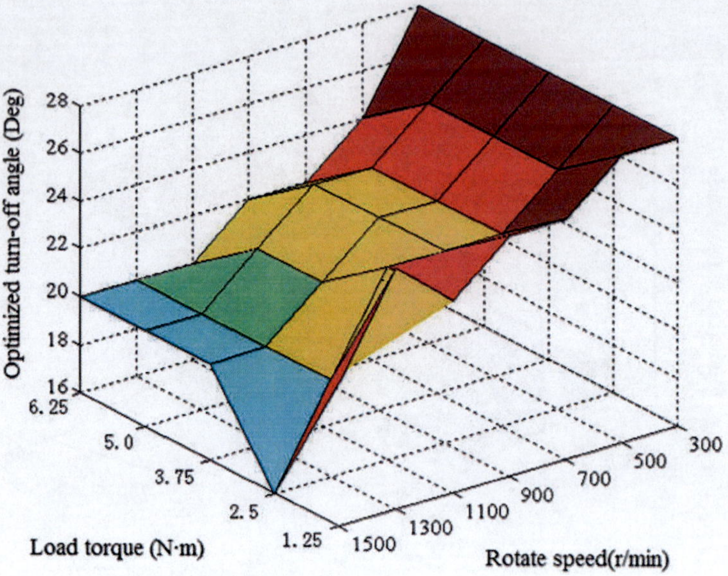

Fig. 4.17 Optimal turn-off angles with maximization of TS

Table 4.3 *PF* when the speed and torque are 1100 r/min and 2.5 N m

Equivalent power coefficient		Turn-on angle				
		−12 deg.	−10 deg.	−8 deg.	−6 deg.	−4 deg.
Turn-off angle	16 deg.	0.485383	0.528785	0.581067	0.625189	0.633456
	18 deg.	0.476317	0.516624	0.565312	0.613381	0.648572
	20 deg.	0.454562	0.500255	0.537373	0.586669	0.610239
	22 deg.	0.416111	0.463739	0.488096	0.535858	0.570908
	24 deg.	0.342369	0.395655	0.442047	0.466652	0.464377
	26 deg.	0.354902	0.363214	0.353717	0.3493	0.331208

angles. The obtained equivalent power data are mathematically calculated to obtain the equivalent power coefficient values under different constraint conditions. Table 4.3 shows the equivalent power coefficient values obtained by system model simulation and mathematical operation under different turn-on and turn-off angles when the motor speed is 1100 r/min and the load torque is 2.5 N m.

The equivalent power coefficient values obtained under different speeds and load torques are expressed in tabular form. Then, according to the objective function formula (4.13), the optimal turn-on angle and turn-off angle values are obtained under each constraint condition to complete the optimization of motor efficiency index.

In order to visually display the optimization results of the motor efficiency index, the optimized turn-on and turn-off angles obtained under various states are

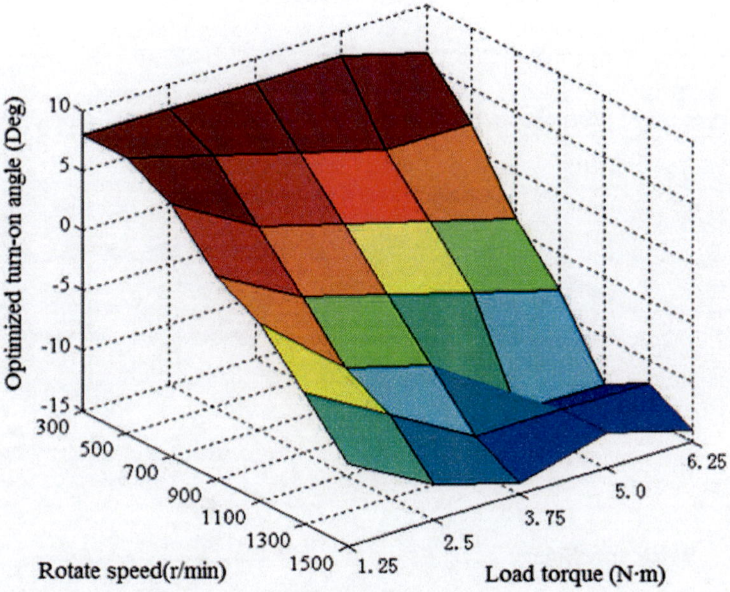

Fig. 4.18 Optimal turn-on angles with maximization of *PF*

expressed in the form of three-dimensional surface. Figures 4.18 and 4.19 respectively show the three-dimensional surface of the optimized turn-on angle and turn-off angle changing with the motor speed and load torque.

It can be seen from Figs. 4.18 and 4.19 that both the turn-on angle and the turn-off angle are not constant after the optimization of torque ripple. In addition, the variation trends of the optimized turn-on angle and turn-off angle are similar to those obtained under the condition of average torque optimization. The optimized turn-on angle presents a trend of gradual decrease with the increase of motor speed. Meanwhile, the reduction amplitude is also different under different load torque conditions. However, the optimal turn-off angle is always maintained within a certain range with the increase of motor speed, because the value of current is the most critical factor affecting the efficiency of the motor under the condition of constant speed and average torque.

4.5.5 Multi-objective Function of the SRM Dynamic Performance

It can be clearly seen from Sects. 4.5.2 to 4.5.4 that the turn-on and turn-off angles obtained by maximizing the average torque, minimizing torque ripple, and maximizing the equivalent power coefficient are different under each constraint

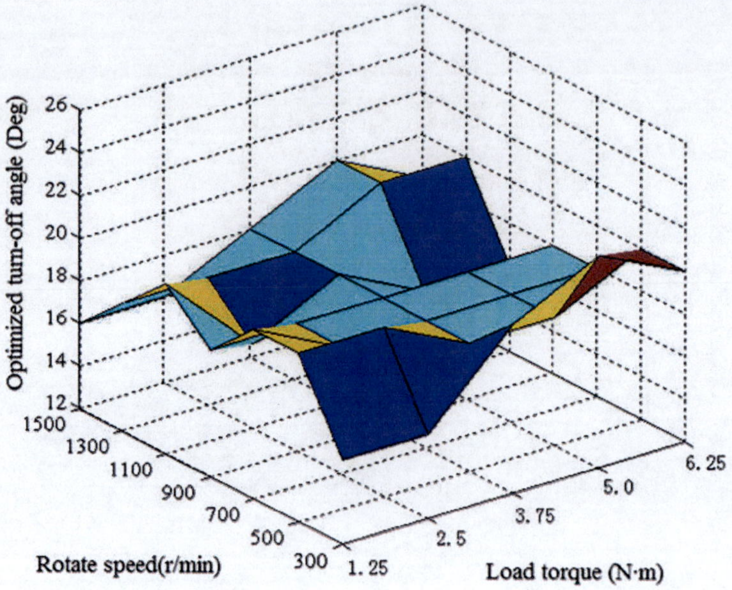

Fig. 4.19 Optimal turn-off angles with maximization of *PF*

condition, that is, the synchronous optimization of the three parameter indexes is impossible by using these three methods. Therefore, in order to improve the dynamic characteristics of SRM drive system to the maximum extent, a multi-index synchronous optimization objective function is proposed, which can improve the RMS current, torque smoothness coefficient, and equivalent power coefficient of the motor through the distribution of weight coefficient.

According to the characteristics of the three indexes, the comprehensive performance indicator of the motor is defined as follows:

$$K(\theta_{on}, \theta_{off}) = \frac{k_1 \cdot (1/I_{rms})}{(1/I_{rms})_{\max}} + \frac{k_2 \cdot TS}{(TS)_{\max}} + \frac{k_3 \cdot PF}{(PF)_{\max}} \tag{4.13}$$

where k_1 means RMS current weight coefficient, k_2 represents torque ripple weight coefficient, and k_3 refers to equivalent power weight coefficient.

Therefore, the synchronous optimization objective function for these three indicators is expressed as follows:

$$K_{opt}(\theta_{on}, \theta_{off}) = \max\{K(\theta_{on}, \theta_{off})\} \tag{4.14}$$

As the core of the driving system of EVs, the output torque of the motor is directly related to the dynamic performance of the vehicle. In addition, as an important indicator of EVs, the driving distance is closely related to the efficiency of the motor. Therefore, the importance of the torque ripple in these three indicators is relatively small. Furthermore, when the speed and load torque are fixed, the current value can reflect the power characteristics from another aspect. Therefore, considering the importance of average torque, torque ripple, and efficiency as well as their relationships, when selecting three weight coefficients, the weight coefficients of RMS current, torque smoothness coefficient, and equivalent power coefficient are respectively selected as 0.35, 0.3, and 0.35 in this chapter. However, it should be noted that the three coefficients can be determined according to the actual requirements, although only one combination is considered in this chapter.

4.5.6 Multi-objective Synchronization Optimization and Analysis

According to the motor comprehensive performance index defined in Sect. 4.5.5 and the three performance index data under different constraint conditions, the comprehensive performance index values corresponding to different turn-on angle and turn-off angle are calculated under each combination between load torque and rotation speed, and the database is established as shown in Table 4.4. Table 4.4 shows the calculated motor comprehensive performance index when the motor speed is 1100 r/min and the load torque is 2.5 N m.

As shown in Table 4.4, based on the three performance index data of each condition, the motor comprehensive performance index database was established under different motor speeds and load torques, and the optimal results is obtained through the previously established multi-index synchronous optimization objective function shown in Eq. (4.15), that is, the optimal turn-on angle and turn-off angle under this constraint condition are obtained.

The optimal turn-on angle and the optimal turn-off angle under each constraint condition is obtained by solving the objective function under different speed and load torque conditions. In order to visually display the optimization results of the motor's comprehensive performance index, the optimized turn-on and turn-off angles obtained under various states are expressed in the form of three-dimensional

Table 4.4 *K* when the speed and torque are 1100 r/min and 2.5 N m

Comprehensive performance index parameters *K*		Turn-on angle				
		−12 deg.	−10 deg.	−8 deg.	−6 deg.	−4 deg.
Turn-off angle	16 deg.	0.673237	0.758042	0.846937	0.90513	0.673237
	18 deg.	0.65148	0.72287	0.803663	0.885704	0.65148
	20 deg.	0.625801	0.706009	0.777373	0.865617	0.625801
	22 deg.	0.595384	0.679804	0.737684	0.83088	0.595384
	24 deg.	0.475686	0.593627	0.687615	0.768593	0.475686
	26 deg.	0.488483	0.505937	0.54951	0.571101	0.488483

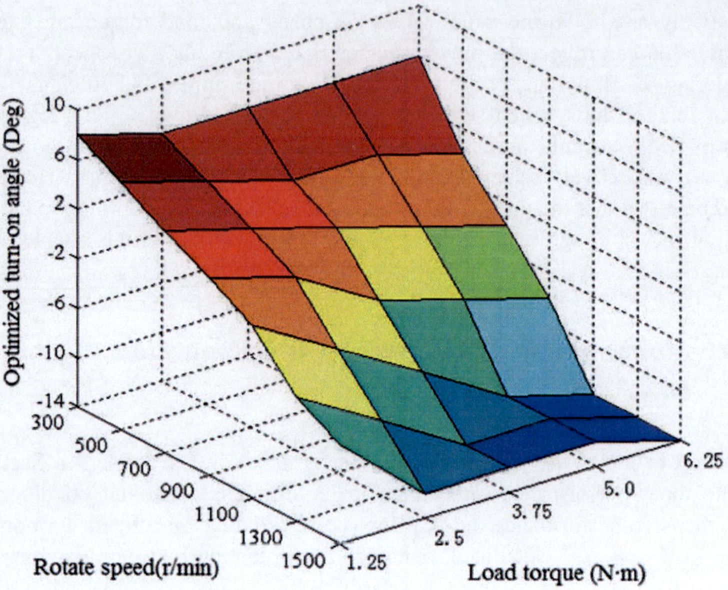

Fig. 4.20 Optimal turn-on angles with maximization of *K*

surface. Figures 4.20 and 4.21 respectively show the three-dimensional curved surfaces of the turn-on angle and turn-off angle changing with the motor speed and load torque as long as the motor comprehensive performance index are optimized.

According to Figs. 4.20 and 4.21, it can be concluded that after the optimization of the comprehensive performance index of the motor, the variation trends of the turn-on angle and the turn-off angle with the motor speed and the load current are illustrated as follows:

(1) With the continuous increase of the motor rotation speed, the turn-on angle decreases gradually as long as the motor comprehensive performance index are optimized.
(2) The influence of load torque on optimized turn-on angle is also obvious. Under the same rotation speed, the optimal turn-on angle decreases with the increase of load torque.
(3) Except for the low speed of the motor at 300 r/min and the high speed at 1500 r/min, the optimized turn-off angle gradually decreases with the increase of the motor speed, and the turn-off angle obviously fluctuates at the speed of 300 and 1500 r/min.
(4) Similarly, the influence of load torque on the optimized turn-off angle is also obvious. Except for the motor at low speed at 300 r/min and high speed at 1500 r/min, the optimized turn-off angle can gradually decrease with the increase of the load torque of the motor.

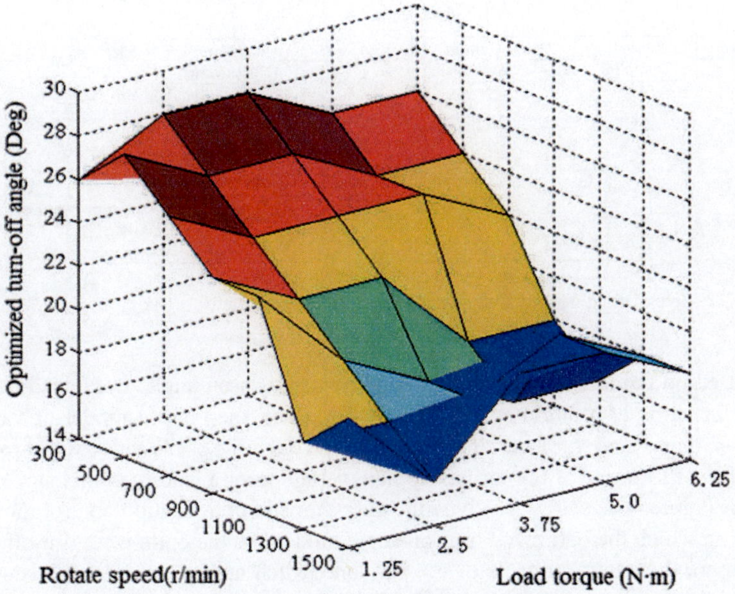

Fig. 4.21 Optimal turn-off angles with maximization of *K*

4.6 Optimized Controller Design for SRM Drive System

As described in Sect. 4.4, the study has deeply analyzed the influences of turn-on angle and turn-off angle on the average torque, torque ripple, and efficiency of the SRM drive system. Therefore, it is of great significance to select the switching angles of the motor drive system and design the high efficiency and stable torque controller correctly to improve the motor drive characteristics, and the comfort and power performance of EVs.

4.6.1 Dynamic Performance Optimized Controller Design

According to the established function relationships between motor performance indexes and switch angles under different rotating speeds and load torques, to satisfy four different objective functions, average torque optimization, torque ripple optimization, motor efficiency optimization, and motor comprehensive performance optimization are carried out, respectively. Then, the optimal turn-on angle and turn-off angle are calculated based on motor speeds and load torques under each objective function. As a result, the functional relationships of the optimal turn-on angle and the optimal turn-off angle in terms of the rotational speeds and load torques are calculated and established, respectively, as shown in Figs. 4.14–4.21.

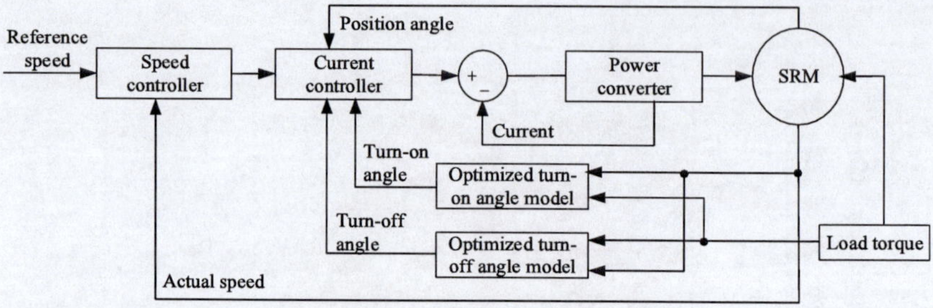

Fig. 4.22 Schematic diagram of the optimized SRM drive system

Based on the relation curves of the optimized turn-on angle, optimized turn-off angle in terms of rotation speeds and load torques, the control model of variable turn-on angle and turn-off angle based on rotation speed and load torque is established to improve the torque ripple, average torque, and the efficiency of the motor. Figure 4.22 shows the dynamic characteristic optimization control model of SRM, in which the optimized turn-on angle model and the optimized turn-off angle model are the control models of the variable switch angles. In order to accurately import the newly established optimization control model into the simulation platform, two Look-up table modules including optimized turn-on angle model and the optimized turn-off angle are built in the MATLAB/Simulink environment. Afterwards, the optimized SRM drive system can be established in this chapter.

In the control process of SRM drive system, according to the current real-time motor speed and load torque values, the controller can search the optimal turn-on angle and turn-off angle in the form of looking up the table, and then gradually reaches the required angle by small incremental steps to make the motor control in the steady performance state.

4.6.2 Comparison Analysis of the Various Optimization Strategies

In order to study the influence of different optimization strategies on the dynamic performance of motor drive system, the optimized SRM drive systems based on the different optimized variable turn-on angle and turn-off angle models are established by means of the four optimization strategies, that is, the average torque optimization, the torque ripple optimization, the motor efficiency optimization, and the multi-index synchronization optimization. Then the four optimized SRM drive systems are simulated to achieve the purpose quantifying the improvement level for the SRM dynamic performance.

On the basis of analyzing the influence of different optimization strategies on the drive system, three indicators, RMS current proportional coefficient, torque smoothness proportional coefficient, and motor efficiency proportional coefficient, are defined in this chapter to study the influence law of different optimization strategies on motor performance indexes.

As for the four optimization strategies, an RMS current proportional coefficient K_{Irms} is introduced in this chapter, which is calculated by means of actual RMS current values and the minimum RMS current value. It should be noted that the minimum RMS current can be obtained after all the combinations between motor speeds and load torque are simulated based on the specific optimization strategy. As a result, the RMS current proportional coefficient K_{Irms} is expressed as follows.

$$K_{Irms} = (I_{rms})_{\min}/I_{rms} \tag{4.15}$$

It can be seen from Formula (4.11) that it is expected that the value of the proportional coefficient is as close to 1 as possible. When the value is closer to 1, it indicates that the optimization strategy improves the average torque better. On the contrary, the closer it is to 0, the less obvious for the optimization strategy to improve the average torque performance.

In the same way, the torque smoothness coefficient data and the maximum torque smoothness coefficient obtained by simulation under each constraint condition are used for proportional calculation to compute the torque smoothness proportional coefficient K_{TSR}. The expression is shown as follows.

$$K_{TSR} = TS/(TS)_{\max} \tag{4.16}$$

It can be seen from Eq. (4.16) that the closer the torque smoothness proportional coefficient is to 1, the more successful the optimization strategy is for optimizing the torque ripple, and the smaller the torque ripple can be at this time. If the value of this coefficient is closer to 0, it means that the torque ripple is larger at this time, and the optimization strategy is less perfect for the optimization of torque ripple.

The expression of the equivalent power proportional coefficient is the same as the description process of the two other proportional coefficients of the RMS current proportional coefficient and the torque smoothness proportional coefficient, which is expressed as follows.

$$K_{PFR} = PF/(PF)_{\max} \tag{4.17}$$

It can also be seen from the expression (4.17) that the closer the equivalent power proportional coefficient is to 1, the more obvious the optimization strategy is to improve the motor efficiency. However, the closer the coefficient value is to 0, the less significant the optimization is for the motor efficiency.

According to the three proportional coefficient expressions, the numerical values of the three proportional coefficients under different optimization strategies are obtained by simulating the model of SRM drive system under various constraints. In order to facilitate direct observation, the simulation data are described in the form of curve. At the same time, the proportional coefficients obtained at different speeds

Table 4.5 Average RMS current proportional coefficient in various strategies

Average of RMS current proportional coefficient		Strategy 1	Strategy 2	Strategy 3	Strategy 4
Rotate speed	300 r/min	1	0.589201	0.984896	0.63187
	500 r/min	1	0.615662	0.985134	0.660391
	700 r/min	1	0.78004	1	0.88004
	900 r/min	1	0.739225	1	0.914024
	1100 r/min	1	0.854446	1	0.949368
	1300 r/min	1	0.892878	0.986416	0.995518
	1500 r/min	1	0.902046	1	0.98741

are averaged and put into a table, where strategies 1 to 4 are the four optimization strategies of average torque optimization, torque ripple optimization, motor efficiency optimization, and multi-index synchronization optimization, respectively.

Table 4.5 shows the data obtained by averaging the RMS current proportional coefficients corresponding to different load torques under the four optimization strategies at different speeds, which can generally reflect the improvement of RMS current in specific motor rotation speeds under different optimization strategies. Figure 4.23 shows the change curves of the RMS current proportional coefficient with load torque at different speeds, which can show the influence of different optimization strategies on RMS current under each constraint condition.

The followings can be seen from Table 4.5 and Fig. 4.23.

(1) The average torque optimization strategy can reduce the RMS current to the greatest extent and achieve the goal of average torque optimization.
(2) The variation range of the proportional coefficient obtained by the torque ripple optimization strategy is 0.589 to 0.902. Although, this optimization strategy has a great influence on RMS current, it is not conducive to improve the average torque of the motor.
(3) According to the motor efficiency optimization strategy, the variation range of the RMS current proportional coefficient is 0.985 to 1, which indicates that this optimization strategy is in good agreement with the average torque optimization strategy. Meanwhile, this is consistent with the previous analysis that the RMS current is the key factor affecting the motor efficiency.
(4) At low speeds (300 and 500 r/min), the RMS current proportional coefficient obtained by the multi-indicator synchronous optimization strategy is small and around 0.65, which indicates that the optimization strategy has a great influence on RMS current at low speeds and is unfavorable for reducing RMS current. However, considering the working process of EVs, the motor basically runs at medium and high speed, while the low speed condition mostly exists when the EV begins to start. In general, under the starting condition, it has low requirements on motor efficiency and high requirements on torque ripple. Therefore, the numerical value of the proportional coefficient at low speed has little influence on the operation of EVs. Under the condition of medium and

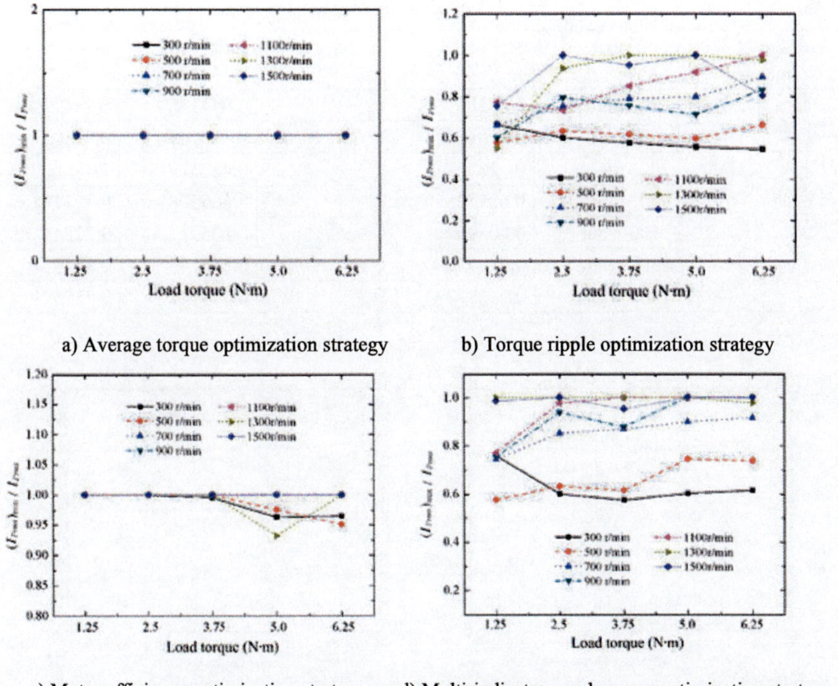

a) Average torque optimization strategy
b) Torque ripple optimization strategy
c) Motor efficiency optimization strategy
d) Multi-indicator synchronous optimization strategy

Fig. 4.23 RMS current proportional coefficient in variable strategies

high speed, the proportion coefficient is basically above 0.90, which indicates that this optimization strategy can well reduce the RMS current and improve the average torque at medium and high speeds.

Similarly, in order to generally reflect the influence of different optimization strategies on the torque ripple at a specific motor speed, the torque smoothness proportional coefficients under the four optimization strategies are calculated. And the average values of torque smoothness proportional coefficients are computed and listed in Table 4.6. Meanwhile, Fig. 4.24 is the curves of the torque smoothness proportional coefficient varying with load torque at different speeds, which can show the specific influence of various optimization strategies on torque smoothness under each constraint condition. Combined the results from Table 4.6 and Fig. 4.24, the following observations can be obtained.

(1) The average torque optimization strategy has much effective influence on the torque smoothness proportional coefficient, which varies from 0.043 to 0.92. It shows that the optimization strategy has a negative influence on the torque ripple at low and medium speeds, which reduces the torque smoothness coefficient and increases the motor's torque ripple.

Table 4.6 Average of *TS* proportional coefficient in various strategies

Average of *TS* proportional coefficient		Strategy 1	Strategy 2	Strategy 3	Strategy 4
Rotate speed	300 r/min	0.042889	1	0.044165	0.926999
	500 r/min	0.120453	1	0.130107	0.965057
	700 r/min	0.428329	1	0.428329	1
	900 r/min	0.665528	1	0.665528	0.892772
	1100 r/min	0.80591	1	0.80591	0.92545
	1300 r/min	0.919938	1	0.836246	0.952914
	1500 r/min	0.919727	1	0.919727	0.955571

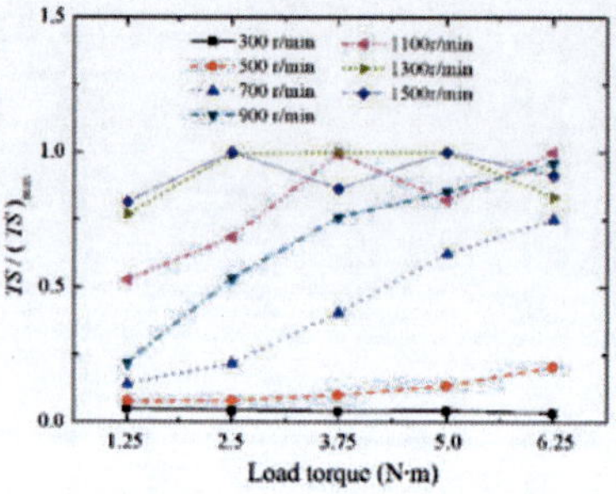

a) Average torque optimization strategy

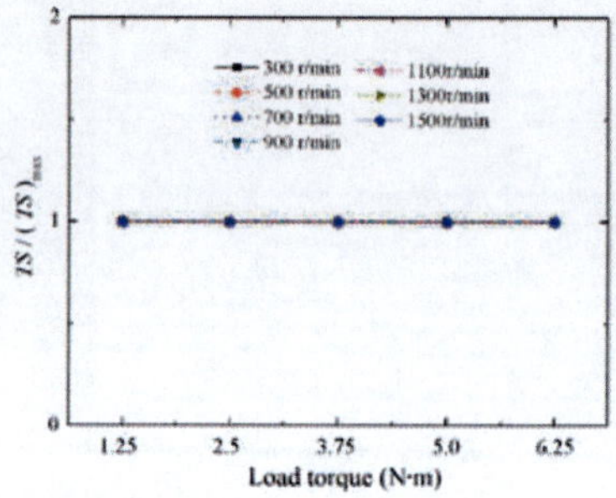

b) Torque ripple optimization strategy

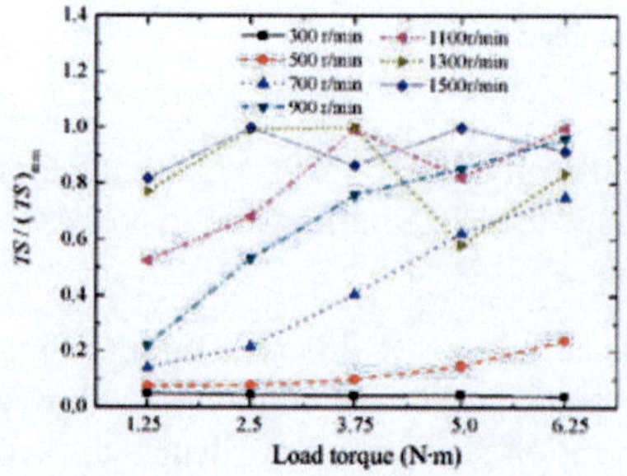

c) Motor efficiency optimization strategy

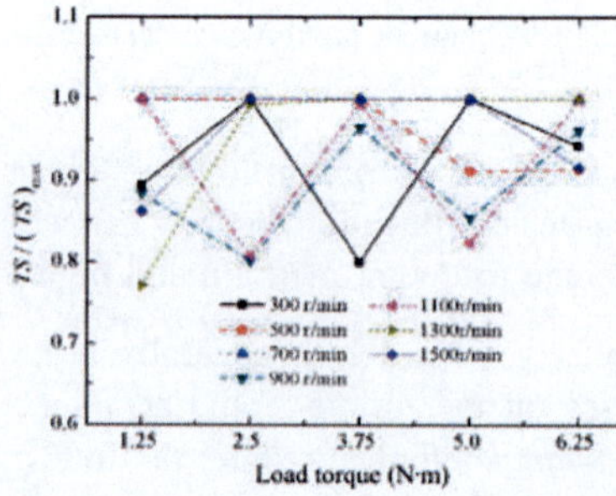

d) Multi-indicator synchronous optimization strategy

Fig. 4.24 *TS* proportional coefficient in various strategies

(2) The torque ripple optimization strategy, focusing on improving torque ripple performance, has the most effective ability to increase the torque smoothness coefficient and reduce the torque ripple for the SRM.
(3) Similar to the average torque optimization strategy, the motor efficiency optimization strategy can greatly increase the torque ripple when the SRM works at low and medium speeds.

Table 4.7 Average of *PF* proportional coefficient in various strategies

Average of *PF* proportional coefficient		Strategy 1	Strategy 2	Strategy 3	Strategy 4
Rotate speed	300 r/min	0.976758	0.608603	1	0.652927
	500 r/min	0.978991	0.628799	1	0.675341
	700 r/min	0.93291	0.847287	1	0.880197
	900 r/min	1	0.73818	1	0.915468
	1100 r/min	0.973024	0.871481	1	0.949391
	1300 r/min	0.993589	0.882763	1	0.985856
	1500 r/min	1	0.903643	1	0.989351

(4) The average value of the torque smoothness proportional coefficient under the multi-index synchronous optimization strategy varies from 0.93 to 1.0, which indicates that the optimization strategy can greatly improve the torque smoothness coefficient and reduce the torque ripple at all rotational speeds.

Table 4.7 and Fig. 4.25 respectively show the variation curves and the average values of equivalent power proportional coefficient with the rotate speeds under different optimization strategies, from which we can analyze the influences of four optimization strategies on the equivalent power coefficient. The followings can be observed.

(1) The variation range of the average equivalent power proportional coefficient is 0.93 to 1.0 under the average torque optimization strategy, which indicates that the optimization strategy can effectively improve the equivalent power coefficient and the efficiency of the SRM.
(2) The torque ripple optimization strategy has a great influence on the equivalent power proportional coefficient. The average value variation range under this strategy is 0.61 to 0.90, which indicates that the optimization strategy has a great influence on the equivalent power coefficient of the motor and is not conducive to improve the efficiency of the motor.
(3) The motor efficiency optimization strategy can maximize the equivalent power coefficient to achieve the purpose of motor efficiency optimization.
(4) Similar to the influence rule of RMS current proportional coefficient, the multi-index synchronous optimization strategy can well improve the equivalent power coefficient of the motor at high speed, although the improvements at low and medium speeds are limited. However, the appropriate efficiency reduction at low speed can also meet the requirements of the operation condition of traction motor for EVs.

To sum up, the four optimization strategies have different effects on motor performance based on the results described above. The average torque optimization strategy and the motor efficiency optimization strategy can improve the average

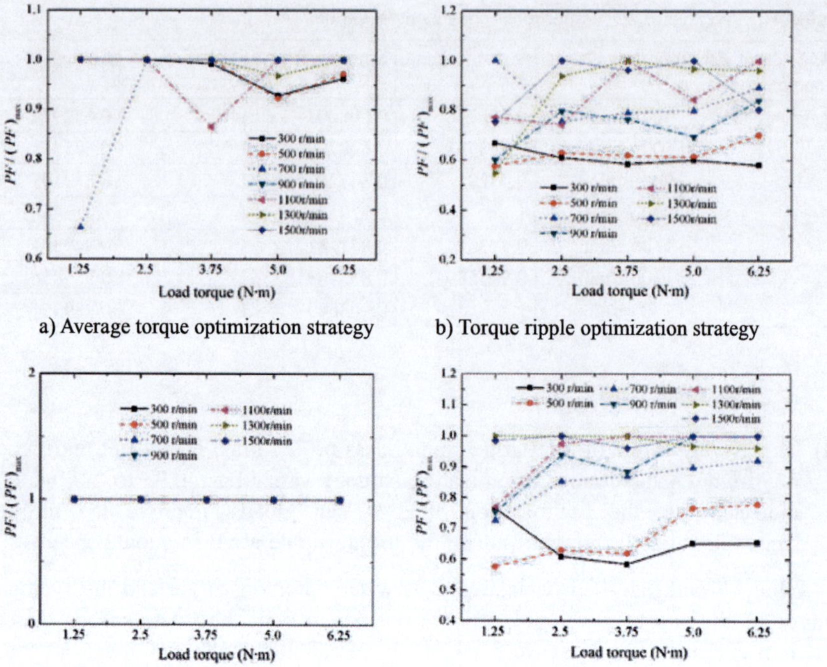

a) Average torque optimization strategy b) Torque ripple optimization strategy

c) Motor efficiency optimization strategy d) Multi-indicator synchronous optimization strategy

Fig. 4.25 Average of *PF* proportional coefficient in various strategies

torque and the motor efficiency very well, but they have great negative influence on the torque ripple. Although the torque ripple optimization strategy can improve the torque smoothness coefficient and reduce the torque ripple of the motor, it is not conducive to improve the average torque and the motor efficiency. By selecting the weight coefficient, the multi-index synchronous optimization strategy can achieve a good balance among the average torque, torque ripple, and motor efficiency, which greatly improves the dynamic characteristics of the SRM and EVs.

4.7 Summary

In this chapter, a nonlinear dynamic model is firstly established for the flux linkage and torque characteristics of the switched reluctance motor under the new winding excitation mode. Then, through the simulation of the dynamic model, the influences of load torque, turn-on angle, and turn-off angle on average torque, torque ripple, and motor efficiency are analyzed. With the goal of improving the comprehensive performance of the motor, a multi-index synchronization optimization objective function is proposed. Furthermore, the mathematical relationships between the synchronization optimization target quantity and the turn-on angle, turn-off angle, load torque and speed are established based on the synchronization optimization. Meanwhile, the multi-index synchronization optimization weight coefficients are introduced, and a variable turn-on and turn-off angle control model based on load torque and motor speed is built. Finally, a comparative analysis of optimization results is carried out for four different optimization strategies, and the results show that the proposed multi-index synchronous optimization controller can improve the average torque, reduce torque ripple, and improve motor efficiency, which will greatly improve the power performance, economy, and comfort index of EVs.

References

1. Zheng H, Jiang J. Study on high-grade torque control strategies for switched reluctance motor. Transactions of China Electrotechnical Society, 2005, 20(9): 24–28.
2. Omekanda AM. A new technique for multidimensional performance optimization of switched reluctance motors for vehicle propulsion. IEEE Transactions on Industry Applications, 2003, 39(3): 672–676.
3. Rahman KM, Schulz SE. High performance fully digital switched reluctance motor controller for vehicle propulsion. IEEE Transactions on Industry Applications, 2002, 38(4): 1062–1071.
4. Inanc N, Ozbulur V. Torque ripple minimization of a switched reluctance motor by using continuous sliding mode control technique. Eletric Power Systems Research, 2003, 66(3): 241–251.
5. Balaji M, Ramkumar S, Kamaraj V. Sensitivity Analysis of Geometrical Parameters of a Switched Reluctance Motor with Modified Pole Shapes. Journal of Electrical Engineering & Technology, 2014, 9(1).
6. He C, Hao C, Qianlong W, et al. Design and control of switched reluctance motor drive for electric vehicles 2016 14th International Conference on Control, Automation, Robotics and Vision (ICARCV). IEEE, 2017.
7. Hwu KI. Dynamic modelling of SRM drive assisted by POWERSYS and SIMULINK. International Review of Electrical Engineering, 2007, 2(2):282–292.
8. Cha HR, Seo JY, Yang H, et al. A Study on Voltage and Switching Angle for Maximum Torque/Efficiency and Minimum Torque Ripple of SRM by using SIMULINK. Journal of Power Electronics, 2001, 1(1): 56–64.
9. Reddy BP, Janaki RV, Kumar KS. Torque ripple minimization of SRM using sense coils and FPGA// 2020 IEEE International Conference on Power Electronics, Smart Grid and Renewable Energy (PESGRE). IEEE, 2020.
10. Sun X, Wu J, Wang S, et al. Analysis of torque ripple and fault-tolerant capability for a 16/10 segmented switched reluctance motor in HEV applications. The international journal for computation and mathematics in electrical and electronic engineering, 2019, 38(6):1725–1737.
11. Buja GS, Maria I. Control Characteristics of the SRM Drives: Operation in the Saturated Region. IEEE Transactions on industrial electronics, 1994, 41(3): 316–325.
12. Wei W, John B Dunlop, Stephen J. Design Optimization of a Switched Reluctance Motor by Electromagnetic and Thermal Finite Element Analysis. IEEE Transactions on Magnetics, 2003, 39(5): 3334–3337.
13. Inderka RB, De Doncker RW AA. High Dynamic Direct Average Torque Control for Switched Reluctance Drives. IEEE Transactions on Industry Applications, 2003, 39(4): 1040–1045.

14. David Cajander, Hoang Le Huy. Design and Optimization of a Torque Controller for a Switched Reluctance Motor Drive for Electric Vehicles by Simulation. Mathematics and Computers in Simulation, 2006, 71: 333–344.
15. Xue XD, Cheng KWE, Lin JK. Optimal Control Method of Motoring Operation for SRM Drives in Electric Vehicles. IEEE Transactions on Vehicular Technology, 2010, 59(3): 1191–1204.

Chapter 5
Design and Control of Regenerative Braking System

5.1 Compound Braking Force Distribution Strategy for EVs

5.1.1 Introduction

Environmental pollution and energy shortage accelerate the development and use of green renewable energy. With the advantages of low pollution, low noise and high energy efficiency, electric vehicles have become the focus of automotive industry research [1–3]. At present, switched reluctance motor (SRM) has the advantages of simple structure, low cost, high reliability, large power generation and high efficiency, such that it is considered as one of the best choices to drive EVs [4–6]. At the same time, the traction system with SRM can recover more energy in the case of regenerative braking. However, the disadvantage of SRM is large torque fluctuation [7, 8], which will directly affect the braking comfort and stability of the vehicle. In addition, SRM has complex characteristics of power generation principle under braking condition, and the energy recovery efficiency and charging current of switched reluctance generator (SRG) imply the driving range and battery life of EVs respectively, such that it is difficult to comprehensively improve the comprehensive braking performance of EVs [9, 10]. In order to increase the driving range of EVs, the braking system of EVs usually combines mechanical braking and regenerative braking [11].

The main difference between EVs and traditional fuel vehicles is the introduction of motor feedback braking in the braking system, which greatly improves energy efficiency, and part of the energy in the power battery comes from the feedback braking of the motor. The compound braking system composed of regenerative braking and mechanical braking can provide regenerative braking force and recover braking energy when the vehicle is decelerating. Therefore, the braking process that the motor participates in can achieve the purpose of energy saving and

Y. Zhu, *The Key Technologies for Powertrain System of Intelligent Vehicles Based on Switched Reluctance Motors*, Recent Advancements in Connected Autonomous Vehicle Technologies 1, https://doi.org/10.1007/978-981-16-4851-9_5

environmental protection, and this is also of great significance for extending the cruising range of EVs [12, 13].

As for the control strategy of compound braking force distribution, firstly, the safety and stability of the braking system should be put in the first place, then the regenerative braking energy recovery ability should be improved. At the same time, the driver's requirements for braking comfort should be met. In addition, the regenerative braking force and mechanical braking force need to be dynamically coordinated controlled under normal braking and emergency braking conditions, such that the introduction of regenerative braking puts forward new requirements for the braking system of EVs. The coordinated control of regenerative braking force and mechanical braking force is the core research content of compound braking technology. The efficiency of braking energy recovery, braking comfort, and feedback current performance largely depend on the coordinated control of compound braking system. In addition, the dynamic response characteristics of mechanical braking and motor regenerative braking are not consistent, which will cause the impact of braking force fluctuation on the comprehensive braking comfort performance in the braking process. Therefore, the control of the compound braking system mainly focuses on the coordinated control strategy of braking force distribution. When designing a compound braking system, the total required braking force is allocated to the regenerative braking force and the mechanical braking force according to the driver's braking intention. When the required braking force is less than the maximum braking force provided by the motor, the required braking force is completely provided by the motor braking force. When the required total braking force exceeds the maximum braking force provided by the motor, the proportion of the regenerative braking force of the motor is dynamically coordinated according to the braking control strategy. When the speed is low enough, the motor will not provide braking force, while the mechanical braking force will provide the required total braking force because of the low efficiency of the regenerative motor under low speed condition. Therefore, the working principle is shown in Fig. 5.1.

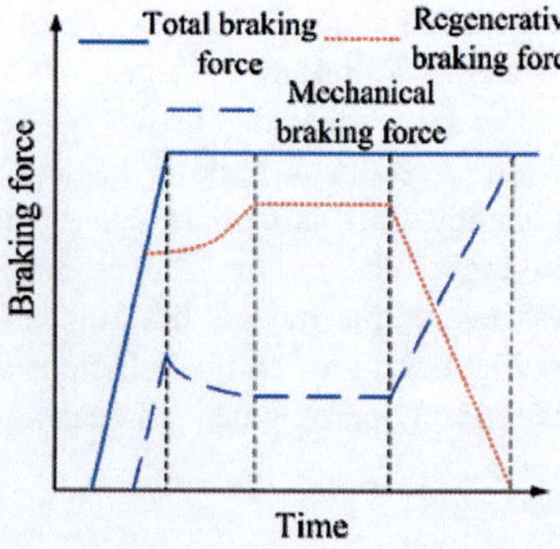

Fig. 5.1 Working principle of compound brake

5.1.2 Compound Braking Structure Forms

At present, the structure of compound braking system of EVs includes parallel structure and series structure. Parallel braking is a kind of braking form which increases the motor feedback braking on the basis of the original mechanical braking system and will not change the original braking form and braking arrangement scheme. It only controls motor regenerative braking torque to realize mechanical electrical compound braking control. The parallel braking form is shown in Fig. 5.2. Parallel braking has the advantages of less control parameters and easy realization, but the disadvantages are low energy recovery efficiency and poor braking comfort. Due to the lack of active braking control in parallel braking, coordinated control between motor braking and mechanical braking cannot be realized. Therefore, parallel braking mode cannot make full use of motor braking force, and the energy recovery is low. Its control schematic diagram is shown as Fig. 5.3.

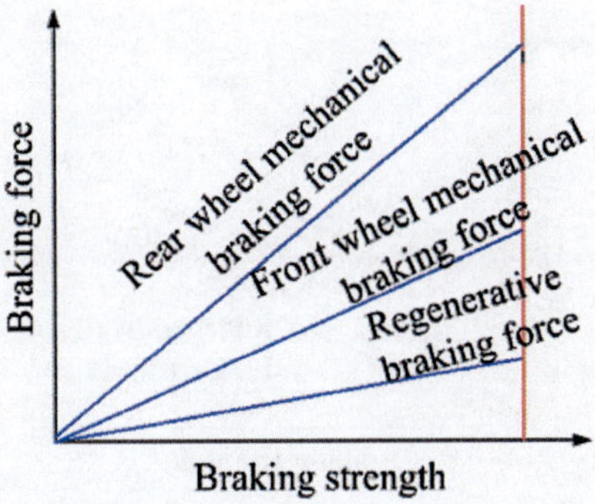

Fig. 5.2 Parallel braking form

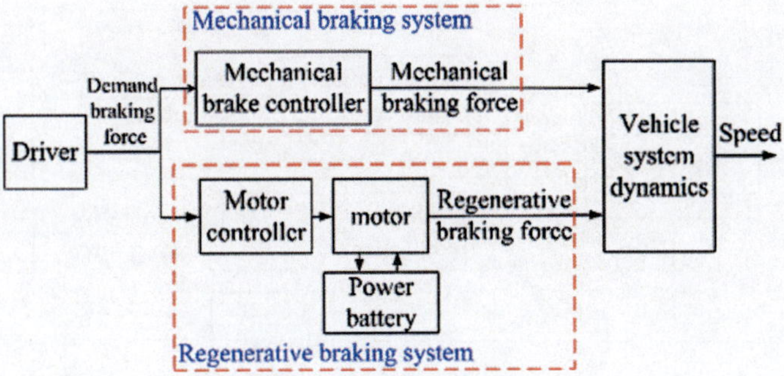

Fig. 5.3 Parallel brake control principle

In the form of series braking, the braking force is preferentially allocated to the motor according to the driver's need to step on the brake pedal. When the demand braking force is less than the maximum braking force output by the motor, the braking system is only provided by the motor; When the demand braking force is greater than the regenerative braking force output by the motor, the mechanical braking force will participate in the braking. The series braking mode is shown in Fig. 5.4. Series braking can give full consideration on the regenerative braking performance of the motor, such that it has high braking energy recovery ability. However, the structure of series braking is complex and the cost is high, such that it needs highly integrated control unit to achieve. The mechanism of series braking is shown in Fig. 5.5. Under the condition of keeping balance between the front and rear axle braking forces, the brake controller calculates the braking force required by the front and rear wheels based on the commands coming from the driver's brake pedal, and distributes the motor regenerative braking force and mechanical friction braking force to the wheels of the front and rear axles according to the braking force distribution control strategy. In order to improve braking performance and braking

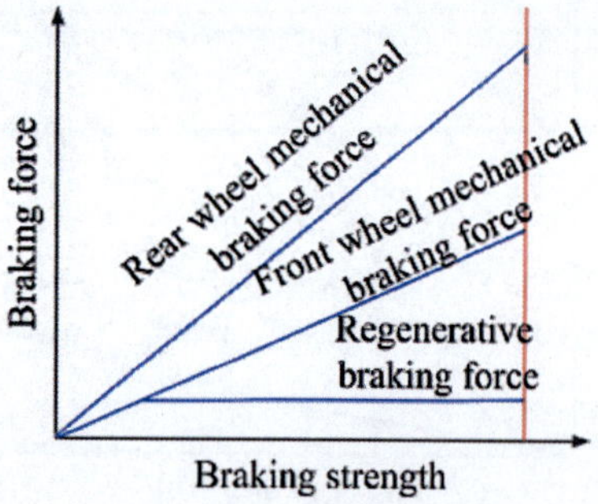

Fig. 5.4 Series braking form

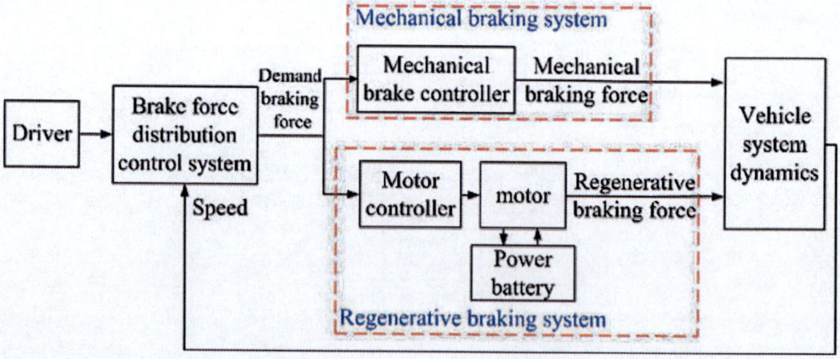

Fig. 5.5 Series brake control principle

energy recovery efficiency at the same time, the series braking structure is selected in this chapter.

5.1.3 Model of Electric Vehicle

5.1.3.1 Vehicle Dynamics Model

In the process of driving, the vehicle will be affected by rolling resistance, air resistance, acceleration resistance, and ramp resistance. Therefore, the external force balance equation of the vehicle in the driving process is

$$F_t = F_f + F_w + F_i + F_j \tag{5.1}$$

where F_t, F_f, F_w, Fi, and F_j are driving force, rolling resistance, air resistance, slope resistance, and acceleration resistance, respectively.

The rolling resistance is expressed as

$$F_f = mgf \cos \alpha \tag{5.2}$$

where m, g, f, and, α are the mass, gravity acceleration, rolling resistance coefficient, and ramp angle of the vehicle, respectively.

The expression of air resistance is

$$F_w = \frac{1}{2} C_D A \rho u^2 \tag{5.3}$$

where C_D, a, ρ, and u are the air resistance coefficient, windward area, air density, and vehicle speed relative to static wind speed (m/s).

The calculation formula of ramp resistance is expressed as follows

$$F_i = mg \sin \alpha \tag{5.4}$$

When the vehicle accelerates on the road, it needs to overcome its own inertia force, namely, acceleration resistance F_j, which is expressed as

$$F_j = \delta m \frac{du}{dt} \tag{5.5}$$

where du/dt is the acceleration of the vehicle and δ is the conversion coefficient of the rotational mass of the vehicle.

When the EV is braking and decelerating, in addition to the rolling resistance, air resistance and ramp resistance, it will also be affected by the front and rear wheel braking force. The calculation formula of the total braking force F_b is as follows

$$F_b = F_{bf} + F_{br} \tag{5.6}$$

where F_{bf} and F_{br} are the front wheel braking force and the rear wheel braking force, respectively.

When the total braking force is only provided by the regenerative braking force of the motor, the total braking force F_b can be also directly expressed as

$$F_b = \frac{T_e i_0 \eta_T}{R} \tag{5.7}$$

where T_e is the output braking torque of SRM drive system, i_0 represents the transmission ratio of main reducer, η_t means the transmission efficiency of transmission system, and R refers to the wheel radius.

Therefore, the dynamic equation of the vehicle in the longitudinal direction during braking condition is

$$m\frac{du}{dt} = mgf\cos\alpha + \frac{1}{2}C_D A \rho u^2 + mg\sin\alpha + F_b \tag{5.8}$$

According to Eqs. (5.7) and (5.8), the vehicle dynamics model is established in the MATLAB/Simulink environment, as shown in Fig. 5.6. According to the initial vehicle speed, the braking deceleration, real-time vehicle speed, and braking distance can be calculated in the process of vehicle deceleration by means of the established vehicle dynamics model.

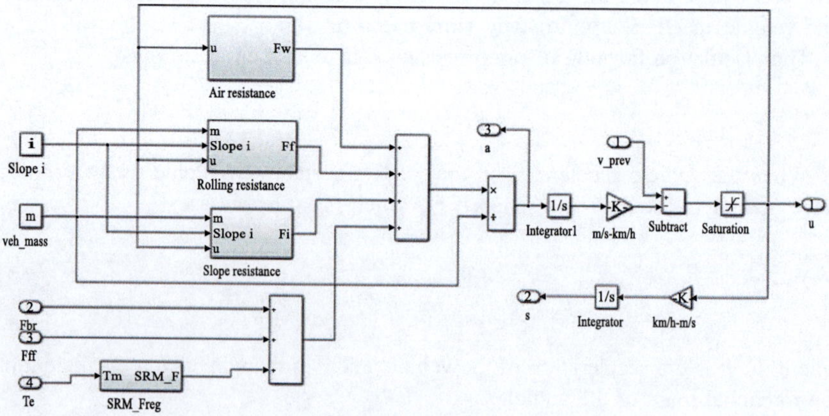

Fig. 5.6 The dynamics model of the electric vehicle

Table 5.1 Parameters of electric vehicle

Parameter	Unit	Value
Vehicle mass	kg	660
Wheel radius	m	0.25
Shaft length	m	1.62
Mass center height	m	0.58
Distance from mass center to center line of front wheel	m	0.76
Distance from mass center to center line of rear wheel	m	0.86
Rolling resistance coefficient		0.014
Air resistance coefficient		0.4
Windward area	m^2	1.4
Main reduction ratio		4.5
Transmission efficiency		0.86
Rated capacity of battery	Ah	100
Voltage of battery	V	72

The main parameters of EVs in this chapter are shown in Table 5.1.

5.1.3.2 Driver Model

In the process of driving, the vehicle is accelerated or decelerated according to other vehicle conditions, road conditions, and pedestrian conditions. In addition, braking performance, braking energy recovery efficiency, and braking comfort should be considered in the design of a compound braking control system. Therefore, the driver model only needs to consider the longitudinal control, that is, the control of the accelerator pedal and brake pedal opening degree. In order to achieve this control, in the MATLAB/Simulink environment, the PID controller is designed based on the initial speed and actual speed of the vehicle to control the driver's pedal, and its mathematical model is

$$\Delta v(t) = v_1(t) - v_2(t) \tag{5.9}$$

$$k(t) = K_p v(t) + K_\mathrm{i} \int_0^t v(t)dt + K_\mathrm{d} \frac{dv}{dt} \tag{5.10}$$

where $v_1(t)$ is the expected speed, $v_2(t)$ refers to the actual speed, and $K(t)$ represents the driver's operation command. K_p, K_i and K_d are proportional coefficient, integral coefficient, and differential coefficient, respectively.

According to the descriptions above, the driver model is built in MATLAB/Simulink environment, as shown in Fig. 5.7.

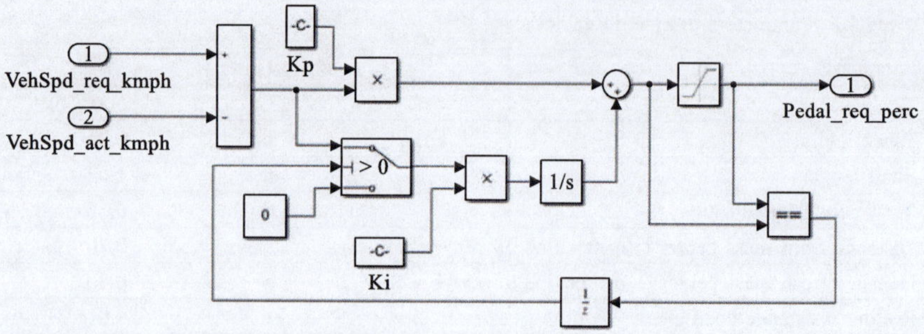

Fig. 5.7 Driver model

5.1.3.3 Power Battery Model

The common modeling methods of power battery include internal resistance model, resistance capacity model, and neural network model. The key point is the coordinated control strategy of compound braking in this chapter. It only needs to recover and store the feedback energy of EVs under braking conditions. In order to simplify the model, this chapter adopts the internal resistance model, which does not require high accuracy of the battery model. Figure 5.8 is the principle of battery internal resistance model.

The ideal output power of the battery is expressed as Eq. (5.11), where, I refers to the discharge current and V_O represents the opening voltage of the battery.

$$P_i = IV_o \tag{5.11}$$

The battery loss should be considered when the battery model is established, which can be expressed as Eq. (5.12) where R is the internal resistance of the battery.

Fig. 5.8 Battery internal resistance model principle

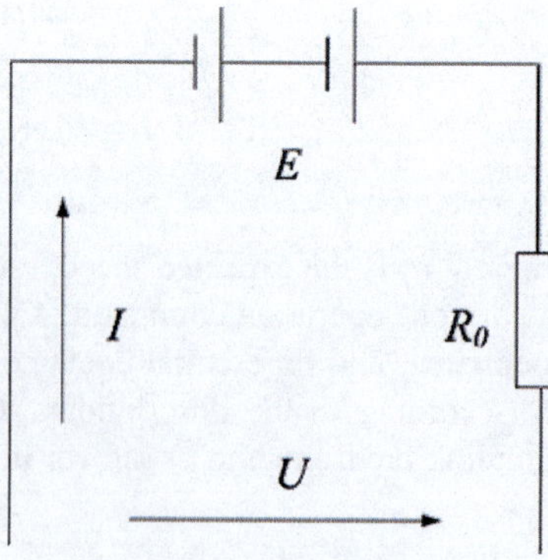

$$P_l = I^2 R \tag{5.12}$$

The actual output power of the battery is

$$P = IV_o - I^2 R \tag{5.13}$$

The discharge current can be obtained from the above three equations, and its expression is

$$I = \frac{V_o - \sqrt{V_o^2 - 4RP}}{2R} \tag{5.14}$$

The common SOC estimation algorithms are open circuit voltage method, current–time integration method, internal resistance method, neural network method, and Kalman filter algorithm. The open circuit voltage method requires a large time cost, and the internal resistance method is more difficult in algorithm. The application cost of neural network and Kalman filter algorithm in the battery management system is relatively high, and a strong control unit computing capability is required. Through comparative analysis, it can be known that the current–time integration method has the advantages of simple algorithm, easy implementation, and high calculation accuracy, such that it is applied to this article. The current–time integral is expressed as Eq. (5.15), where, SOC_0 refers to the initial SOC value, C presents the rated capacity of the battery, and I is the discharge current of the battery.

$$\mathrm{SOC} = \mathrm{SOC}_0 + \frac{1}{C}\int_0^t I dt \tag{5.15}$$

5.1.4 Braking Force Distribution Scheme with Compound Braking Condition

EVs can be affected by air resistance, rolling resistance, inertial force and ground braking force when they decelerate on a level road. Air resistance, rolling resistance, and inertial force are relatively small compared to ground braking force. Therefore, the influence of the aforementioned braking force on braking performance can be ignored. Assuming that the adhesion coefficient of the road surface is φ, the stress condition of the vehicle during braking is shown in Fig. 5.9.

Through the force analysis of the vehicle under braking conditions, the opposite force of the ground on the front and rear wheels is expressed as

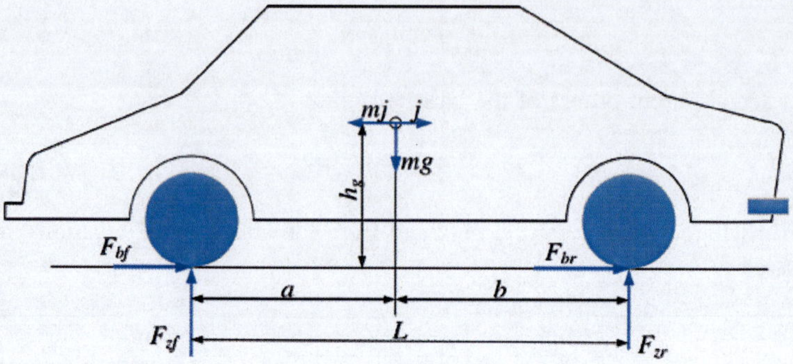

Fig. 5.9 Force analysis of vehicle braking condition

$$\begin{cases} F_{zf} = \dfrac{mg}{L}\left(b + \dfrac{j}{g}h_g\right) \\ F_{zr} = \dfrac{mg}{L}\left(a - \dfrac{j}{g}h_g\right) \end{cases} \tag{5.16}$$

where F_{zf} and F_{zr} refer to the normal opposite forces facing the front and rear wheels respectively under braking condition; a and b are the distances from the mass center of the vehicle to the front and rear axles, respectively; L is the wheelbase of the vehicle; H_g is the height of the mass center; $j = du/dt$ is the braking deceleration.

It should be noted that the braking intensity can be defined as $z = j/g$, which means the distance of the driver pressing the brake pedal.

5.1.4.1 Ideal Control Scheme for Braking Force Distribution

When the vehicle is braked on the road with the road adhesion coefficient of φ, according to formula (5.16), under the condition of the front and rear wheels locking at the same time, the maximum friction braking forces that the road can provide to the front and rear wheels are expressed as

$$\begin{cases} F_{zf} = \dfrac{mg}{L}(b + \varphi h_g) \\ F_{zr} = \dfrac{mg}{L}(a - \varphi h_g) \end{cases} \tag{5.17}$$

When braking, according to the influences from the distribution of front and rear brake force, road adhesion coefficient, and load, there may be three situations in the braking process with the increase of brake pedal stroke, which are as follows. (1) The front wheel is locked, then the rear wheel is locked; (2) The rear wheels were first locked, then the front wheels were locked; (3) The front and rear wheels are locked simultaneously.

In both the braking conditions (1) and (2), the adhesion condition of the road surface is not fully utilized, and the dangerous braking situation is easy to occur. Situation (3) can effectively avoid the side slip of the rear axle, can make full use of the road adhesion conditions and effectively exert the performance of the vehicle braking system.

When the vehicle is braking, the front and rear wheels could be locked at the same time, which is the most ideal braking condition. At this time, the total braking force from the front and rear wheels should be equal to the ground adhesion, and the braking force of the front and rear wheels should be equal to the respective ground adhesion, which can be expressed as Eq. (5.18) where F_{bf} and F_{br} are braking forces of front and rear wheels.

$$\begin{cases} F_{bf} + F_{br} = F_b = \varphi mg \\ F_{bf} = \varphi F_{zf} \\ F_{br} = \varphi F_{zr} \end{cases} \tag{5.18}$$

By combining Eqs. (5.18) and (5.19), it can be further obtained

$$\begin{cases} F_{bf} + F_{br} = \varphi mg \\ \dfrac{F_{bf}}{F_{br}} = \dfrac{b + \varphi h_g}{a - \varphi h_g} \end{cases} \tag{5.19}$$

Further simplification, the ideal front and rear brake force distribution relationship is obtained as follows

$$F_{br} = \frac{1}{2}\left[\frac{mg}{h_g}\sqrt{b^2 + \frac{4h_g L}{mg}F_{bf}} - \left(\frac{mgb}{h_g} + 2F_{bf}\right)\right] \tag{5.20}$$

According to Eq. (5.20), the model is built in MATLAB/Simulink environment according to the ideal front and rear wheel braking distribution control scheme, as shown in Fig. 5.10. In addition, the ideal front and rear wheel braking force distribution curve, I curve, is drawn as shown in Fig. 5.11.

The ideal braking force distribution control scheme is designed based on the front and rear wheels locking at the same time. At this situation, the vehicle can effectively use the adhesion conditions of the road, and the braking stability and braking efficiency of the vehicle are the best. However, due to the influences of complex factors such as cost, technology, and actual working conditions, it is difficult to distribute the braking force according to the ideal braking force curve. In the braking process of EVs, with the increase of the driver's pedal stroke, it is more common that one of front wheels and rear wheels is locked before the other one.

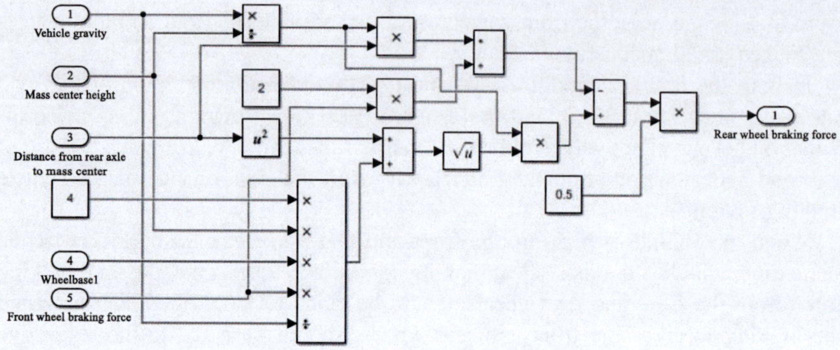

Fig. 5.10 Ideal braking force control scheme model

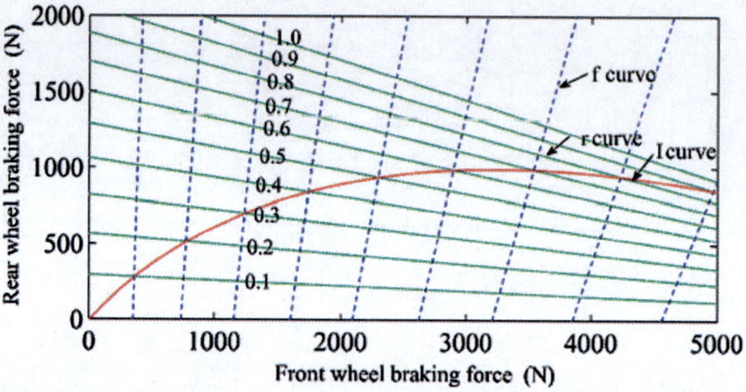

Fig. 5.11 Braking force distribution curve

When the front wheels are locked first, the vehicle will lose the steering ability, but it is still in a stable braking condition. At this time, the relationship between the front and rear wheel braking forces is as follows

$$F_{br} = F_{bf}\left(\frac{L - \varphi h_g}{\varphi h_g}\right) - mg\frac{b}{h_g} \tag{5.21}$$

When the rear wheels of the vehicle are locked first, the rear axle of the vehicle is prone to side slip, and the vehicle is in an unstable braking condition. At this time, the braking force relationship between the front and rear wheels is

$$F_{br} = \frac{\varphi}{L + \varphi h_g}\left(mga - h_g F_{bf}\right) \tag{5.22}$$

According to formulas (5.21) and (5.22), two common braking condition curves can be drawn as shown in Fig. 5.11, where *f* curve group is the front and rear wheel braking force relationship curve based on different road adhesion coefficient under the condition of front wheel locked first, and *r* curve group is the front and rear wheel braking force relationship curve based on different road adhesion coefficient under the condition of rear wheel locked first.

5.1.4.2 Braking Force Distribution Control Scheme Constrained by ECE Regulations

The Economic Commission for Europe proposed regulation No. 13 for two-axle motor vehicles on the road, which clearly stipulates the distribution of braking force between the front and rear wheels of the vehicle. This regulation is called the ECE brake regulation. According to the ECE braking regulations, when the vehicle brakes on the road surface within the range of 0.2–0.8, to improve the braking stability and braking efficiency of the vehicle on the premise of ensuring the braking safety, the braking intensity of the vehicle must meet the following relationship.

$$z \geq 0.1 + 0.85(\varphi - 0.2) \tag{5.23}$$

In order to simplify the structure of the vehicle brake system, the front and rear wheel braking force distribution relationship is set to a fixed value. Usually, the ratio between the front wheel braking force and the total braking force is called the braking force distribution coefficient, which is represented by β

$$\beta = \frac{F_{bf}}{F_b} \tag{5.24}$$

The total braking force is the sum of the front and rear wheel braking forces, and the relationship between the front and rear wheel braking forces is expressed as

$$F_{bf} = \frac{\beta}{1 - \beta} F_{br} \tag{5.25}$$

According to formula (5.23), the braking force distribution relationship can be obtained as

$$\beta \leq \frac{(b + zh_g)(z + 0.07)}{0.85zL} \tag{5.26}$$

The ECE braking regulations stipulate the maximum braking force distribution coefficient. And the Eq. (5.26) can satisfy the front and rear wheel braking relations stipulated by the ECE braking regulations, that is, the boundary curve relationship of the ECE regulations is

$$\frac{h_g(F_{\mathrm{bf}}+F_{\mathrm{br}})^2}{mgL}+\frac{(b+0.07h_g)(F_{\mathrm{bf}}+F_{\mathrm{br}})}{L}+\frac{0.07mgb}{L}-0.85F_{\mathrm{bf}}=0 \tag{5.27}$$

Through formula (5.27), the maximum front wheel braking force restricted by ECE braking regulations can be obtained, and its trajectory curve is the curve of ECE braking safety regulations. The ECE curve can distribute the maximum total braking force to the front wheels. Under the compound braking coordination control strategy, the motor can give priority to regenerative braking, such that it can effectively improve the new braking energy recovery.

Generally, in order to simplify the design of the brake system, the front and rear wheel braking force fixed ratio lines are used to represent the maximum front wheel braking force distribution curve restricted by ECE braking safety regulations. According to formula (5.26), the maximum front wheel braking force distribution coefficient $\beta_{\max}$ can be obtained, and the expression is

$$\beta_{\max}=\frac{2\sqrt{0.07bh_g}+b+0.07h_g}{0.85L} \tag{5.28}$$

The braking intensity z on the curve of ECE safety regulation is expressed as

$$z=\sqrt{0.07b/h_g} \tag{5.29}$$

According to Eqs. (5.27) and (5.28), the ECE braking regulation boundary curve and $\beta_{\max}$ line can be drawn, as shown in Fig. 5.12.

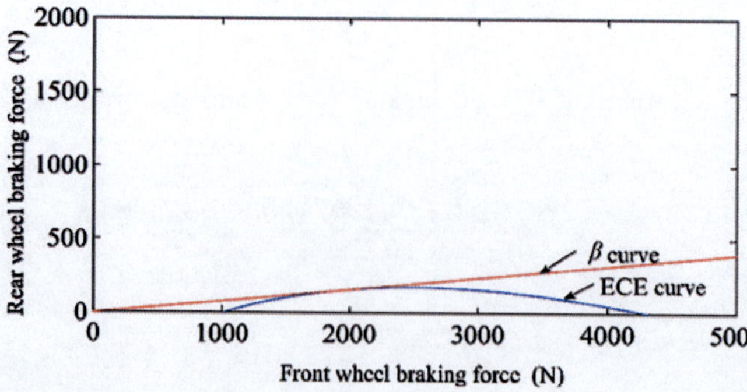

Fig. 5.12 ECE regulation curve and $\beta_{\max}$ curve

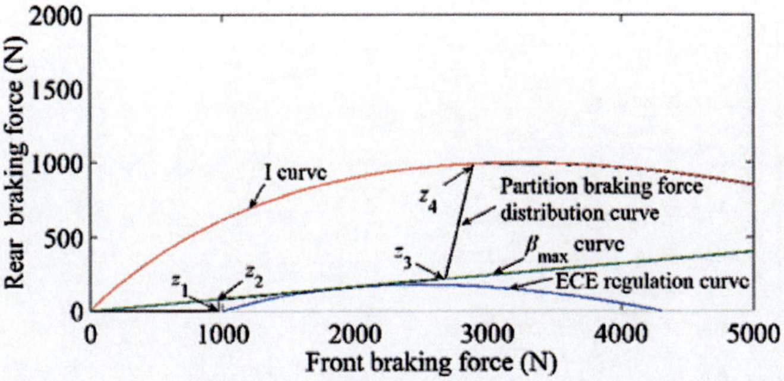

Fig. 5.13 Segmented braking force control scheme

5.1.4.3 Segmented Braking Force Distribution Control Scheme

Several typical braking force distribution control schemes have their own advantages and disadvantages. Although the ideal braking force distribution control scheme can ensure the stability of vehicle braking, the braking system is too complex and accuracy requirements for the control system are too high, such that it is difficult to realize. The ECE braking regulation control scheme can recover the vehicle braking energy to the greatest extent based on the maximum front wheel braking force distribution control strategy, but the control system is complicated and easily exceeds the constraints of ECE regulations. Therefore, when designing a vehicle braking system program, it is necessary to consider braking safety to ensure braking performance, and also increasing braking energy recovery. Based on the ideal braking force distribution scheme and the maximum front wheel braking force distribution control scheme restricted by ECE regulations, a segmented braking force control scheme is proposed in this chapter, in which the front and rear wheel braking forces are divided into five parts. The distribution curve is shown in Fig. 5.13.

Based on the braking intensity, the scheme is divided into five parts between the ideal braking force curve and β_{max} line, and its expression is shown as Eq. (5.30). This scheme can comprehensively consider the motor braking torque according to different braking conditions.

$$\begin{cases} \begin{cases} F_{bf} = mgz \\ F_{br} = 0 \end{cases}, & 0 < z \leq z_1 \\ \begin{cases} F_{bf} = mgz_1 \\ F_{br} = mgz - F_{bf} \end{cases}, & z_1 < z \leq z_2 \\ \begin{cases} F_{bf} = \beta_{\max} mgz \\ F_{br} = (1 - \beta_{\max}) mgz \end{cases}, & z_2 < z \leq z_3 \\ \begin{cases} F_{bf} = \varphi \frac{mg}{L}(b + zh_g) \\ F_{br} = mgz - F_{bf} \end{cases}, & z_3 < z \leq z_4 \\ \begin{cases} F_{bf} = z\frac{mg}{L}(b + zh_g) \\ F_{br} = \frac{1}{2}\left[\frac{mg}{h_g}\sqrt{b^2 + \frac{4h_g L}{mg}F_{bf}} - \left(\frac{mgb}{h_g} + 2F_{bf}\right)\right] \end{cases}, & z > z_4 \end{cases} \tag{5.30}$$

OA segment: The braking intensity corresponding to point A is 0.06, which is a low-intensity braking condition, which can effectively ensure the safety performance of braking. Therefore, the required total braking force can be assigned to the motor first, which can maximize the regenerative braking energy.

AB segment: With the increase of braking intensity, in order to ensure the braking safety performance and meet the ECE law, the rear wheel braking force part participates in the braking process, but it is mainly motor feedback braking.

BC segment: Point B is the intersection of the fixed front wheel braking force curve and the $\beta_{\max}$ line. When the braking intensity is greater than z_2, in order to obtain more braking energy, the braking force distribution curve should be on the $\beta_{\max}$ line under the premise of ensuring braking safety.

CD segment: Point C is the intersection of f curve and $\beta_{\max}$ curve under the road adhesion coefficient is 0.6. When the braking intensity is greater than Z_3, it is easy to exceed the maximum friction provided by the ground, and the front wheel may lock and drag. Therefore, when the braking intensity increases, the distribution of braking force will follow the f curve with the ground adhesion coefficient of 0.6, and the proportion of rear wheel braking force will increase.

DF section: Point D is the intersection of f curve and ideal braking force curve at the road adhesion coefficient is 0.6. When the braking intensity is greater than 0.4, to ensure the braking performance, the front and rear wheel braking force should be distributed along the ideal braking force distribution curve.

When the braking intensity exceeds 0.7, it belongs to the emergency braking condition. Under the premise of ensuring the braking safety, the motor no longer participates in the braking, and the required braking force is completely provided by the mechanical braking force.

5.1.4.4 Motor Braking Torque Correction

During the braking process, the segmented braking force distribution control scheme can recover the braking energy as much as possible on the premise of ensuring the braking safety and braking performance of the EVs, but the braking energy recovery performance is also affected by the vehicle speed and the working state of the battery. In addition, when the brake pedal stroke gradually increases, in order to improve braking comfort and avoid the instability of the braking system, the motor regenerative braking torque needs to be gradually withdrawn; When the emergency braking condition, the total braking force needs to be completely provided by the mechanical friction braking force, which helps to improve the braking safety performance of EVs. Therefore, the influence factors of vehicle speed, SOC, and braking intensity are introduced to correct the braking torque of the motor during the braking process.

Speed influence factor k_1: When the speed is very low, the torque provided by the motor is small, so the recoverable braking energy is very small. At this time, in order to ensure the braking comfort, the motor should gradually withdraw from regenerative braking. Therefore, the influence factor of vehicle speed k_1 is proposed to modify the regenerative braking torque of motor, which is expressed as Eq. (5.31) where the threshold values of vehicle speed are u_l and u_h.

$$k_1 = \begin{cases} 0, & u < u_l \\ \frac{u - u_l}{u_h - u}, & u_l \leq u \leq u_h \\ 1, & u > u_h \end{cases} \tag{5.31}$$

SOC influence factor k_2: Vehicle braking energy recovery will be affected by SOC of battery state of charge. When the SOC of battery is greater than a certain value, if the motor continues to carry out regenerative braking, the battery life will be shortened and the battery will be damaged. Therefore, in order to prolong the service life and improve the braking smoothness, the motor regenerative braking torque should be gradually withdrawn. The SOC influence factor k_2 is proposed to modify the motor regenerative braking torque, which is expressed as Eq. (5.32) where the set SOC thresholds are $\mathrm{SOC_l}$ and $\mathrm{SOC_h}$, respectively.

$$k_2 = \begin{cases} 1, & \mathrm{SOC} < \mathrm{SOC_l} \\ \frac{\mathrm{SOC} - \mathrm{SOC_l}}{\mathrm{SOC_h} - \mathrm{SOC}}, & \mathrm{SOC_l} \leq \mathrm{SOC} \leq \mathrm{SOC_h} \\ 0, & \mathrm{SOC} > \mathrm{SOC_h} \end{cases} \tag{5.32}$$

Braking intensity influence factor k_3: In order to ensure the smooth withdrawal of the regenerative braking force of motor under high-intensity braking condition and improve braking comfort, the braking intensity influence factor k_3 is proposed to correct the regenerative braking torque, which is expressed as Eq. (5.33) where the set braking intensity thresholds are z_l and z_h, respectively.

$$k_3 = \begin{cases} 1, & z < z_l \\ \frac{z - z_l}{z_h - z}, & z_l \le z \le z_h \\ 0, & z > z_h \end{cases} \tag{5.33}$$

Considering the influence factors of vehicle speed, SOC, and braking intensity, the regenerative braking torque output by the motor is computed as Eq. (5.34), where, T_e is the braking torque output by SRM drive system.

$$T_{reg} = k_1 k_2 k_3 T_e \tag{5.34}$$

5.1.4.5 Braking Force Distribution Control Strategy

Electric vehicles driven by the front axle are studied in this paper. The regenerative braking force provided by the motor only acts on the front wheels, and the rear wheel braking force is always provided by mechanical braking force. When the driver steps on the brake pedal, combined with the impact of braking intensity, vehicle speed, and SOC of battery on braking performance, the braking system will first distribute the total braking force to the front axle based on the segmented braking force distribution control scheme. If it is judged as an emergency braking condition, the motor will not participate in braking; If the front axle braking force can provide the required total braking force, the mechanical braking force of the rear wheels will not participate in braking; If the front axle braking force cannot meet the demand of the total braking force, the total braking force is first provided by the front axle, and then the remaining part of the total braking force is provided by the mechanical braking force of the rear axle.

Considering that the motor can only provide limited regenerative braking force, when the total braking torque required is less than the regenerative braking torque provided by the motor, the braking force required by the front wheel is only provided by the motor, which only has the braking condition participated by the motor, so the braking energy recovery efficiency of the braking system of EVs is the largest; When the required total braking torque is greater than the torque provided by the motor, the motor provides the braking torque with optimal comprehensive braking performance, and the remaining front wheel braking force is provided by the mechanical braking torque. At this time, the mechanical braking torque is the difference between the total braking torque required by the front wheel and the regenerative braking torque, and its expression is

$$T_{ff} = T_{bf} - T_{reg} \tag{5.35}$$

In conclusion, the logic block diagram of braking force distribution control for EVs is shown in Fig. 5.14.

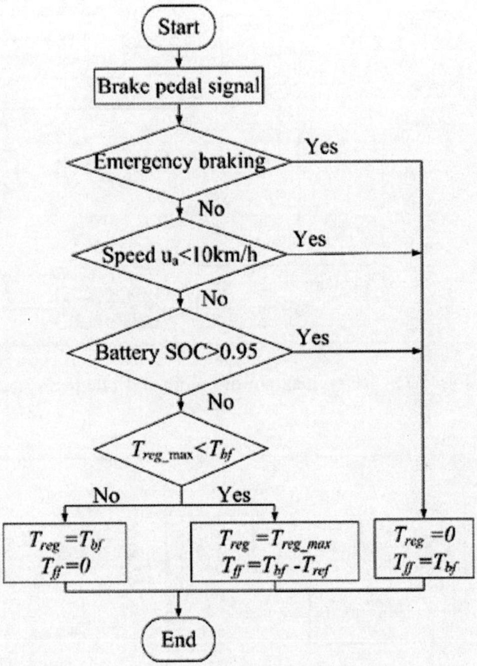

Fig. 5.14 Compound braking force control process

5.1.5 Compound Braking System for Electric Vehicles

Based on the established EV model, the proposed compound braking coordination control scheme and the braking force distribution control strategy, combined with the established SRM drive system model, the compound braking system model is established, as shown in Fig. 5.15. The model is composed of mechanical braking system and regenerative braking system. The required braking force is distributed to the mechanical braking system and regenerative braking system under the action of braking force distribution controller. The EV decelerates under the action of mechanical braking force and regenerative braking force, and the braking energy is recovered and stored in the power battery.

The compound braking system model of EV is established in the MATLAB/Simulink environment, as shown in Fig. 5.16.

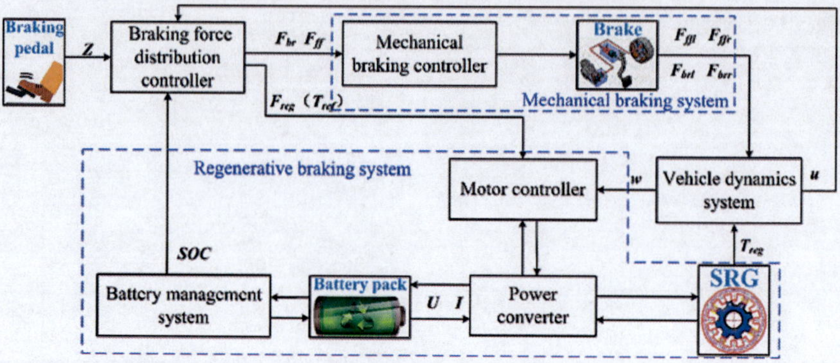

Fig. 5.15 Block diagram of compound braking system

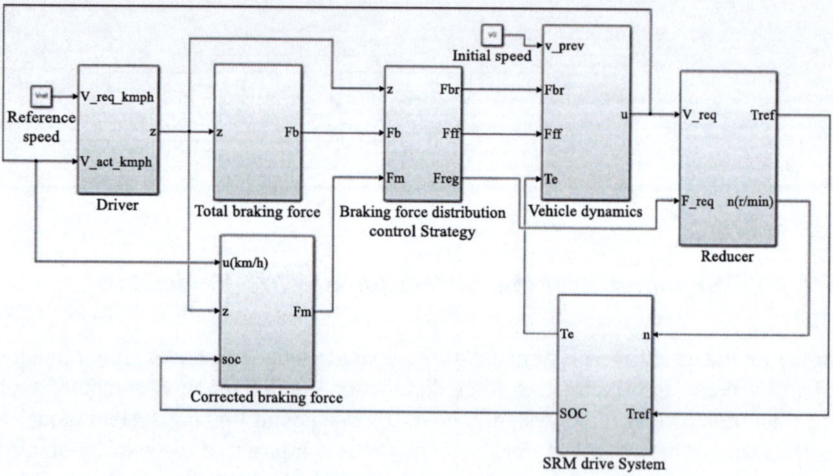

Fig. 5.16 Compound braking system model of EVs

5.2 Optimal Control of EV Braking System Under Sliding Condition

5.2.1 Introduction

When the vehicle decelerates, it will be in the sliding condition or braking condition, in which the sliding condition is the vehicle deceleration braking with only motor participation. In the urban and mountainous road conditions, the sliding deceleration state of the vehicle is more common, and the energy recovery efficiency of the EV is greatly improved to extend the range of the EV when only the motor participates in the sliding braking condition [14, 15].

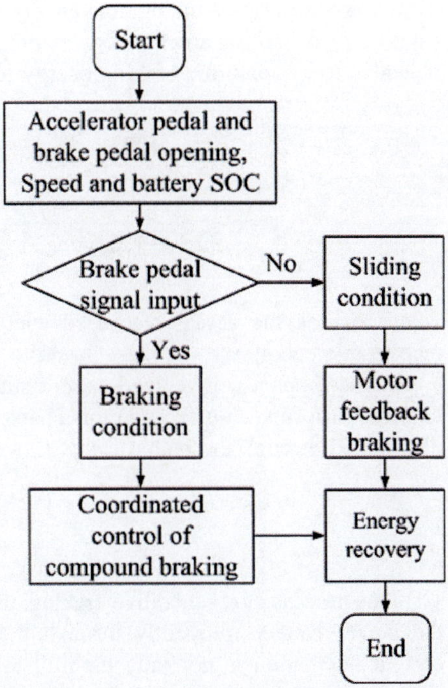

Fig. 5.17 Deceleration condition judgment of vehicle

The energy recovery of feedback braking of EVs is performed only when they are in sliding or braking condition. Braking condition is a driving state in which the driver controls the brake pedal according to the actual operation condition of EVs, and the vehicle is in deceleration braking. The sliding condition is that the driver does not control the accelerator pedal and brake pedal [16, 17]. As shown in Fig. 5.17, according to the opening of the accelerator pedal, the opening of the brake pedal, the vehicle speed and the SOC of the battery, it is judged that the EV is in the sliding condition or the braking condition. If the input signal of the accelerator pedal and the brake pedal is zero and the vehicle speed is greater than the automatic starting speed, it is judged as the sliding condition; if there is the input signal of the brake pedal, it is judged as the braking condition.

5.2.2 Optimization Objectives

In the deceleration process of EVs under sliding condition, the braking energy recovery performance of EVs will affect the driving range of vehicle. Meanwhile, the riding comfort of occupants and the lifetime of power battery shall also be considered when the motor is involved in the feedback braking.

In order to measure the braking energy recovery capability of EVs under sliding condition, the braking energy recovery efficiency η_{reg} is selected as the optimization objective to evaluate the braking energy recovery performance, the expression is

$$\begin{aligned}\eta_{reg} &= \frac{E_{reg}}{E_{bk}} \times 100\% \\ &= \frac{\int_0^t UIdt}{\frac{1}{2}mu^2 - \int_0^t fmgudt - \int_0^t \frac{C_D A}{21.15} \cdot (3.6u)^2 \cdot udt} \times 100\%\end{aligned} \tag{5.36}$$

where E_{reg} is the actual recovered energy of SRM and E_{bk} is the maximum recoverable energy.

Braking impact j_r is defined as an optimization objective to assess the braking comfort performance of EVs in this chapter [18], and the braking smoothness j_s is determined through the reciprocal of j_r, which is expressed as

$$j_s = \frac{1}{j_r} = \frac{1}{d^2u/d^2t} \tag{5.37}$$

In the process of regenerative braking, the phase current will flow into and out of the power battery frequently through the main switch. Nevertheless, the large current fluctuation is not only harmful to the main switch, but also reduces the lifetime of the power battery. Consequently, the current fluctuation i_r should be decreased. The current smoothness i_s is calculated to evaluate current fluctuation in this chapter and expressed as

$$i_s = \frac{1}{i_r} = \frac{1}{\left[\frac{1}{t}\int_0^t (i(t) - i_{ave})^2 dt\right]^{1/2}} \tag{5.38}$$

where $i(t)$ denotes bus current, i_{ave} represents average bus current.

The braking energy recovery efficiency, braking smoothness, and current smoothness defined above are chosen as optimization objectives to enhance the working distance, braking comfort performance, and battery lifetime for EVs, respectively. Therefore, these objectives values should be supposed to be higher to improve the braking performance of EVs.

5.2.3 Multi-Objective Optimization of Braking System in Sliding Condition

The SRM drive system that drives the EV works in a wide speed range, the high-speed range is a single pulse control mode, and the medium and low-speed range is a current chopping control mode. In addition, the dynamic characteristics of the motor under sliding condition will be affected by the turn-on and turn-off angles, and the dynamic characteristics of EVs will also be affected by the reference braking force. Therefore, the influence of the turn-on angle θ_{on}, the turn-off angle θ_{off}, and the reference braking force F_{ref} on the driving range, riding comfort, and battery lifetime is studied in this paper, as shown in Fig. 5.18.

In order to improve the efficiency of EVs braking energy recovery, improve braking comfort, and extend battery lifetime, a multi-objective synchronization optimization strategy based on weight factors is proposed. Compared with the single-objective optimization strategy, the multi-objective optimization strategy can simultaneously meet the requirements of three objectives and can maximize the comprehensive performance of braking. The weight factors assigned under the multi-objective optimization strategy can simultaneously improve the comprehensive braking performance of EVs under sliding braking conditions. The objective function is defined, and its expression is shown in Eq. (5.39) where, $\eta_{\max}$, $j_{s\max}$, and, $i_{s\max}$ represent the maximum values of all sampled data acquired from braking energy recovery efficiency, braking smoothness, and current smoothness, respectively; ω_1, ω_2, and ω_3 are the weight factors of the three objectives.

$$K\left(\theta_{on}, \theta_{off}, F_{ref}\right) = \omega_1 \frac{\eta_{reg}}{\eta_{\max}} + \omega_2 \frac{j_s}{j_{s\max}} + \omega_3 \frac{i_s}{i_{s\max}} \tag{5.39}$$

Hence, the optimization function is confirmed based on the above objective function, which is expressed as

$$K_{opt}\left(\theta_{on}, \theta_{off}, F_{ref}\right) = \max\left\{\omega_1 \frac{\eta_{reg}}{\eta_{\max}} + \omega_2 \frac{j_s}{j_{s\max}} + \omega_3 \frac{i_s}{i_{s\max}}\right\} \tag{5.40}$$

Considering the needs of comprehensive braking performance of EVs, braking energy recovery efficiency is the most important objective to improve the range of EVs. The optimization objective s of braking smoothness and current smoothness are equally important to improve the comprehensive braking performance of EVs,

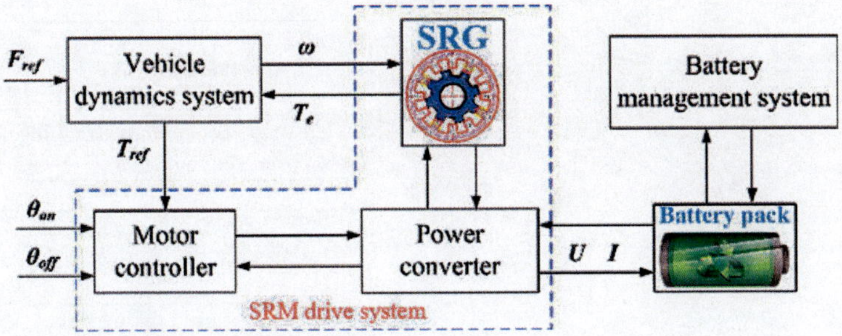

Fig. 5.18 Braking system under sliding condition

but the focus should be on improving the working range. Therefore, in this paper, the weight factors of the three optimization objectives of braking energy recovery efficiency, braking smoothness, and current smoothness are selected as 0.6, 0.2, and 0.2, respectively. In addition, the weight factors are different under different optimization requirements.

The braking system of EV under sliding condition is simulated with different combinations of turn-on angle, turn-off angle, and reference braking force. The objective function value database under different parameter combinations is obtained based on Eq. (5.39), and then the optimal turn-on angle, turn-off angle, and reference braking force of the braking system under different vehicle speeds are obtained from Eq. (5.40). The comprehensive braking performance database based on vehicle speed is established, and the multi-objective optimization process of braking system under sliding condition is shown in Fig. 5.19.

Figures 5.20, 5.21, and 5.22 show the optimal reference braking force, turn-on angle and turn-off angle of the EVs braking system based on the multi-objective optimization strategy under different vehicle speeds. It can be seen from Fig. 5.20 that the reference braking force first increases and then decreases with the decrease of vehicle speed. When the vehicle speed is from 30 to 40 km/h, the reference braking force reaches the maximum value and basically remains at about 330 N, which is basically the same as the variation trend of motor external characteristic curve; In Figs. 5.21 and 5.22, the turn-on angle and turn-off angle tend to increase with the decrease of vehicle speed. In addition, based on the reference braking force, turn-on angle and turn-off angle of the multi-objective optimization strategy,

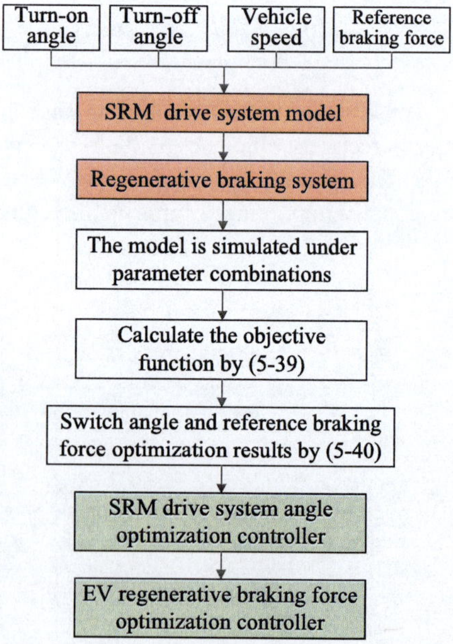

Fig. 5.19 Multi-objective optimization process of braking system under sliding condition

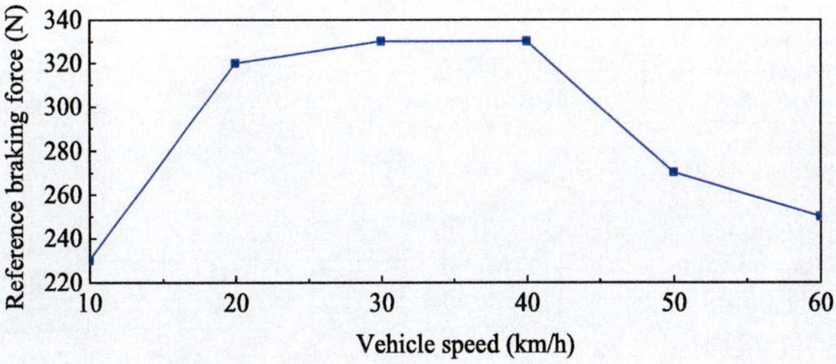

Fig. 5.20 Optimal reference braking force

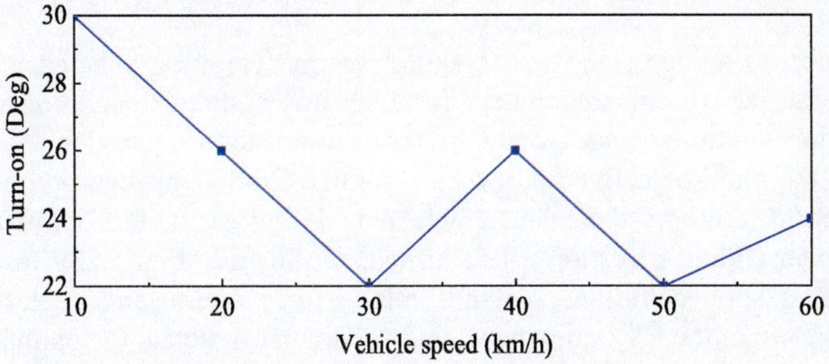

Fig. 5.21 Optimal turn-on angle

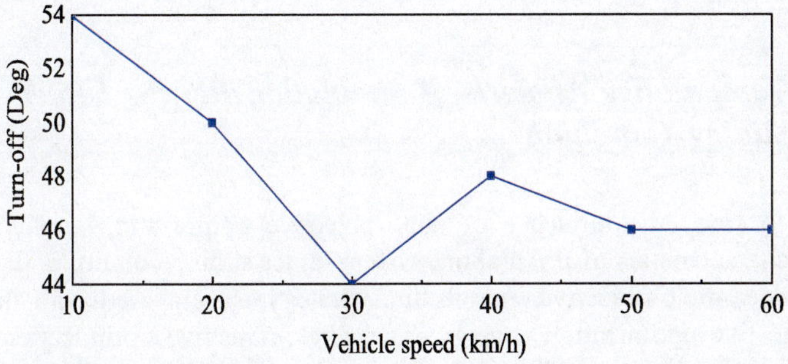

Fig. 5.22 Optimal turn-off angle

the braking force optimization controller and angle optimization controller are established. The optimal demand braking force and switching angle can be obtained under the action of the two controllers through the real-time speed of EVs. The multi-objective optimization strategy can effectively balance the three objectives of braking energy recovery efficiency, braking comfort, and braking efficiency current smoothness.

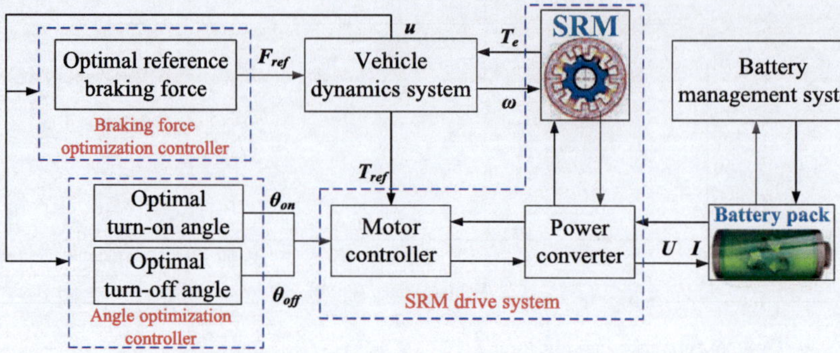

Fig. 5.23 Optimal braking system under sliding condition

In order to obtain the best braking energy recovery efficiency, braking smoothness and current smoothness for EVs under sliding condition, an angle optimization controller and a braking force optimization controller are designed based on the multi-objective optimization results. Combining two optimized controllers, SRM drive system and vehicle dynamics system, the optimal braking system under sliding condition is established, as shown in Fig. 5.23. Based on the angle optimization controller and the braking force optimization controller, the braking system of the EV under the sliding condition can obtain the optimal turn-on angle, turn-off angle, and reference braking force through the real-time vehicle speed, which realizes the goal of improving the comprehensive braking performance of the braking system under the sliding condition of EVs.

5.2.4 *Comparative Analysis of Simulation Results Under Sliding Condition*

In order to study the influence of multi-objective optimization strategy on the dynamic characteristics of EV braking system under sliding condition, the braking system under multi-objective optimization strategy is compared with that under single-objective optimization strategy. The single-objective optimization strategy includes efficiency optimization strategy, braking comfort optimization strategy, and current smoothness optimization strategy. Based on the single-objective optimization strategy and the multi-objective optimization strategy, the dynamic optimization model of EV braking system is established.

5.2.4.1 Comparative Analysis of Different Optimal Control Strategies

In order to acquire the influence of regenerative braking system on braking performance under sliding braking condition, the dynamic performance of EVs under four strategies are compared and analyzed, including multi-objective optimization strategy, efficiency optimization strategy, braking comfort optimization strategy, and current smoothness optimization strategy. Compared with the multi-objective optimization strategy, the efficiency optimization strategy only considers the braking energy recovery efficiency, that is, the three weight factors of the objective function are selected as 1, 0, and 0, respectively. Similarly, the three weight factors are fixed at 0, 1, 0 and 0, 0, 1 under the braking comfort optimization strategy and the current smoothness optimization strategy, respectively.

The efficiency factor $K_{\eta reg}$ is expressed by the ratio of the optimal braking energy recovery efficiency to the maximum braking energy recovery efficiency, and its expression is

$$\left\{ K_{\eta_{reg}} = \frac{\eta_{opt}}{\eta_{\max}} \right. \tag{5.41}$$

where η_{opt} is the optimal braking energy recovery efficiency and $\eta_{\max}$ is the maximum braking energy recovery efficiency.

Braking smoothness factor K_{js} is defined as the ratio of optimal braking smoothness and maximum braking smoothness, and the expression is

$$\left\{ K_{js} = \frac{j_{sopt}}{j_{s\max}} \right. \tag{5.42}$$

where j_{sopt} is the optimal braking smoothness, and $j_{s\max}$ is the maximum braking smoothness.

Similarly, the current smoothness factor K_{is} is defined as the ratio of the optimal current smoothness to the maximum current smoothness, and its expression is

$$\left\{ K_{i_s} = \frac{i_{sopt}}{i_{s\max}} \right. \tag{5.43}$$

where i_{sopt} is the optimal current smoothness, and $i_{s\max}$ is the maximum current smoothness.

From Eqs. (5.41), (5.42), and (5.43), the analysis shows that the closer the factor value is to 1, the better the comprehensive braking performance of regenerative braking system can be improved under sliding braking condition of EVs. In addition, $K\eta_{reg}$, Kj_s, and Ki_s are all kept at 1 obviously under efficiency optimization strategy, braking comfort optimization strategy, and current smoothness optimization strategy, respectively.

When the EV is running at 60 km/h, the multi-objective optimization strategy is compared with the efficiency optimization strategy, brake comfort optimization strategy, and current smoothness optimization strategy. The efficiency factor, braking smoothness factor, and current smoothness factor are compared and analyzed. Figures 5.24, 5.25, and 5.26 respectively show the curves of efficiency factor, brake comfort factor, and current smoothness factor with the change of vehicle speed under the four optimization strategies.

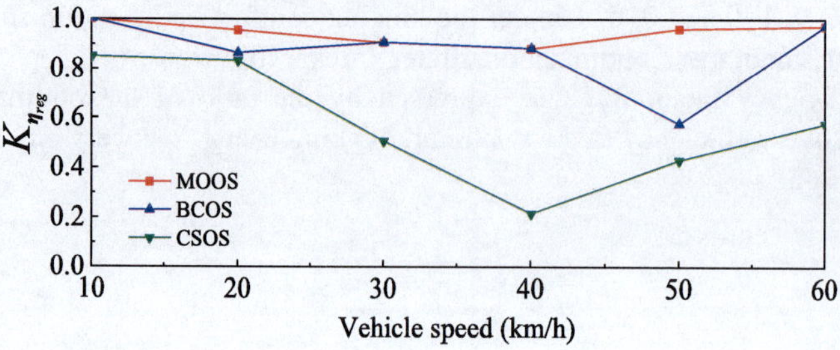

Fig. 5.24 Efficiency factor

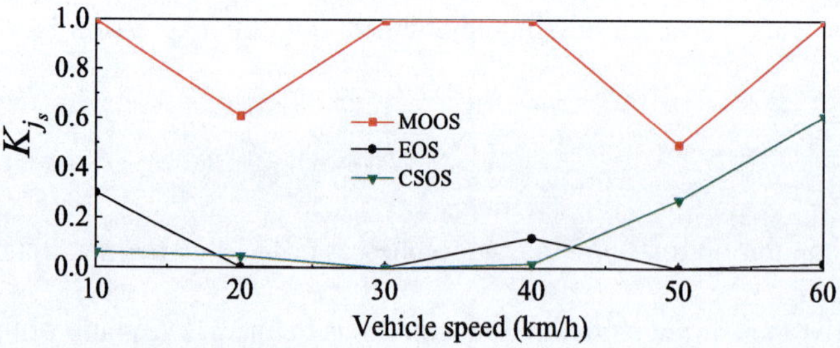

Fig. 5.25 Braking smoothness factor

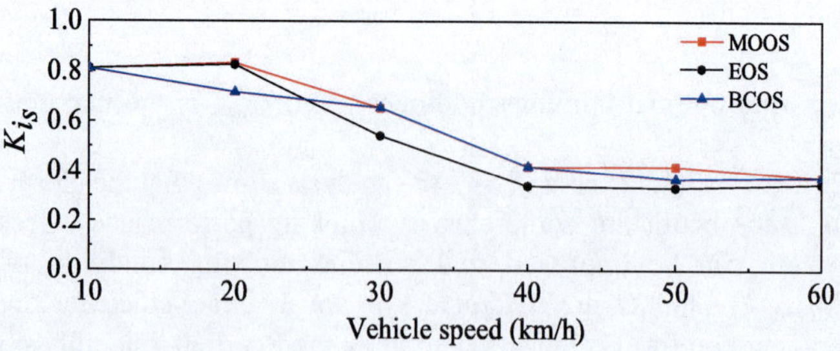

Fig. 5.26 Current smoothness factor

As show in Fig. 5.24, the efficiency factors are all above 0.9 under the multi-objective optimization strategy. The efficiency factors vary from 0.2 to 0.8 in a large range under the current smoothness optimization strategy, which is significantly smaller than those under the multi-objective optimization strategy. At the same time, the efficiency factors under the braking comfort optimization strategy are also smaller than those under the multi-objective optimization strategy. It can be seen from Fig. 5.25 that the braking comfort factors are below 0.3 under efficiency optimization strategy and below 0.2 under current smoothness optimization strategy when the vehicle speed is less than 45 km/h. However, the braking comfort factors are all above 0.6 under multi-objective optimization strategy. The change trend of current smoothness factors is basically the same under three optimization strategies. However, the current smoothness factors are the largest under multi-objective optimization strategy as shown in Fig. 5.26. As a result, the regenerative braking system under the multi-objective optimization strategy can efficiently improve the braking energy recovery efficiency, enhance braking comfort, and reduce the current fluctuation under sliding braking condition of EVs.

5.2.4.2 Dynamic Characteristic Analysis of Braking System

To further assess the influence of the regenerative braking system on the braking performance and obtain the dynamic characteristics of EVs under the multi-objective optimization strategy, the actual recovered energy E_{reg}, state of charge (SOC), braking distance, and vehicle speed are simulated and analyzed under four optimization strategies. Considering the work ability of the vehicle in this paper, two typical initial vehicle speeds are selected as 60 and 30 km/h, which represent the deceleration condition of EVs from high and low speed. The initial SOC of the battery under both conditions is set to 0.6. The simulation results at both speeds are shown in Figs. 5.27 and 5.28.

As shown in Fig. 5.27a, under the initial vehicle speed of 60 km/h, the E_{reg} is around 54.1 kJ, 50.9 kJ, 44.2 kJ, and 5.1 kJ in the efficiency optimization strategy, multi-objective optimization strategy, braking comfort optimization strategy, and current smoothness optimization strategy, respectively. Although the E_{reg} in the multi-objective optimization strategy is somewhat lower than that in the efficiency optimization strategy, it is far greater than that in the current smoothness optimization strategy and the braking comfort optimization strategy. In addition, the change trend of SOC is also the same under the four optimization strategies, which is shown in Fig. 5.27b. The growth rate of SOC is 0.317%, 0.301%, 0.261%, and 0.028% under the efficiency optimization strategy, multi-objective optimization strategy, braking comfort optimization strategy, and current smoothness optimization strategy, respectively. It can be seen from Fig. 5.27c that the braking distance is the shortest under the efficiency optimization strategy, which is 201.6 m. And the braking distance is 249.4 m under the multi-objective optimization strategy, which is 57.5 and 112.3 m shorter than that under the braking comfort optimization strategy and current smoothness optimization strategy. The vehicle deceleration

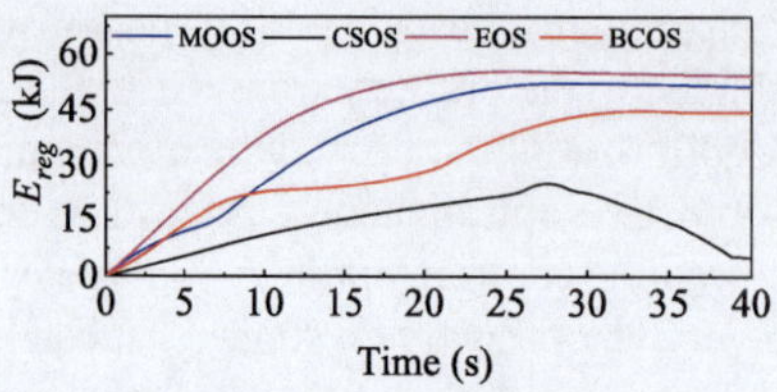

(a) Actual recovered energy

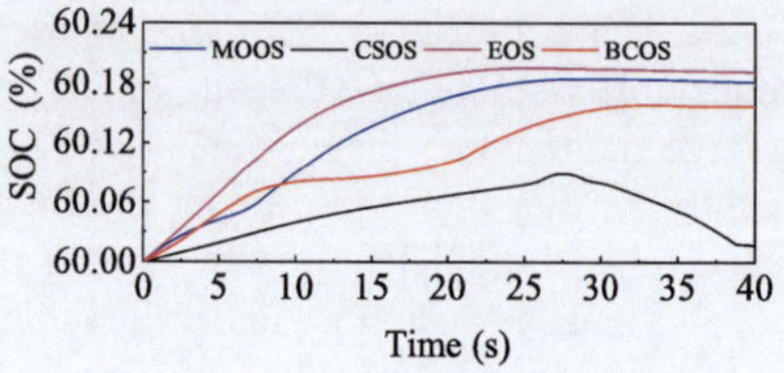

(b) SOC

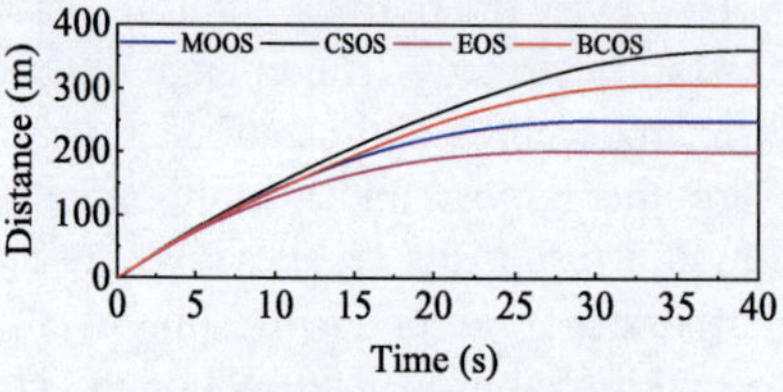

(c) Sliding braking distance

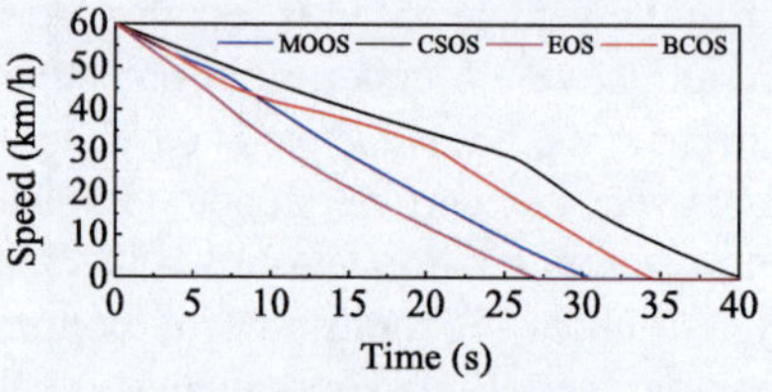

(d) Vehicle speed

Fig. 5.27 Simulation results at the initial speed of 60 km/h

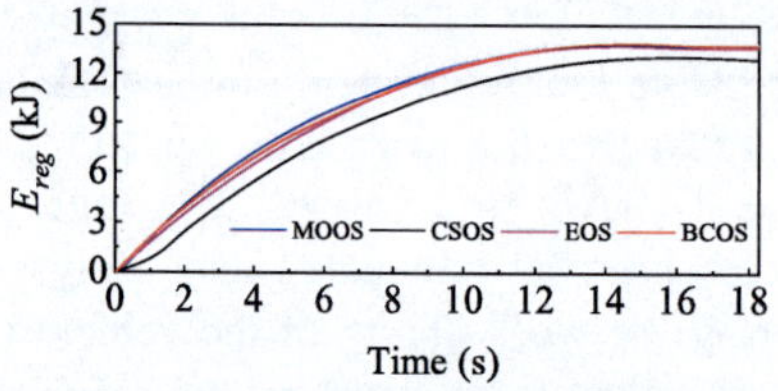

(a) Actual recovered energy

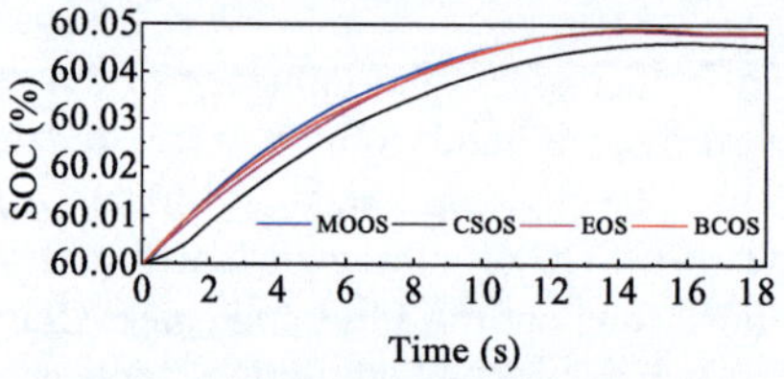

(b) SOC

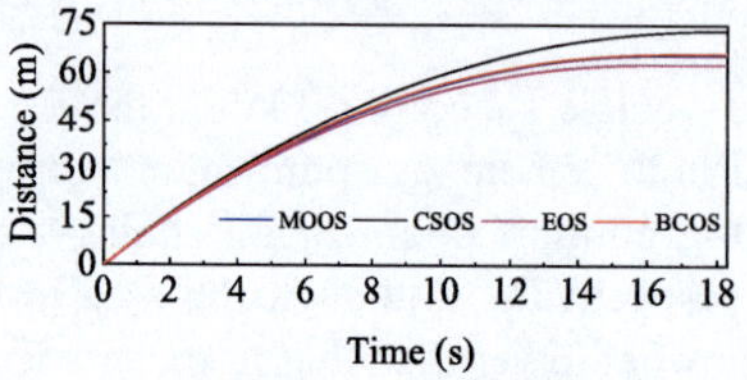

(c) Sliding braking distance

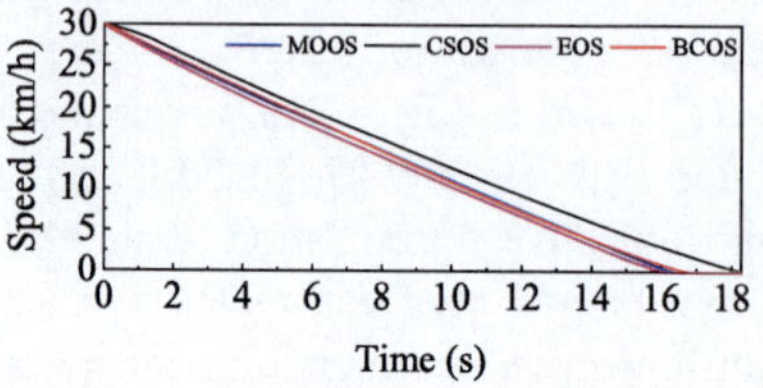

(d) Vehicle speed

Fig. 5.28 Simulation results at the initial speed of 30 km/h

time is 27.20 s, 30.48 s, 34.58 s, and 39.64 s under the four strategies, respectively, as shown in Fig. 5.27d. In addition, the change of E_{reg}, SOC, braking distance, and vehicle speed under the initial speed of 30 km/h is the same as that under the initial speed of 60 km/h in Fig. 5.28. Compared with the initial speed of 60 km/h, the E_{reg}, SOC, braking distance, and vehicle speed are greatly reduced under the initial speed of 30 km/h. Therefore, the braking performance of the regenerative braking system in the low-speed segment is relatively inferior to that in the high-speed segment under sliding braking condition of EVs.

Furthermore, in order to evaluate the braking comfort and current smoothness of regenerative braking system under sliding braking condition, the braking deceleration and bus current are compared and analyzed under the multi-objective optimization strategy and efficiency optimization strategy in Figs. 5.29 and 5.30. When the initial vehicle speed is 60 km/h, it can be seen from Fig. 5.29 that the braking deceleration changes in the larger value range of 0.4–0.9 m/s^2 under the efficiency optimization strategy, while the braking deceleration under the multi-objective optimization strategy changes in the smaller value range of 0.3–0.4 m/s^2 before 27 s. Besides, when the initial vehicle speed is 30 km/h, the fluctuation of braking deceleration under the multi-objective optimization strategy is also significantly smaller than that under the efficiency optimization strategy. Compared to the efficiency optimization strategy, it can be observed that the current fluctuation under the multi-objective strategy is obviously smaller than that under the efficiency optimization strategy at both initial speeds of 60 and 30 km/h, as show in Fig. 5.30.

Combine the analysis of the above figure, compared to other optimization strategies, the regenerative braking system under the multi-objective optimization strategy can efficiently balance energy recovery, braking comfort, and current smoothness, which will greatly increase the driving range, improve braking comfort performance, and extend battery lifetime of EVs, respectively.

In order to further analyze the braking performance of the regenerative braking system under sliding braking condition of EVs, the actual recovered energy increase

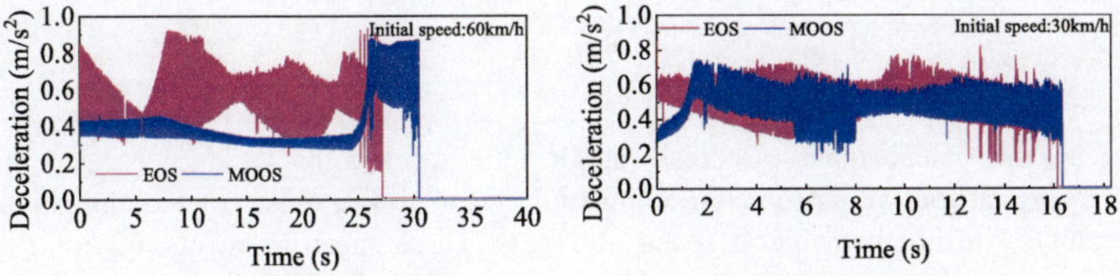

(a) Initial speed 60 km/h (b) Initial speed 30 km/h

Fig. 5.29 Braking deceleration

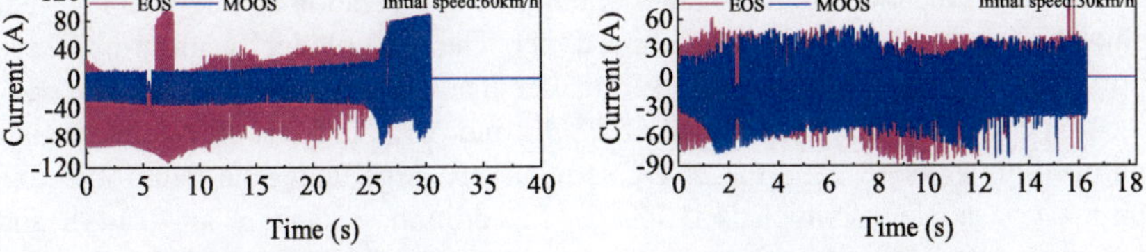

(a) Initial speed 60 km/h (b) Initial speed 30 km/h

Fig. 5.30 Braking deceleration and bus current

Table 5.2 Comparison of dynamic characteristics of braking systems in various speed segments

		60–50 km/h	40–30 km/h	20–10 km/h
Multi-objective optimization strategy	ΔE_{reg} (kJ)	13.85	8.51	4.69
	ΔSOC (%)	0.049	0.030	0.0166
	a_v	0.1498	0.0638	0.0577
	c_v	15.26	13.96	25.22
Efficiency optimization strategy	ΔE_{reg} (kJ)	14.59	11.57	4.74
	ΔSOC (%)	0.0514	0.0409	0.0167
	a_v	0.1618	0.0924	0.0690
	c_v	24.45	40.01	26.72
Braking comfort optimization strategy	ΔE_{reg} (kJ)	12.69	6.56	4.68
	ΔSOC (%)	0.0449	0.0233	0.0165
	a_v	0.1112	0.0169	0.0453
	c_v	19.93	20.11	27.44
Current smoothness optimization strategy	ΔE_{reg} (kJ)	8.13	5.93	4.51
	ΔSOC (%)	0.0288	0.0210	0.0159
	a_v	0.1527	0.0935	0.0861
	c_v	7.8799	11.68	22.71

ΔE_{reg}, the state of charge increase ΔSOC, the acceleration variance a_v, and the charging current variance c_v are compared and analyzed under four optimization strategies in different vehicle speed segments. Three speed segments, 60 km/h to 50 km/h, 40 km/h to 30 km/h, and 20 km/h to 10 km/h, are selected, which represent high-speed, medium-speed, and low-speed, respectively. The comparative results are shown in Table 5.2.

When the vehicle decelerates from 60 km/h to 50 km/h, the ΔE_{reg} is 13.85 kJ, 14.59 kJ, 12.69 kJ, and 8.13 kJ under the multi-objective optimization strategy, efficiency optimization strategy, braking comfort optimization strategy, and current smoothness optimization strategy, respectively. The ΔE_{reg} under the multi-objective optimization strategy is only 0.74 kJ smaller than that under the efficiency optimization strategy. The change trend of ΔSOC and ΔE_{reg} is the same under the four optimization strategies, and the ΔSOC is relatively large under the multi-objective optimization strategy, which is 0.0490%. In addition, a_v and c_v are 0.1498 and 15.26 under the multi-objective optimization strategy, which are much smaller than that under the efficiency optimization strategy. In the deceleration segment from

40 km/h to 30 km/h, the ΔE_{reg} and ΔSOC under the multi-objective optimization strategy are only slightly lower than that the efficiency optimization strategy. Meanwhile, a_v and c_v are relatively large under the efficiency optimization strategy, and the values are 0.0924 and 40.1. The a_v and c_v are smaller under the multi-objective optimization strategy, and their values are 0.0638 and 13.96. In addition, ΔE_{reg} is 6.56 and 5.53 kJ under braking comfort optimization strategy and current smoothness optimization strategy, such that the value is smaller than that under multi-objective optimization strategy. In the low-speed segment from 20 km/h to 10 km/h, the ΔE_{reg}, ΔSOC, a_v, and c_v are 4.69 kJ, 0.0166%, 0.0577, and 25.22 under the multi-objective optimization strategy, respectively. On the other hand, ΔE_{reg} in the high-speed segment, medium-speed segment, and low-speed segment is 13.85 kJ, 8.51 kJ, and 4.69 kJ under the multi-objective optimization strategy, respectively. In addition, the a_v of high-speed segment, medium-speed segment, and low-speed segment are 0.1498, 0.0638, and 0.0577 under the multi-objective optimization strategy, respectively, which is a decreasing trend. And compared to the medium-speed segment and low-speed segment, the c_v of the high-speed segment is the smallest under the multi-objective optimization strategy, which is 15.26. Moreover, the change trend of ΔE_{reg}, a_v, and c_v is also roughly the same under other optimization strategies.

Based on the above analysis, compared to single-objective optimization strategies, the regenerative braking system under the multi-objective optimization strategy can well recover the braking energy, suppress the deceleration variance, and suppress the current variance through the distribution of weight factors, which will increase the driving range, improve braking comfort performance, and extend battery lifetime of EVs, respectively. Furthermore, the braking energy recovery and current smoothness of the regenerative braking system in the high-speed segment are obviously greater than those in the low-speed segment and medium-speed segment, while the braking comfort performance is the worst in the high-speed segment.

5.3 Optimal Control of Braking System Under Braking Condition

5.3.1 *Introduction*

Compound braking is a more common brake of electric vehicle, which is jointly participated by motor and machinery. The driver steps on the brake pedal to recover the kinetic energy of the vehicle according to the compound braking control strategy.

When the vehicle is in the braking condition, the target priority of the vehicle braking safety performance is higher than the braking energy recovery. Therefore, in the emergency braking condition, the motor does not participate in the braking, and the total braking force is only provided by the mechanical friction braking force [19–21]. Firstly, the coordinated control of compound braking system of EV under braking condition is to determine the real-time total braking force according to the driver's input to the brake pedal; Secondly, the distributed braking force is determined according to the designed coordinated control strategy of compound braking control; Then, combined with the established control model of SRM drive system and considering the state of charge (SOC) of power battery, the turn-on angle, turn-off angle, and reference braking force of EV under braking condition are determined; Finally, according to the real-time vehicle speed and braking intensity, the braking energy recovery efficiency, braking comfort, and current fluctuation are optimized by multi-objective optimization strategy. The coordinated control flow chart of EV compound braking system under braking condition is shown in Fig. 5.31.

Compared with the optimization objective of EVs under sliding condition, the braking recovery energy, braking comfort, and current fluctuation also need to be considered under braking condition to improve the range, ride comfort, and battery lifetime of EVs, respectively. Therefore, the optimization objective compound braking system of EVs under braking condition is also braking energy recovery efficiency, braking smoothness, and current smoothness.

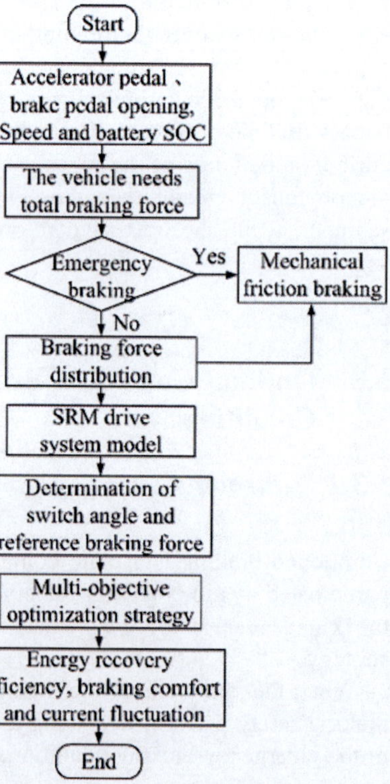

Fig. 5.31 Flow chart of coordinated control of braking system under braking condition

5.3.2 Influence Analysis of Control Parameters

The comprehensive performance of compound braking system of EVs under braking condition is related to the turn-on angle, turn-off angle, reference braking force, vehicle speed, and braking intensity. Therefore, the influences of turn-on angle, turn-off angle and reference braking force on the optimization objectives under different vehicle speeds and braking intensities are studied in this paper.

5.3.2.1 Influence of Turn-On Angle

Figures 5.32, 5.33, and 5.34 show the relationship between the turn-on angle and the optimization objectives of compound braking system of EVs under different constraint conditions. Among them, the braking intensity of 0.1, 0.3, and 0.5 represent the compound braking condition of EVs under low, medium, and high braking intensity, respectively. The fixed turn-off angle is 44° and the reference braking force is 360 N. The relationship between turn-on angle and braking energy recovery efficiency, braking smoothness, and current smoothness is studied under different vehicle speeds.

Under the condition of low-intensity braking, as shown in Fig. 5.32, the braking energy recovery efficiency changes within a certain range with the increase of turn-on angle, and the braking smoothness and current smoothness tend to increase with the increase of turn-on angle. EVs have high energy recovery efficiency under

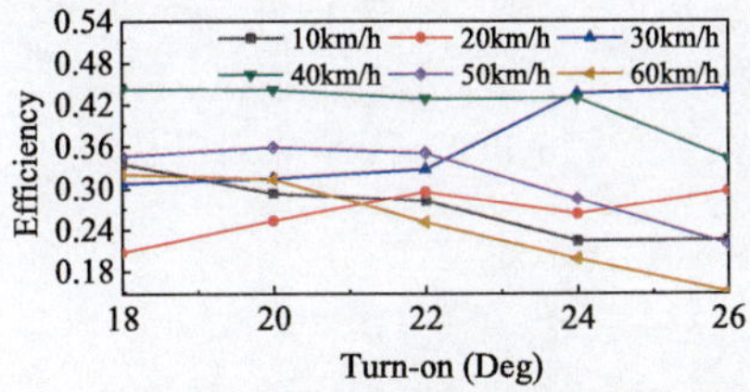

(a) Braking energy recovery efficiency

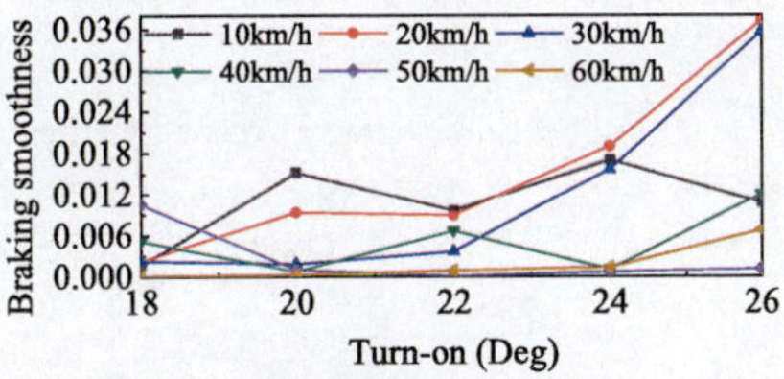

(b) Braking smoothness

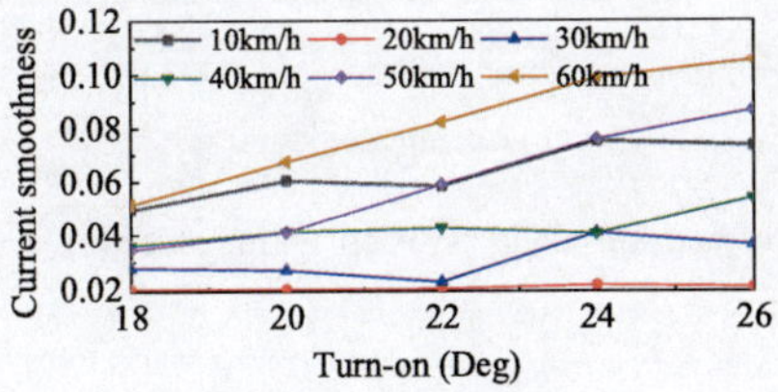

(c) Current smoothness

Fig. 5.32 The influence of turn-on angle on each optimization objective under low-intensity braking

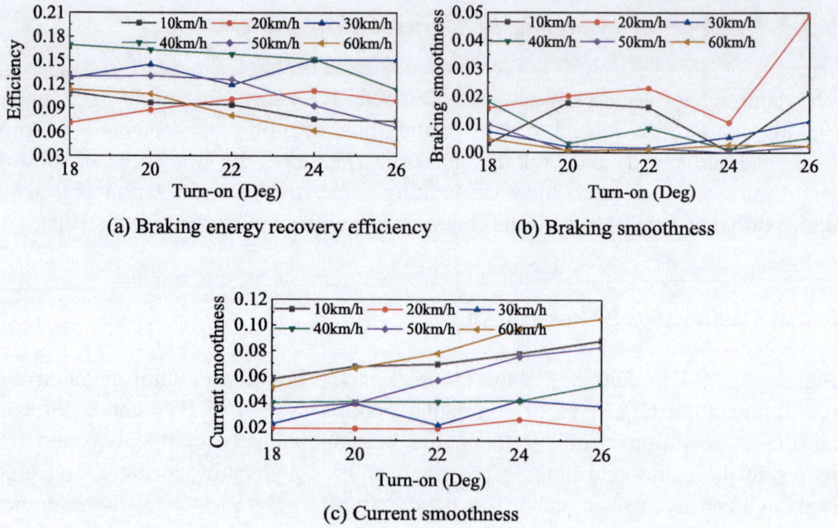

(a) Braking energy recovery efficiency (b) Braking smoothness

(c) Current smoothness

Fig. 5.33 The influence of turn-on angle on each optimization objective under medium-intensity braking

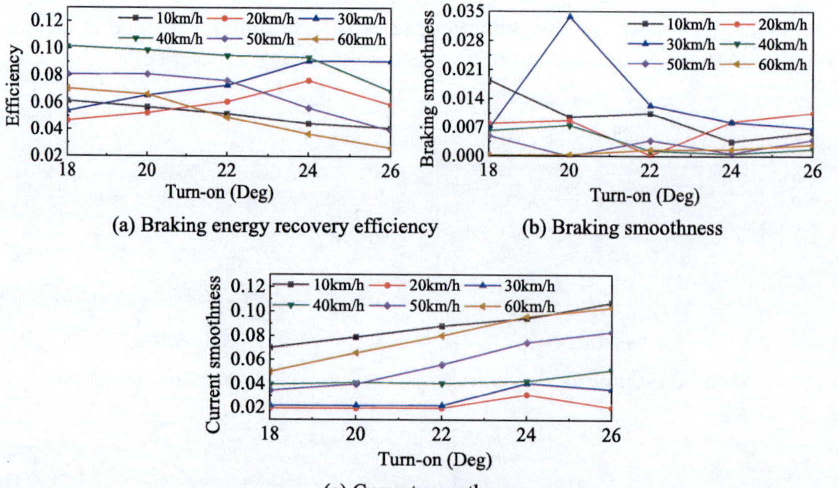

(a) Braking energy recovery efficiency (b) Braking smoothness

(c) Current smoothness

Fig. 5.34 The influence of turn-on angle on each optimization objective under high-intensity braking

conditions of vehicle speeds ranging from 30 to 40 km/h. The braking smoothness tends to increase as the vehicle speed decreases, while the current smoothness decreases as the vehicle speed decreases.

Under the condition of medium-intensity braking, as shown in Fig. 5.33, the braking energy recovery efficiency varies from 0.06 to 0.17 with the increase of the turn-on angle under different vehicle speeds, and the braking energy recovery efficiency is higher in the intermediate vehicle speed range. The braking smoothness tends to increase with the turn-on angle increases. The braking smoothness at medium and high speeds is smaller than that at medium and low speeds; With the increase of the turn-on angle, the current smoothness tends to increase. Within the range of 20–60 km/h, the current smoothness at high speeds is greater than that at low speeds.

Under the condition of high-intensity braking, as shown in Fig. 5.34, the braking energy recovery efficiency varies from 0 to 0.1 under different constraint conditions, and the braking energy recovery efficiency is relatively large in the medium speed range, but the braking energy recovery efficiency is significantly reduced compared with that under the condition of medium and low-intensity; Compared with the low and medium-intensity braking conditions, the braking smoothness and current smoothness change trends under high-intensity braking conditions are roughly the same.

Based on the above analysis, the braking smoothness and current smoothness can be increased by increasing the turn-on angle, thereby achieving the purpose of improving braking comfort and prolonging the lifetime of the power battery. Therefore, the determination of the turn-on angle becomes the key.

5.3.2.2 Influence of Turn-Off Angle

Figures 5.35, 5.36, and 5.37 show the relationship between the turn-off angle and the optimization objectives of compound braking system of EVs under different constraint condition, in which the fixed turn on angle and reference braking force are 22° and 360 N respectively. The relationship between the turn-off angle and braking energy recovery efficiency, braking smoothness, and current smoothness under different speed conditions is researched.

Under the condition of low-intensity braking, as shown in Fig. 5.35, with the increase of turn-off angle, and the braking energy recovery efficiency, the braking energy recovery efficiency increases is higher when the vehicle speed is between 30 and 40 km/h; The results show that the braking smoothness decreases with the increase of the turn off angle in the range of 0–0.02, and the braking smoothness decreases with the increase of the vehicle speed; The current smoothness decreases with the increase of the turn-off angle in the range of 0.02–0.15, and the current smoothness in the medium and high speed range is obviously larger than that in the medium and low speed range.

Under the condition of medium-intensity braking, as shown in Fig. 5.36, the change trend of braking energy recovery efficiency, braking smoothness and current smoothness with turn-off angle is basically the same as that under low-intensity braking condition. With the increase of turn-off angle, the braking energy recovery

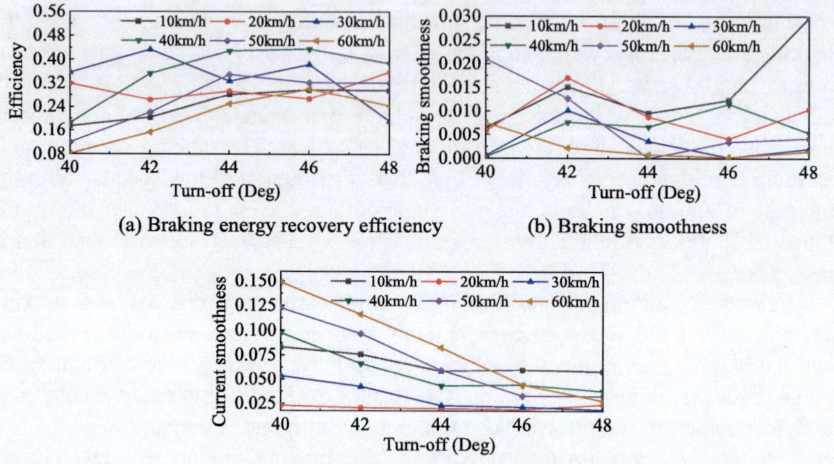

(a) Braking energy recovery efficiency (b) Braking smoothness

(c) Current smoothness

Fig. 5.35 The influence of turn-off angle on each optimization objective under low-intensity braking

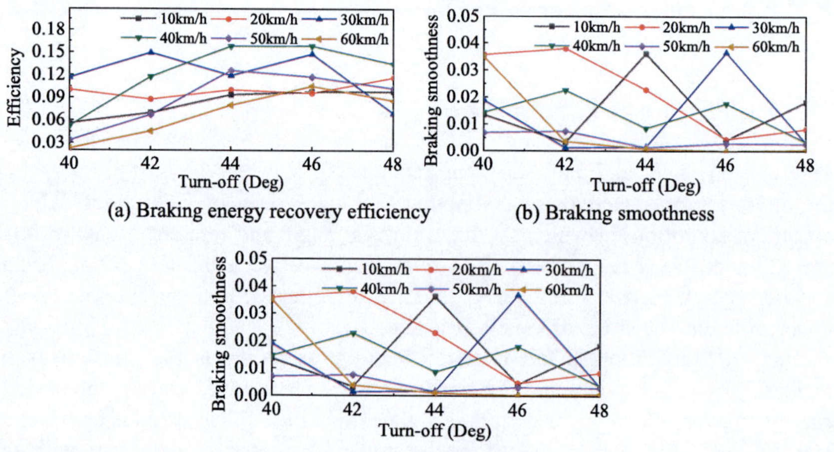

(a) Braking energy recovery efficiency (b) Braking smoothness

(c) Current smoothness

Fig. 5.36 The influence of turn-off angle on each optimization objective under medium-intensity braking

efficiency increases in the range of low vehicle speed, and the braking smoothness and current smoothness decrease.

Under the condition of high-intensity braking, as shown in Fig. 5.37, the change trend of three optimization objectives with turn-off angle is basically the same as that under medium and low-intensity braking condition, but the braking energy

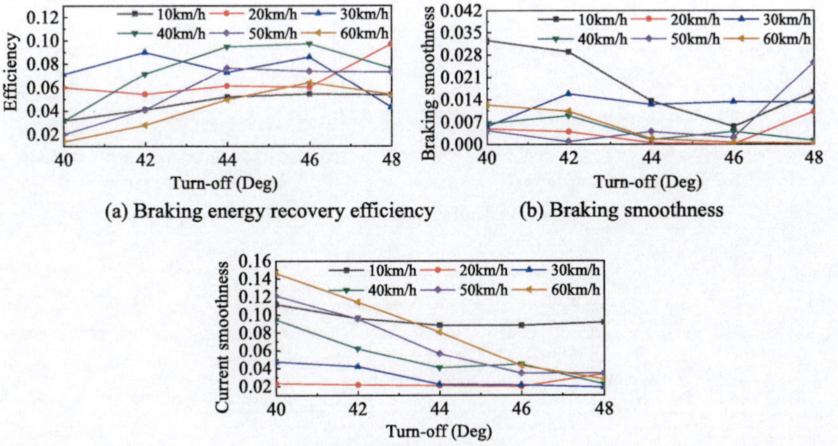

(a) Braking energy recovery efficiency (b) Braking smoothness

(c) Current smoothness

Fig. 5.37 The influence of turn-off angle on each optimization objective under high-intensity braking

recovery efficiency under high-intensity braking condition is significantly lower than that under medium and low-intensity braking condition.

Based on the above analysis, the braking energy recovery efficiency can be improved by increasing the turn-off angle to meet the purpose of extending the driving range of EVs. However, with the increase of the turn-off angle, the braking smoothness and current smoothness will decrease. Therefore, to improve the efficiency of braking energy recovery under the premise of considering braking smoothness and current smoothness, the determination of the turn-off angle becomes the key.

5.3.2.3 Influence of Reference Braking Force

Figures 5.38, 5.39, and 5.40 show the relationship between the reference braking force and the optimization objectives under different constraint conditions of compound braking system of EVs under braking condition, in which the fixed turn-on angle and turn-off angle are 22° and 44°, respectively. The relationship between the reference braking force and braking energy recovery efficiency, braking smoothness, and current smoothness under different speed conditions is studied, and briefly analyze it.

Under the condition of low-intensity braking, as shown in Fig. 5.38, with the increase of reference braking force, and the braking energy recovery efficiency, the braking energy recovery efficiency increases is relatively large when the vehicle speed is between 30 and 50 km/h; The results show that the braking smoothness varies from 0 to 0.07 under different vehicle speeds, and the braking smoothness is

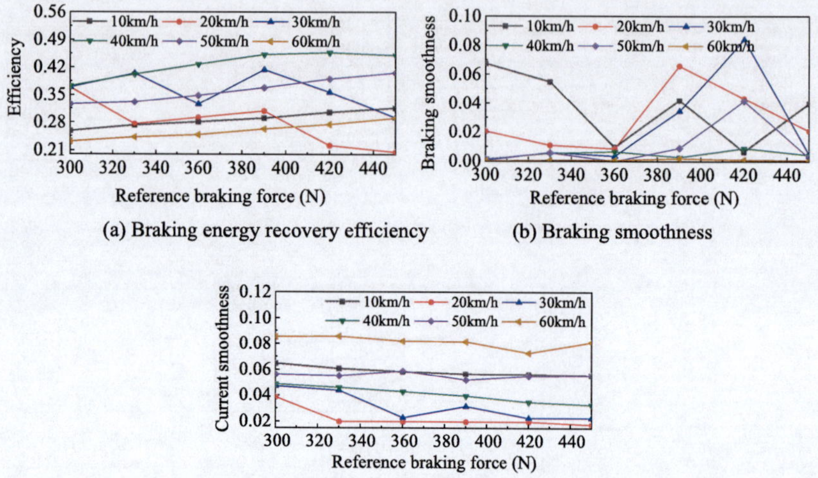

(a) Braking energy recovery efficiency (b) Braking smoothness

(c) Current smoothness

Fig. 5.38 The influence of reference braking force on each optimization objective under low-intensity

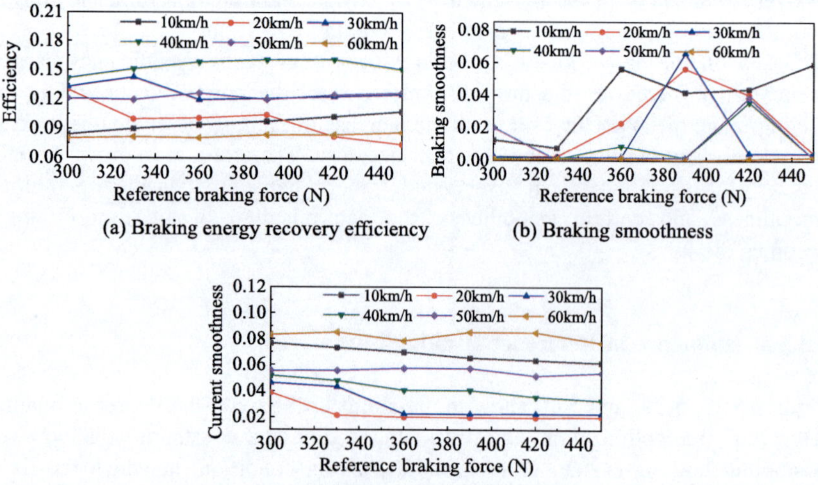

(a) Braking energy recovery efficiency (b) Braking smoothness

(c) Current smoothness

Fig. 5.39 The influence of reference braking force on each optimization objective under medium-intensity braking

relatively large in the range of medium and low speed; With the increase of the reference braking force, the change of the current smoothness is small. Except for the 10 km/h vehicle speed, the current smoothness in the high-speed range is larger than the current smoothness in the low, medium and low speed range.

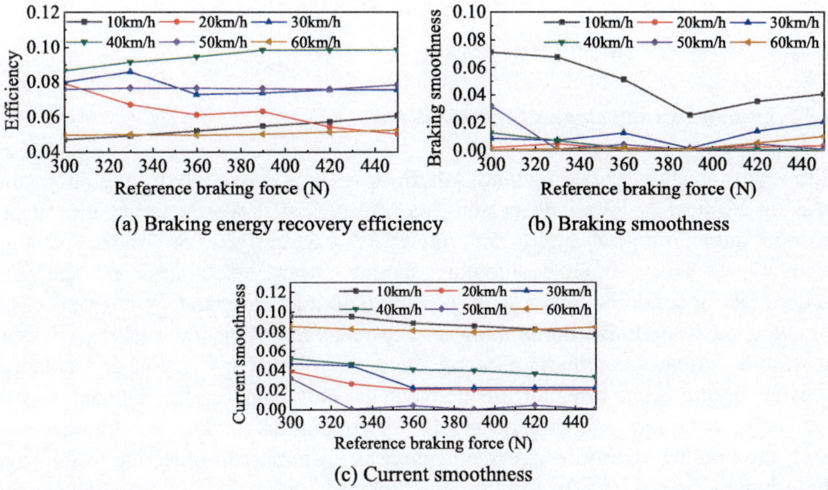

Fig. 5.40 The influence of reference braking force on each optimization objective under high-intensity braking

Under the condition of medium-intensity braking, as shown in Fig. 5.39, with the increase of reference braking force, the braking energy recovery efficiency basically remains unchanged, and the braking energy recovery effect is the best under the intermediate speed of 30–50 km/h; With the increase of the reference braking force, the braking smoothness changes in a large range from 0 to 0.07, and the braking smoothness in the medium and high-speed range is smaller; Similar to the low intensity braking condition, the current smoothness in the medium-intensity braking condition decreases with the decrease of vehicle speed.

Under the condition of high-intensity braking, as shown in Fig. 5.40, the change trend of braking energy recovery efficiency, braking smoothness, and current smoothness with reference braking force is basically the same as that under medium and low-intensity braking condition, but the braking energy recovery efficiency under high-intensity braking condition is much smaller than that under medium and low-intensity braking condition; The braking smoothness under high-intensity braking condition is also significantly lower than that under medium and low-intensity braking condition; Compared with the medium and low-intensity braking condition, the current smoothness changes under the high-intensity braking condition are smaller.

Based on the above analysis, it is necessary to comprehensively consider the braking energy recovery efficiency, braking smoothness, and current smoothness conditions to improve the comprehensive braking performance of EVs, and the determination of the reference braking force becomes the key.

5.3.3 *Multi-objective Optimization of Braking System in Braking Condition*

5.3.3.1 Control Parameter Optimization

Similar to the sliding braking condition, in order to improve the driving range and braking comfort of EVs, and prolong the lifetime of power batteries, the effects turn-on angle, turn-off angle, and reference braking force on braking energy recovery efficiency, braking smoothness, and current smoothness are analyzed under braking condition, and a multi-objective optimization strategy is proposed to simultaneously optimize the braking energy recovery efficiency, braking smoothness, and current smoothness. Among them, the objective function and comprehensive optimization function are defined, as shown in Eqs. (5.39) and (5.40). Similarly, considering comprehensively improving the braking performance of EVs, the braking energy recovery efficiency is an important objective to improve the driving range of EVs. In addition, the braking comfort performance and current smoothing performance should be taken into account, so the weight factors of the three optimization objectives are 0.6, 0.2, and 0.2, respectively.

Compared with braking system of EVs in sliding condition, the comprehensive performance of compound braking system in braking condition is also affected by turn-on angle, turn-off angle, and reference braking force, but these control parameters are affected by vehicle speed and braking intensity at the same time. Therefore, based on the multi-objective optimization strategy, different parameter combinations are determined by the orthogonal experimental design method, and the optimal turn-on angle, turn-off angle, and reference braking force under different speed and braking intensity are obtained by Eqs. (5.39) and (5.40). At the same time, the comprehensive braking performance database is established based on vehicle speed and braking intensity, and the multi-objective optimization process of the composite braking system under braking conditions is shown in Fig. 5.41. According to the optimization function obtained above in this paper, the multi-objective optimization framework of regenerative braking system under braking condition of EVs is designed and shown in Fig. 5.41. Firstly, in order to optimize the above three objectives simultaneously in real time under braking condition, according to the orthogonal optimization design method, the multi-objective optimization database is established based on different combinations of turn-on angle, turn-off angle, and reference braking force. Secondly, the parameters obtained by the orthogonal experimental design method are simulated in the regenerative braking system. Then, the objective function is defined by three weight factors, as shown in Eq. (5.39), and the sensitivity analysis and range analysis of the control parameters are carried out by orthogonal optimization experiment method, and the optimization function is defined based on the objective function, as shown in Eq. (5.40). Finally, the comprehensive objective K and the optimization values are obtained through the processing and analysis of the

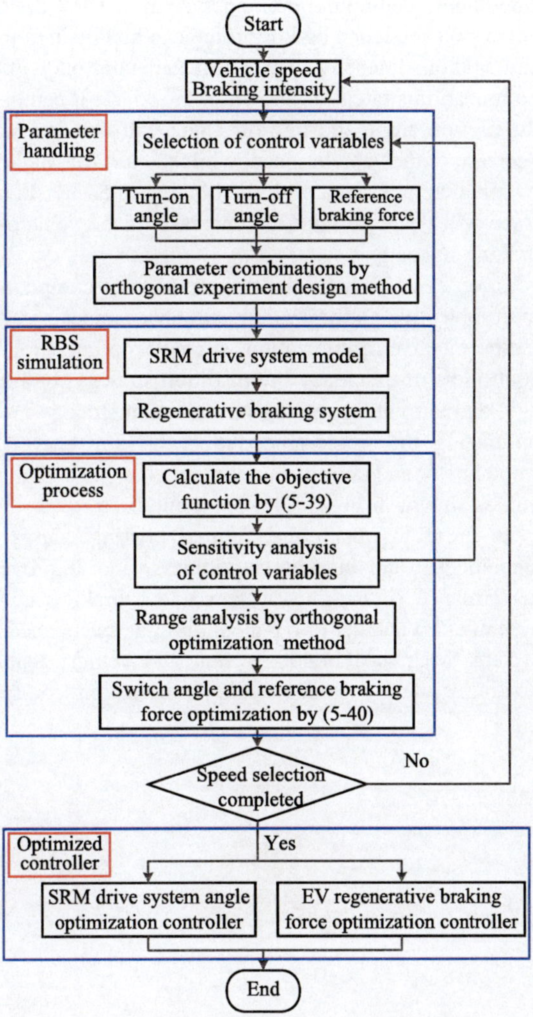

Fig. 5.41 Multi-objective optimization process of compound braking system under braking conditions

database, and angle optimization controller and braking force optimization controller are designed through multi-objective optimization results.

5.3.3.2 Optimization of Controller Parameters

An angle optimization controller and a braking force optimization controller are established based on the multi-objective optimization results of the compound brake system under braking condition. In order to achieve the purpose of optimizing the braking energy recovery efficiency, braking smoothness, and current smoothness compound braking system of EVs, the optimal turn-on angle, turn-off angle, and reference braking force can be obtained through real-time vehicle speed and braking intensity. The optimized controller obtained by the multi-objective optimization strategy is shown in Fig. 5.42. It can be analyzed from the figure that the turn-on angle and turn-off angle will first increase and then decrease with the decrease of the vehicle speed, while the turn-on angle and turn-off angle will change within a certain range with the braking intensity; In addition, the reference braking force will decrease with the decrease of the vehicle speed and the increase of the braking intensity.

In order to evaluate the performance of compound braking system of EV with multi-objective optimization strategy under braking condition, besides the multi-objective optimization strategy optimization controller, the optimization controllers of efficiency optimization strategy, braking comfort optimization strategy, and current smoothing optimization strategy are also established. The goal of optimizing the single-objective of braking energy recovery efficiency, braking smoothness, and current smoothness is realized, and comparative analysis is carried out, as shown in Figs. 5.43, 5.44, and 5.45.

In order to obtain the optimal braking energy recovery efficiency, braking smoothness, and current smoothness, and improve the comprehensive braking performance of braking system under braking condition, a compound braking system of EVs under braking condition is established based on regenerative braking system, mechanical braking system and, vehicle dynamics system. The regenerative braking system includes angle optimization controller, reference braking force

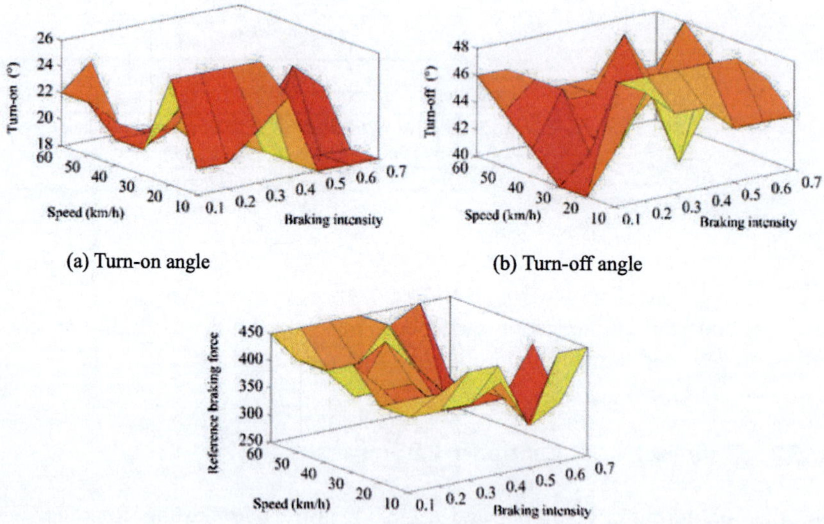

(a) Turn-on angle

(b) Turn-off angle

(c) reference braking force

Fig. 5.42 Optimization controller under multi-objective optimization strategy

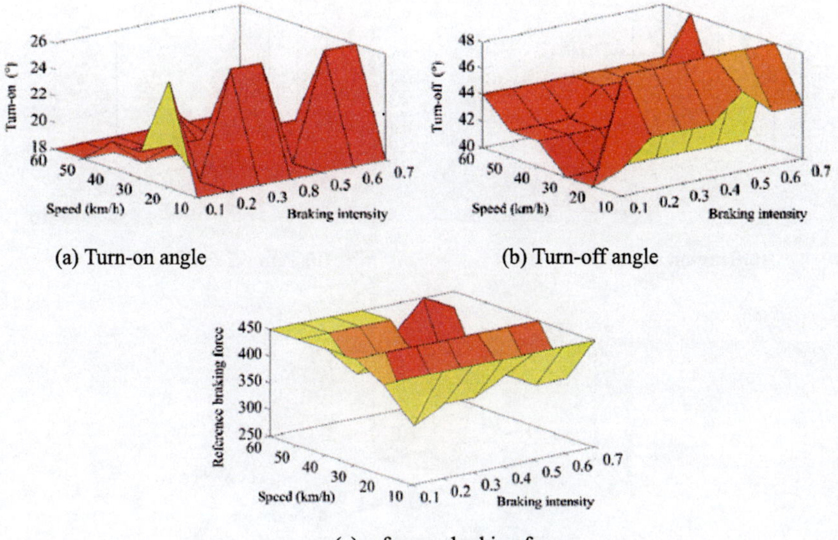

(a) Turn-on angle (b) Turn-off angle

(c) reference braking force

Fig. 5.43 Optimization controller under the efficiency optimization strategy

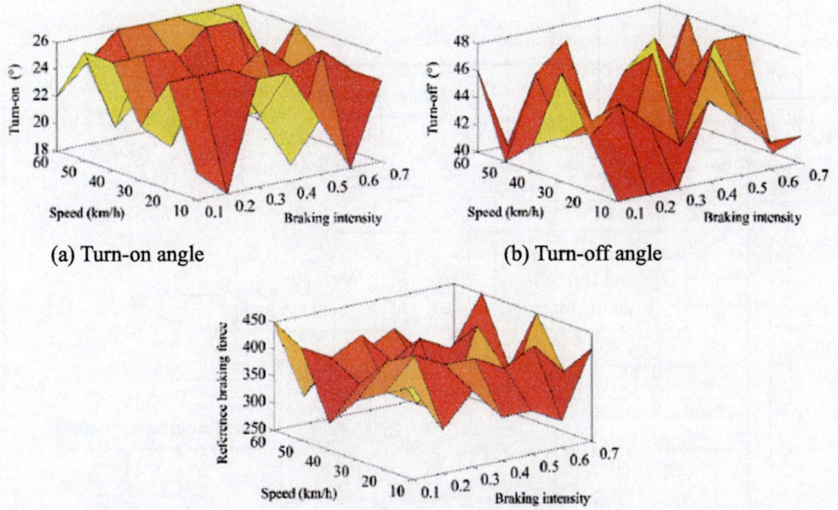

(a) Turn-on angle (b) Turn-off angle

(c) reference braking force

Fig. 5.44 Optimization controller under the braking comfort optimization strategy

optimization controller, and SRM drive system, as shown in Fig. 5.46. Based on the angle optimization controller and reference braking force optimization controller, the braking system can obtain the optimal turn-on angle, turn-off angle, and

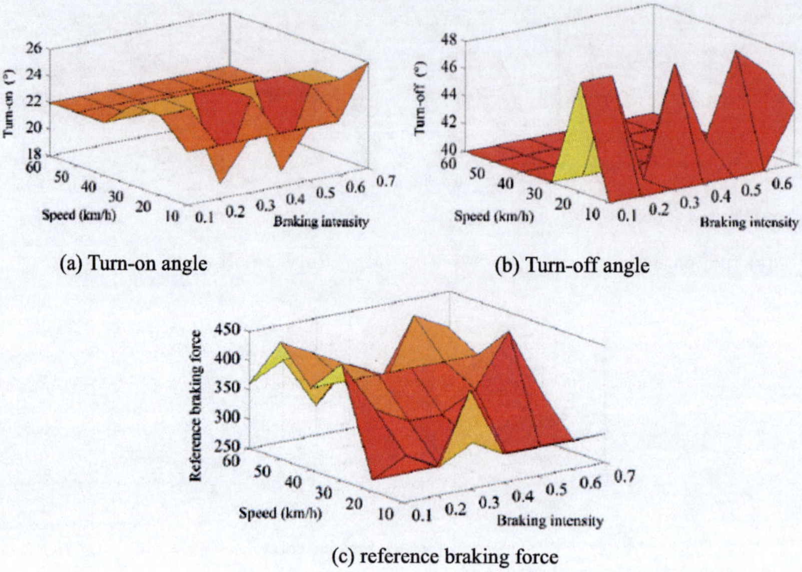

(a) Turn-on angle
(b) Turn-off angle
(c) reference braking force

Fig. 5.45 Optimization controller under the current smoothness optimization strategy

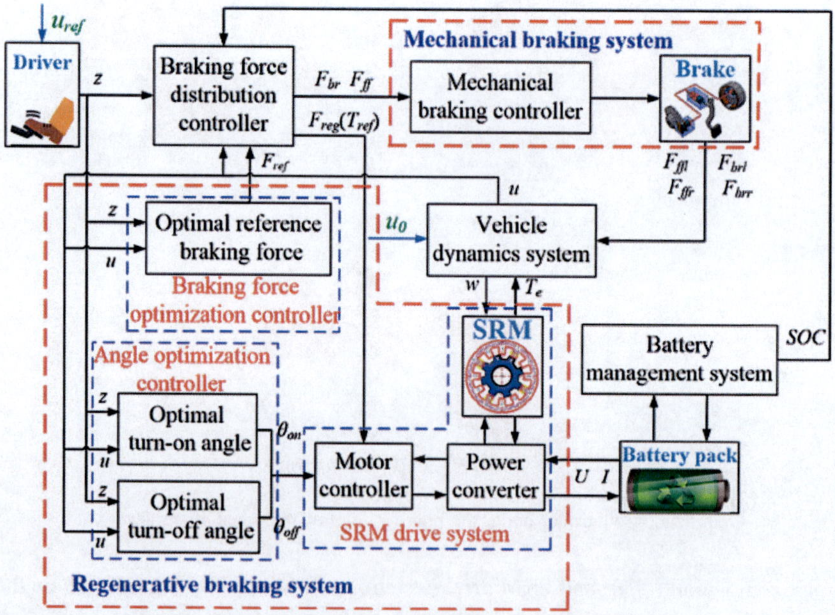

Fig. 5.46 Optimal compound braking system under braking conditions

reference braking force through real-time vehicle speed and braking intensity. The purpose of optimizing the comprehensive performance of the braking system under the multi-objective optimization strategy is achieved.

5.3.4 Comparative Analysis of Simulation Results Under Braking Condition

In order to explore the influence of multi-objective optimization strategy on the dynamic characteristics compound braking system of EVs under braking condition, in this paper, the compound braking system of EVs under multi-objective optimization strategy is compared with that under efficiency optimization strategy, braking comfort optimization strategy, and current smoothing optimization strategy.

5.3.4.1 Comparative Analysis of Different Optimal Control Strategies

The comparative analysis mode compound braking system of EVs under braking condition is similar to that under sliding condition. In order to explore the dynamic characteristics of compound braking system under several optimization strategies, the efficiency factor $K_{\eta reg}$, braking smoothness factor K_{js} and current smoothing factor K_{is} are also compared and analyzed. The EV decelerates and stops at the initial speed of 60 km/h. In addition, the dynamic characteristics of the compound braking system under high, medium and low-intensity braking conditions are analyzed, as shown in Figs. 5.47, 5.48, and 5.49.

Under the condition of low-intensity braking, as shown in Fig. 5.47. The efficiency factors are all above 0.8 under multi-objective optimization strategy, while the efficiency factors are mostly below 0.6 under braking comfort optimization strategy and current smoothing optimization strategy. At the speed of 30 km/h, the braking comfort factor under multi-objective optimization strategy is equal to that of the other two optimization strategies. In addition, the braking smoothness factor under the multi-objective optimization strategy is significantly greater than that under the other two optimization strategies; During the deceleration process of EVs, the current smoothness factor from 0.3 to 0.65 under the three optimization strategies varies, but the current smoothness factor under the multi-objective optimization strategy is the largest.

Under the condition of medium-intensity braking, as shown in Fig. 5.48. The efficiency factor under the multi-objective optimization strategy is above 0.9, the efficiency factor under the braking comfort optimization strategy varies from 0.3 to 0.8, and the efficiency coefficient is less than 0.4 when the vehicle speed ranges from 30 to 60 km/h. In addition, the efficiency factor is less than 0.7 under the current smoothness optimization strategy, which is significantly less than that under the multi-objective optimization strategy. Compared with the other two

Fig. 5.47 Three factors under low-intensity braking condition

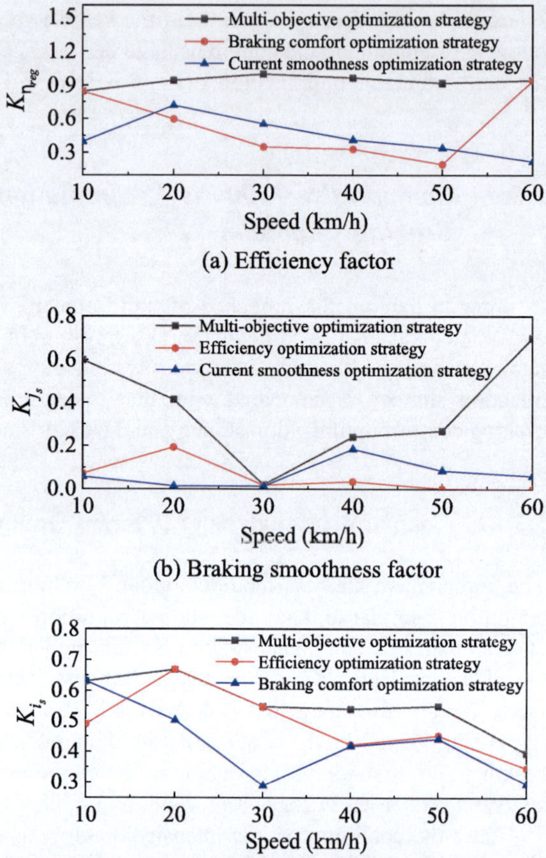

(a) Efficiency factor

(b) Braking smoothness factor

(c) Current smoothness factor

optimization strategies, the braking smoothness factor under the multi-objective optimization strategy is the largest, and the braking smoothness factor changes from 0 to 0.1 under the efficiency optimization strategy and the current smoothness optimization strategy; The change trend of the current smoothing factor under the three optimization strategies is basically the same, but the current smoothing factor under the multi-objective optimization strategy is the largest.

Under the condition of high-intensity braking, as shown in Fig. 5.49. The efficiency factor, braking smoothness factor, and current smoothness factor under the multi-objective optimization strategy are significantly larger than those under the other two optimization strategies. Based on the above analysis, under the braking condition, the compound braking system under the multi-objective optimization strategy can effectively improve the braking energy recovery efficiency, braking comfort, and current smoothness.

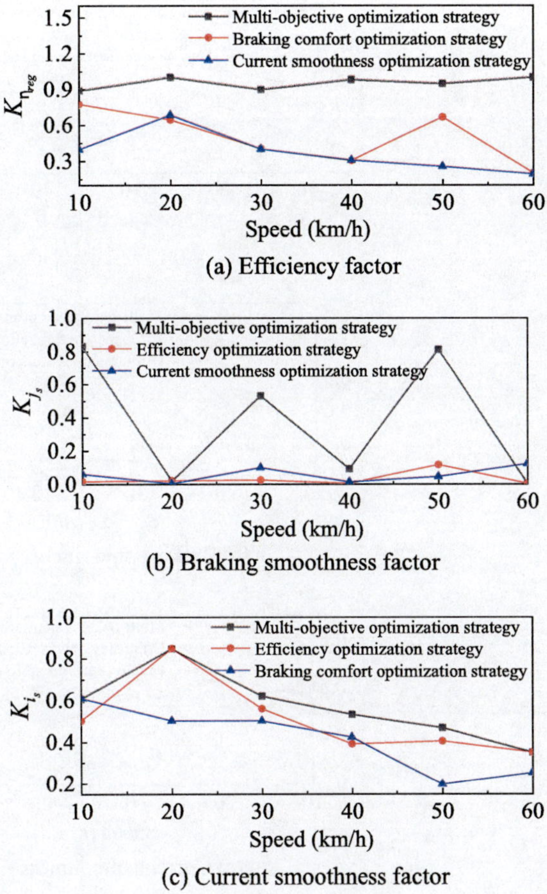

(a) Efficiency factor

(b) Braking smoothness factor

(c) Current smoothness factor

Fig. 5.48 Three factors under medium-intensity braking condition

5.3.4.2 Dynamic Characteristic Analysis of Compound Braking System

In order to verify the effectiveness of the proposed compound braking coordinated control scheme under the multi-objective optimization strategy, the initial speed of 60 and 30 km/h in this paper represent the high-speed and low-speed braking condition for simulation research respectively, and the braking force distribution is explored based on the established compound braking system of EVs, as shown in Fig. 5.50.

When the initial speed is 60 km/h, the total braking force decreases gradually in the deceleration braking process of EVs. According to the compound braking control scheme, the total braking force is distributed to the regenerative braking force of the front wheel, the mechanical braking force, and the braking force of the rear wheel. Before 3.5 s, the total braking force is preferentially distributed to the

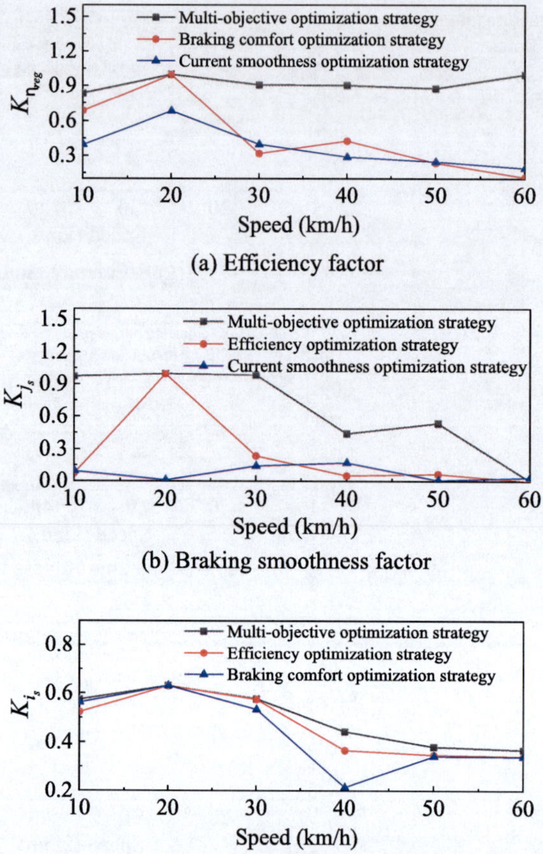

(a) Efficiency factor

(b) Braking smoothness factor

(c) Current smoothness factor

Fig. 5.49 Three factors under high-intensity braking condition

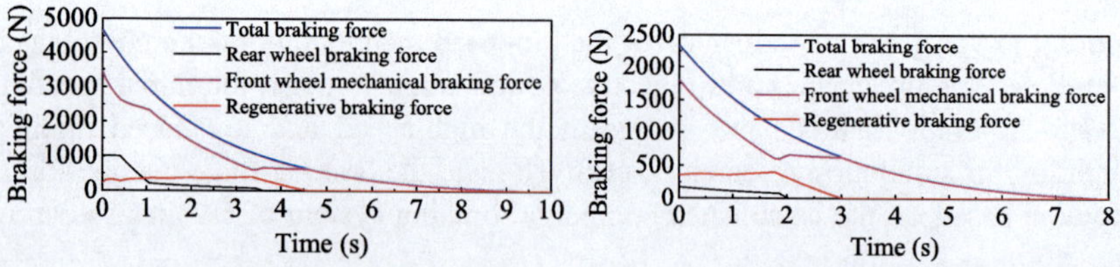

(a) Initial speed 60 km/h (b) Initial speed 30 km/h

Fig. 5.50 Braking force distribution

regenerative braking force of the front wheel, and the insufficient part is supplemented by the mechanical braking force of the front wheel and the braking force of the rear wheel; After 3.5 s, as the vehicle speed is gradually lower than 8 km/h, the regenerative braking force of the motor will gradually withdraw and decrease with the total braking force; After 3.6 s, the rear wheel braking force is no longer involved in the braking, the regenerative braking force of the motor decreases, and the remaining total braking force is provided by the front wheel mechanical braking force, which has an increasing trend; After 4.5 s, the motor no longer provides the braking force, and all the braking force is provided by the front wheel mechanical braking force until the EV decelerates and stops.

When the initial speed is 30 km/h, the change trend of braking force is basically the same as that of the initial 60 km/h. After 2 s, the front wheel mechanical braking force does not participate in the braking, the regenerative braking force gradually decreases, and the rear wheel braking force has an increasing trend; After 3 s, the total braking force required by the EV is all provided by the front wheel mechanical braking force.

In order to further evaluate the dynamic characteristics of the compound braking system model under the multi-objective optimization strategy, the actual energy recovery, SOC, braking distance, and vehicle speed under the multi-objective optimization strategy are compared with those under the other three single-objective optimization strategies. Among them, the initial SOC is set to 60%, and the simulation results are shown in Figs. 5.51 and 5.52.

When the initial vehicle speed is 60 km/h, as shown in Fig. 5.51, the actual recovered energy and SOC under several optimization strategies increase gradually before 4.5 s, and keep stable after 4.5 s. After that, the actual recovered energy and SOC under the efficiency optimization strategy are the largest, and the actual

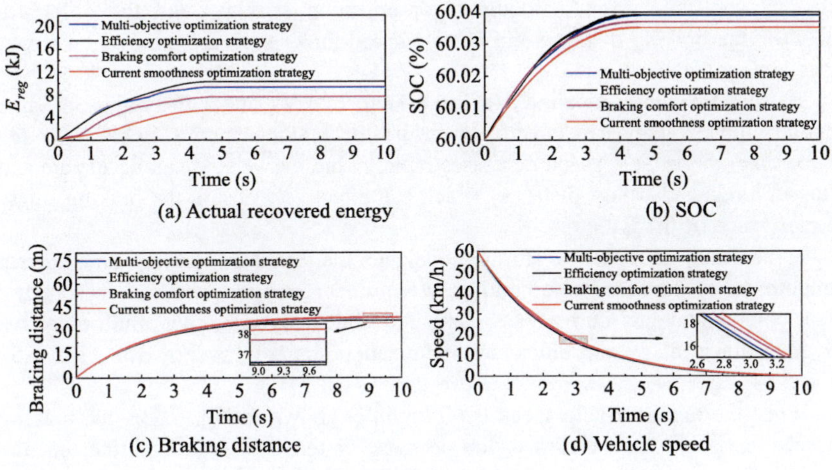

Fig. 5.51 Simulation results at the initial speed of 60 km/h

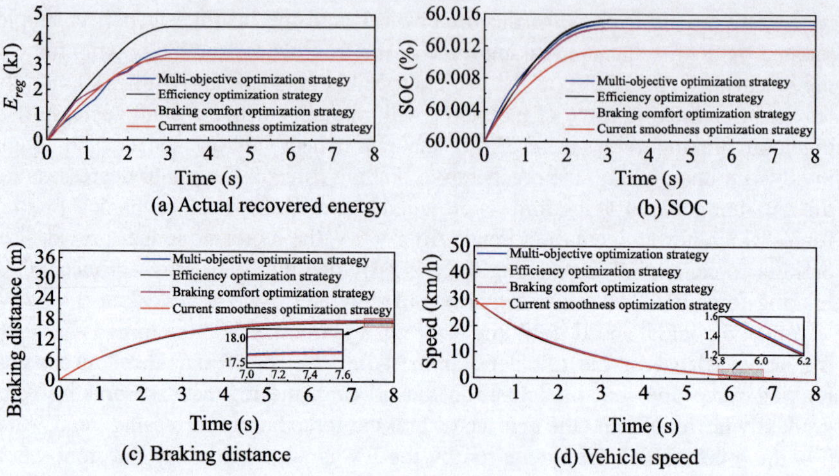

Fig. 5.52 Simulation results at the initial speed of 30 km/h

recovered energy under the multi-objective optimization strategy is only 1.1 kJ less than that under the efficiency optimization strategy, and much larger than that under the braking comfort optimization strategy and the current smoothness optimization strategy; The results show that the difference between braking distance and vehicle speed is small under several optimization strategies.

When the initial vehicle speed is 30 km/h, as shown in Fig. 5.52, the actual energy recovery, SOC, braking distance, and vehicle speed change trend under several optimization strategies are the same as those under the initial vehicle speed of 60 km/h. The actual recovered energy and SOC under the multi-objective optimization strategy are larger than those under the braking comfort optimization strategy and the current smoothness optimization strategy, and the difference between the braking distance and vehicle speed under several optimization strategies is very small.

To sum up, the compound braking system of EVs under the multi-objective optimization strategy can effectively ensure the braking energy recovery, at the same time, the vehicle speed decreases rapidly, and the vehicle can decelerate and stop in a relatively short distance, which can effectively ensure the braking safety performance of the EVs.

In addition, in order to further evaluate the braking comfort and current smoothness of the compound braking system under braking condition, the braking deceleration and bus current are compared and analyzed under multi-objective optimization strategy and efficiency optimization strategy, as shown in Figs. 5.53 and 5.54.

When the initial vehicle speed is 60 km/h, as shown in Fig. 5.53a, the braking deceleration also decreases with the decrease of the total braking force, but the deceleration variation range under the multi-objective optimization strategy is

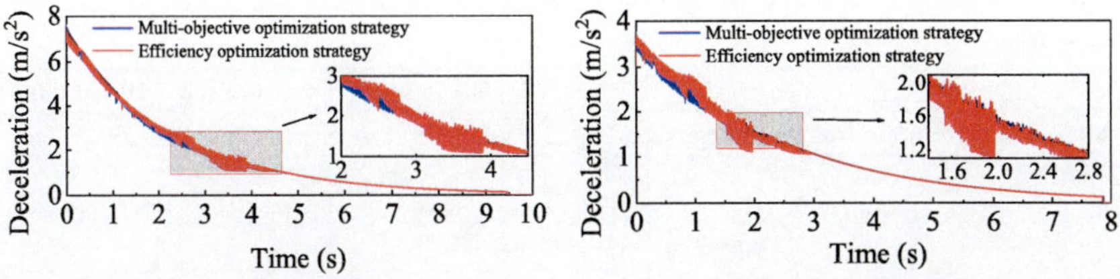

(a) Initial speed 60 km/h

(b) Initial speed 30 km/h

Fig. 5.53 Braking deceleration

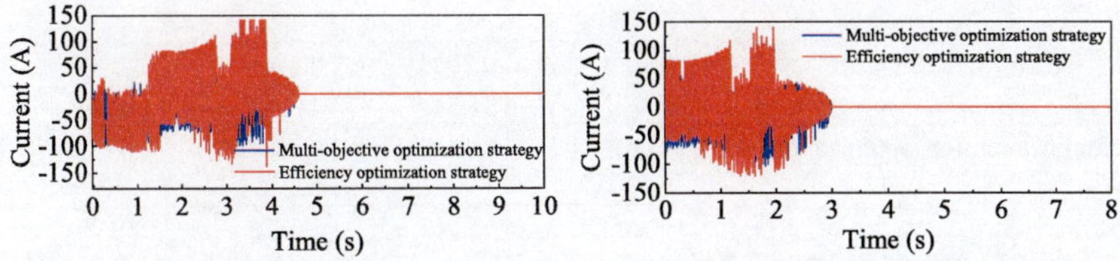

(a) Initial speed 60 km/h

(b) Initial speed 30 km/h

Fig. 5.54 Bus current

relatively smaller than that under the efficiency optimization strategy; The deceleration with initial speed of 30 km/h is similar, as shown in Fig. 5.53b.

As shown in Fig. 5.54, after the initial speed of 60 km/h for 4.5 s and 30 km/h for 3 s, the motor does not participate in braking, so the bus current is 0. Under the condition of motor participating in braking, the current fluctuation range under efficiency optimization strategy is much larger than that under multi-objective optimization strategy.

In order to further analyze the braking performance of the compound braking system of EVs at different speed ranges under braking condition, the actual energy recovery increment ΔE_{reg}, the state of charge increment ΔSOC, braking deceleration variance a_v and charging current variance c_v are selected for comparative analysis under different optimization strategies at high-speed range of 60–50 km/h, medium-speed range of 40–30 km/h and low-speed range of 20–10 km/h, respectively. The comparison results are shown in Table 5.3.

On the one hand, the ΔE_{reg} and ΔSOC of medium and high-speed segment are larger than those of low-speed segment, and the current smoothness of high-speed segment is better than that of medium and low-speed segment. On the other hand, in the high-speed segment, the ΔE_{reg} and ΔSOC under the multi-objective optimization strategy are 12.33 kJ and 0.0895% respectively, which are much larger than those under the braking comfort optimization strategy and the current smoothness optimization strategy; The a_v under multi-objective optimization strategy, efficiency optimization strategy, and current smoothness optimization strategy are 0.2226,

Table 5.3 Comparison results of dynamic characteristics of braking systems in different speed segments

		60–50 km/h	40–30 km/h	20–10 km/h
Multi-objective optimization strategy	ΔE_{reg} (kJ)	12.33	9.46	4.59
	ΔSOC (%)	0.0895	0.0515	0.0190
	a_v	0.2226	0.2551	0.2608
	c_v	18.01	21.03	30.52
Efficiency optimization strategy	ΔE_{reg} (kJ)	12.67	9.89	4.91
	ΔSOC (%)	0.0919	0.0629	0.0206
	a_v	0.2933	0.2868	0.2833
	c_v	31.51	21.29	31.28
Braking comfort optimization strategy	ΔE_{reg} (kJ)	7.34	8.49	4.32
	ΔSOC (%)	0.0542	0.0328	0.0175
	a_v	0.1670	0.1622	0.2516
	c_v	22.61	23.50	32.43
Current smoothness optimization strategy	ΔE_{reg} (kJ)	5.14	6.87	4.51
	ΔSOC (%)	0.0631	0.0452	0.0121
	a_v	0.2473	0.2762	0.3074
	c_v	7.66	12.90	20.1017

0.2933, and 0.2473 respectively. The c_v under multi-objective optimization strategy, efficiency optimization strategy, and braking comfort optimization strategy are 18.01, 31.51, and 22.61, respectively. Therefore, the braking comfort and current smoothness performance are relatively better under multi-objective optimization strategy. Similarly, ΔE_{reg}, ΔSOC, a_v, and c_v are the same in the medium and low-speed segments.

Based on the above analysis, compared with the single-objective optimization strategy, the multi-objective optimization strategy under braking condition can effectively balance the braking recovery energy efficiency, braking comfort, and current smoothness through the weight coefficient, which can improve the driving range of EVs, increase the braking comfort performance and extend the lifetime of power batteries.

Fig. 5.55 Schematic diagram of PIL experiment

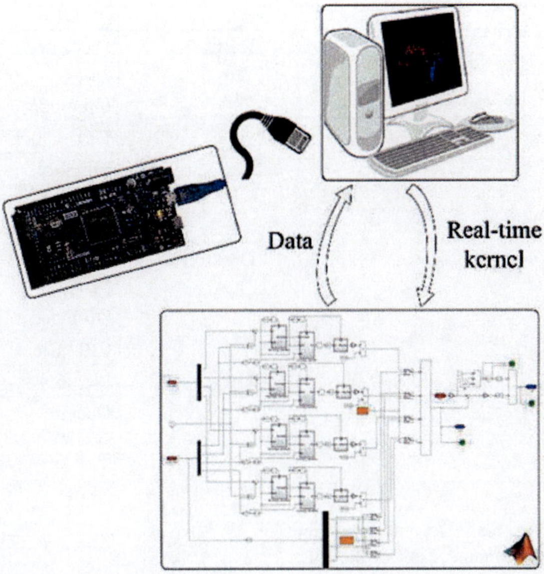

5.3.5 *Processor in Loop Test Verification*

To verify the real-time ability of the braking torque control strategy, the processor in loop (PIL) experiment, which mean the processor is in the real world while the others in the simulation loop, is carried out with the Simulink Desktop Realtime Toolbox and Raspberry Pi 3 B + development board. A real-time kernel for executing Simulink models is provided by the toolbox on a Windows desktop and the braking torque control strategy is flashed to the Raspberry Pi. The schematic diagram of this PIL experiment is shown in Fig. 5.55. Some key features of the test hardware are listed in Table 5.4. Figure 5.56 shows the real-time motor torque and current by the proposed control strategy comparing with the reference values. The test results show that the proposed braking torque control strategy can improve the comprehensive braking performance of EVs under sliding braking condition.

5.4 Summary

This chapter mainly studies the coordinated control strategy compound braking of EVs. Firstly, the EV model is established based on the vehicle dynamics model, driver model, and power battery model; secondly, the new piecewise braking force distribution control scheme is proposed based on the ideal braking force control scheme and the braking force control scheme constrained by ECE regulations; Then, considering the influence of vehicle speed, battery SOC, and braking

Table 5.4 Key specs of the PIL hardware

	Features	
Raspberry Pi	Processor	Broadcom BCM2837B0
	CPU Op-mode	64bits
	CPU Clock	1.4 GHz
	Memory	1 GB LPDDR2 SDRAM
	Input power	5 V/2.5A DC via microUSB
Desktop	Processor	Intel i7-8750H
	CPU Op-mode	64bits
	CPU Clock	2.2 GHz
	Memory	16 GB
	Operate system	Windows 10

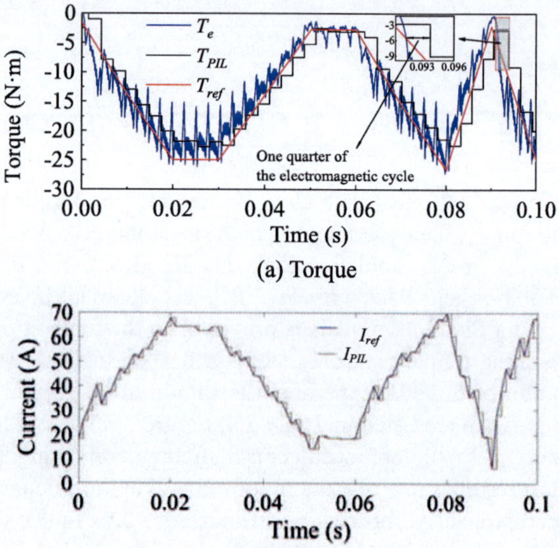

(a) Torque

(b) Current

Fig. 5.56 Torque and current calculation result of the PIL test

intensity on the regenerative braking force of the motor, the braking torque of the motor is modified, and the front wheel braking force distribution control strategy is proposed on this basis. Finally, combined with the SRM drive system, the compound braking system model of the EVs is established in the MATLAB/Simulink environment.

In this chapter, the braking system control of EVs under sliding condition is also studied. Firstly, the optimization objective is determined based on the braking demand of the EV under the sliding condition; secondly, the multi-objective optimization strategy is proposed for the braking system of EVs under the sliding condition; Then, the angle optimization controller and the braking force optimization controller are designed based on the multi-objective optimization strategy; Finally, the comparison and analysis of the braking system under the single-objective optimization strategy and the dynamic analysis of the braking system are carried out, and the multi-objective optimization strategy is obtained. It can effectively improve the comprehensive braking performance braking system of EVs under sliding condition.

The compound braking system of EVs under braking condition is studied and analyzed in this chapter. Firstly, the optimization objective is determined based on the braking condition of EVs, and the influence of control parameters on braking performance is explored based on the proposed multi-objective optimization strategy. Secondly, the angle optimization controller and braking force optimization controller are designed based on several optimization control strategies; Then, compared with the compound braking system under the single-objective optimization strategy and the dynamic analysis of the compound braking system, it is concluded that the multi-objective optimization strategy can effectively improve the comprehensive braking performance braking system of the EV under braking condition. Finally, the real-time effectiveness of the braking control strategy is further verified by the processor in the loop test.

References

1. Huang Y, Wang H, Khajepour A, He H, and Ji J. Model predictive control power management strategies for HEVs: a review. J. Power Sources, 2017, 341: 91–106.
2. Boldea I, Tutelea L, Parsa L, and Dorrell. D. Automotive electric propulsion systems with reduced or no permanent magnets: an overview. IEEE Trans. Ind. Electron, 2014, 61(10): 5696–5711.
3. Tang X, Yang W, Hu X, and Zhang D. A novel simplified model for torsional vibration analysis of a series-parallel hybrid electric vehicle. Mech. Syst. Signal. Process, 2017, 85: 329–338.
4. Sikder C, Husain I, and Sozer Y. Switched reluctance generator control for optimal power generation with current regulation. IEEE Trans. Ind. Appl., 2014, 50(1): 307–316.
5. Rahman KM, Fahimi B, Suresh G, Rajarathnam AV, and Ehsani M. Advantages of switched reluctance motor applications to EV and HEV: design and control issues. IEEE Trans. Ind. Appl., 2000, 36(1): 111–121.
6. Saad NH, El-Sattar AA, and Metally ME. Artificial neural controller for torque ripple control and maximum power extraction for wind system driven by switched reluctance generator. Ain Shams Eng. J., 2018, 9(4): 2255–2264.
7. Ding W, Liu G, and Li P. A hybrid control strategy of hybrid-excitation switched reluctance motor for torque ripple reduction and constant power extension. IEEE Trans. Ind. Electron., 2020, 22(1): 38–48.
8. Zhu Y, Yang C, Zhao G, and Wang D. Optimization of switched reluctance motor drive system in electric vehicles application. journal of motor and control. Electr. Mach. Contrl., 2018, 22(10): 25–34.

9. Namazi MM, Saghaiannejad SM, Rashidi A, and Ahn JW. Maximum power recovery of regenerative braking in electric vehicles based on switched reluctance drive. J. Electr. Eng. Technol, 2018, 13(2): 800–811.
10. Yahia H, Liouane N, and Dhifaoui R. Differential evolution method-based output power optimisation of switched reluctance generator for wind turbine applications. IET Renew. Power Gen., 2014, 8(7): 759–806.
11. Zhang JZ, lv C, and Li YT. Hybrid propulsion and hybrid braking technologies of electrified vehicles: status and prospect. J. of Automotive Safety and Energy, 2014, 5(3): 209–223.
12. Gang L, Lin, et al. Simulation Study on Regenerative Braking System of Pure Electric Vehicle with Compound Power Supply. AEMCME 2019.
13. Xie YB, Wang SC. Research on Regenerative Braking Control Strategy and Simulink Simulation for 4WD Electric Vehicle. IOP Conference Series Materials Science and Engineering, 2018, 398.
14. Wang F, Lin Z, and Mao X. Regenerative braking of hybrid power system in coasting mode. JSCUT, 2009, 37(7): 62–68
15. Park, J.Y. Apparatus and method for guiding driver to coasting of in eco-friendly vehicle. 2016.
16. Abbas H, Kim Y and Siegel J. Synthesis of pontryagin's maximum principle analysis for speed profile optimization of all-electric vehicles. J. DYN. SYST-T ASME vol. 2019, 141(7).
17. Satvik S and Shubham G. Quantification of regenerative braking energy in a two-wheeler incorporating various duty cycles. 2020 International Conference on Computational Performance Evaluation, Jul. 2020.
18. Liu H, He X, Chu L, et al. Study on control strategy of regenerative braking for electric bus based on braking comfort. International Conference on Electronic & Mechanical Engineering & Information Technology. IEEE, 2011.
19. Yuan Y, Zhang J, Li Y, et al. A Novel Regenerative Electrohydraulic Brake System: Development and Hardware-in-Loop Tests. IEEE Transactions on Vehicular Technology, 2018: 11440–11452.
20. Bunyaeva EV, Skorik VG, Vlas'Evskii SV, et al. A method for improving the energy efficiency of an alternating current electric locomotive in the regenerative braking mode. Russian Electrical Engineering, 2016, 87(2): 73–76.
21. Liu W, Qi H, Liu X, et al. Evaluation of regenerative braking based on single-pedal control for electric vehicles. Frontiers of Mechanical Engineering, 2020, 15(1): 166–179.

Chapter 6
Performance Matching Design for the Vehicle Drive System

6.1 Introduction

In general, EVs are composed of electric drive and control system, transmission system, steering system, and braking system. The electric drive system is the heart of the EV, and it is also the biggest difference from the vehicles with internal combustion engines. Normally, the electric drive and control system consists of power source, motor, and motor speed controller, which can efficiently convert the energy stored in the battery into the kinetic energy of the vehicle, and realize the recovery of braking energy at the same time. In the field of drive motors for EVs, the selection of drive motors is restricted by the vehicle design goals, the vehicle constraints, and the energy sources. Therefore, choosing and studying the most suitable motor for the EV is currently an important research topic [1–4].

Compared to traditional fuel vehicles, EVs can greatly reduce energy consumption and also reduce exhaust emissions. However, due to the nonlinearity and complexity of the vehicle driving conditions and load characteristics, the correct selection and performance improvement of drive motors focusing on requirements of EV is greatly stringent. Therefore, the parameter matching design considering the driving motor and EV drive system is a very complicated problem that needs to be solved urgently [5]. Many scholars have carried out extensive research on the parameter matching between drive motors and the drive system of EVs or hybrid vehicles [6, 7].

By analyzing the parameter matching problems of EV drive system and the drive motors, it can be seen that there are two main parameter matching methods that are most widely used at present [8–11]. One is to consider the dynamic performance indicators of the vehicles as design requirements to improve the output characteristics of the drive motor, achieving the parameter matching between drive motors and the drive system. The other is to match the work ranges or points of the drive motor under specific vehicle driving conditions with the high-efficiency requirements of the vehicles, improving the dynamic and economy performance of the

Y. Zhu, *The Key Technologies for Powertrain System of Intelligent Vehicles Based on Switched Reluctance Motors*, Recent Advancements in Connected Autonomous Vehicle Technologies 1, https://doi.org/10.1007/978-981-16-4851-9_6

electric drive system. These two methods describe the parameter matching from the aspects of dynamic performance and motor efficiency, while ignore the driving range that is an important indicator for EVs. In addition, the parameter characteristics of the battery under specific drive motor and vehicle drive system play a vital role in the power performance and economic performance of EVs, such that they should be taken more attention for designing the drive system of EVs.

Therefore, based on the drive cycle conditions of EVs, the parameters matching of the SRM with two-phase excitation mode, the battery, and the EV drive system are carried out in this chapter, with the goal of improving the power performance and driving range of the vehicle. It is of great significance in theoretical and engineering areas to complete the parameters matching of the SRM, the battery and the EV drive system for improving the vehicle performance.

6.2 Design Requirements of the EV Drive System

6.2.1 Vehicle System Restrictions

The parameter indicators for evaluating vehicle dynamic performance mainly include maximum vehicle speed, acceleration time, and maximum gradient. These performance indicators can provide the basic requirements for the design of traction motors for electric vehicles. In addition, the operating conditions of EV are complex, such that vehicles' traction motors are usually required to meet the requirements, which include frequent starting and stopping, acceleration and deceleration, as well as wide speed range. On the other hand, to ensure vehicle comfort and driving safety, traction motors also need to meet various harsh weather and environmental conditions, as well as various complex road conditions. Therefore, compared to general industry motors, traction motors used in EV have stricter performance requirements, and the differences between traction motor and industry motor are shown in Table 6.1 [12].

6.2.2 Dynamic Load Characteristic of the Vehicle

Various external forces acting on the vehicle during the driving process include driving force and driving resistance. The driving force of the EV is obtained from the torque of the motor, which is transmitted to the driving wheels through the drivetrain. The link between driving force and driving resistance should conform to the vehicle driving equation, which is shown as

Table 6.1 Comparisons between EV traction motor and normal industry motor

Indicators	Normal industry motor	EV traction motor
Size	Generally unlimited	Layout space restrictions
Working environment	Low vibration and moderate temperature	Severe vibration, large temperature changes
Cooling method	Air cooling	Liquid cooling
Reliability requirements	Lower	Higher
Speed change	Smaller	frequently
Acceleration performance	No request	High request
Power density	Lower	Higher
Efficiency requirements	Working range	All range
Cost performance	General	Extremely high
Speed range	Narrow	wide
Low vibration requirements	Rated speed point	All working conditions

$$F_t = \sum F \tag{6.1}$$

where F_t represents vehicle driving force, and $\sum F$ denotes the sum of the vehicle driving resistance.

When a vehicle is driving on the road, it must overcome the rolling resistance from the ground and the air resistance from the air. Meanwhile, the vehicle also needs to overcome acceleration resistance and gradient resistance under certain driving conditions.

(1) Rolling resistance F_f

When the wheel is rolling, the vertical and tangential interaction forces are generated, and the tire deforms in the contact area between the tire and the road surface. The elastic hysteresis loss caused by the internal friction of the tire appears as a resistance couple that hinders the rolling of the wheel. The rolling resistance is expressed as

$$F_f = mgf \tag{6.2}$$

where F_f is rolling resistance of the vehicle, f means rolling resistance coefficient, and m refers to the total mass of the vehicle.

(2) Air resistance F_w

Air resistance is the component of the air force in the driving direction when the vehicle is driving. Air resistance is divided into friction resistance and pressure resistance, which is shown as

$$F_w = C_D A \rho u_r^2 \tag{6.3}$$

where F_w denotes the air resistance of the vehicle, C_D is coefficient of air resistance, A represents windward area, ρ refers to air density, and u_r means the relative speed of the vehicle.

(3) Ramp resistance F_i

When the vehicle is driving uphill, the component force of the vehicle gravity along the ramp direction is the ramp resistance, which is expressed as

$$F_i = mg \sin \alpha \tag{6.4}$$

where F_i is the resistance of the vehicle on the ramp and α means ramp angle.

(4) Acceleration resistance F_j

The vehicle needs to overcome the inertial force generated when the vehicle accelerates. The expression of acceleration resistance is shown as

$$F_j = \delta ma \tag{6.5}$$

where F_j is acceleration resistance of the vehicle, δ means conversion coefficient of vehicle rotating mass, and a denotes the acceleration degree of the vehicle.

Among the resistances introduced above, air resistance and rolling resistance exist under any driving conditions, while acceleration resistance and ramp resistance only exist under certain working conditions. To intuitively describe the effect of each resistance, taking a small EV with the curb weight of 500 kg as an example, the resistance torque can be calculated according to the Eqs. (6.1–6.5), and the results are shown in Fig. 6.1.

6.2.3 *Design of the EV Drive System*

The motor is the core component of the EV drive system. Its typical torque-speed characteristic curve is composed of a constant torque area and a constant power area. The left side of the rated speed is the constant torque area, and the right side of the rated speed is the constant power area, as shown in Fig. 6.2.

The power of the traction motor mounted on an EV is usually selected initially based on the expected maximum speed of the vehicle. Although the maximum speed is only an indicator of the vehicle dynamic performance, it reflects the vehicle climbing and acceleration capabilities. The selected motor power should not be less

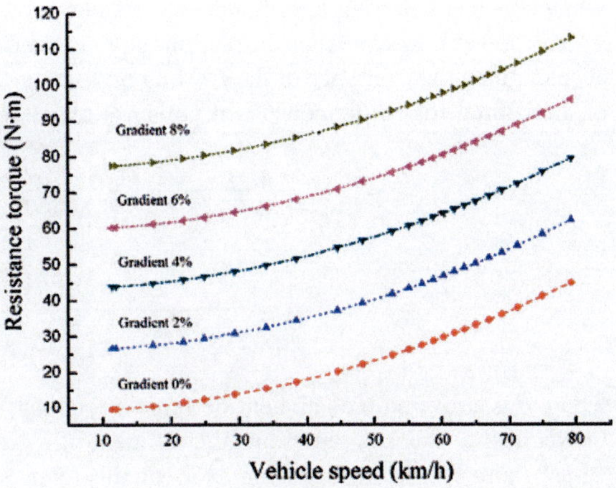

Fig. 6.1 Resistance torque of the vehicle under various gradient

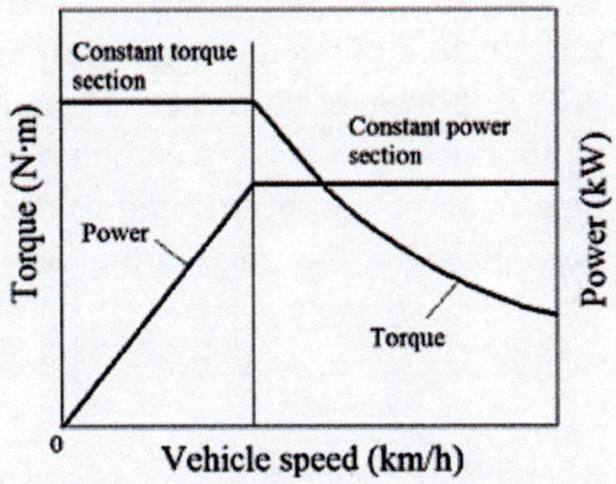

Fig. 6.2 Typical torque-speed profile of electric motor for EVs

than the sum of the driving resistance when the vehicle is running at the highest speed on a flat road. For the vehicle driving at the highest speed, the equation of the power is shown as

$$\sum_{V_{\max}} P = \frac{1}{\eta_T}\left(\frac{Mgf}{3600}V_{\max} + \frac{C_D A}{76140}V_{\max}^3\right) \tag{6.6}$$

where M is the mass of the whole vehicle, f denotes rolling friction coefficient, C_D represents coefficient of air resistance, $V_{\max}$ refers to the maximum speed of EV, A means windward area, and η_T is the mechanical efficiency of the drivetrain.

When the power characteristic of the traction motor is selected according to the climbing ability of the vehicle, the consumed power when climbing a certain slope at a certain driving speed is expressed as

$$\sum_{i} P = \frac{1}{\eta_T}\left(\frac{Mgf}{3600}V_a + \frac{C_D A}{76140}V_a^3 + \frac{Mgi}{3600}V_a\right) \tag{6.7}$$

where V_a is EV driving speed and i means slope.

The power characteristics of traction motors also can be selected according to the acceleration characteristics of the EV on horizontal roads. when the EV accelerating on horizontal roads, the consumed power is expressed as

$$\sum_a P = \frac{1}{\eta_T}\left(\frac{Mgf}{3600}V_a + \frac{C_D A}{76140}V_a^3 + \frac{\delta M}{3600}\frac{dv}{dt}V_a\right) \tag{6.8}$$

$$\delta = 1 + \frac{1}{M}\frac{\Sigma I_w}{R^2} + \frac{1}{M}\frac{I_f i_g^2 i_0^2 \eta_t}{R^2} \tag{6.9}$$

where δ is conversion coefficient of vehicle rotating mass, I_w means the moment of inertia of the wheel, I_f represents the moment of inertia of the flywheel, R refers to wheel radius, i_0 is transmission ratio of the main reducer, and i_g represents the transmission ratio of the gearbox.

The traction motor drive system of EVs should meet the requirements of acceleration, climbing, and maximum speed, which are important design indexes of vehicle power performance. Compared to traditional engine vehicles, the traction motor drive system has a certain overload capacity, which reflects the backup power of EV, and is closely related to the power performance of the vehicle. Hence, the peak power of the traction motor is expressed as

$$P_{e\max} = \lambda P_e \tag{6.10}$$

where P_{emax} is peak power of traction motor, P_e denotes the rated power of the traction motor, and λ means overload factor.

6.3 Dynamic Characteristic Analysis of the SRM

As the heart of EVs, torque and power characteristics of the motor drive system directly determine the various performances of EV, including acceleration performance, climbing ability, and maximum vehicle speed. In addition, when an EV is running under various driving conditions, the required torque and power directly depend on the vehicle driving resistance that the vehicle has to overcome.

Based on the new type SRM with two-phase excitation mode mentioned in Chap. 2 and the established optimized control model of the SRM drive system described in Chap. 4, the selected SRM drive system is simulated and analyzed, and the torque characteristic curve and the motor output power characteristic curve of the SRM are obtained, as shown in Figs. 6.3 and 6.4.

Figures 6.3 and 6.4 describe the torque characteristics and power characteristics of the new SRM drive system. The characteristic curves reflect that the prototype characteristics meet the requirements of ideal characteristics, and the motor starting

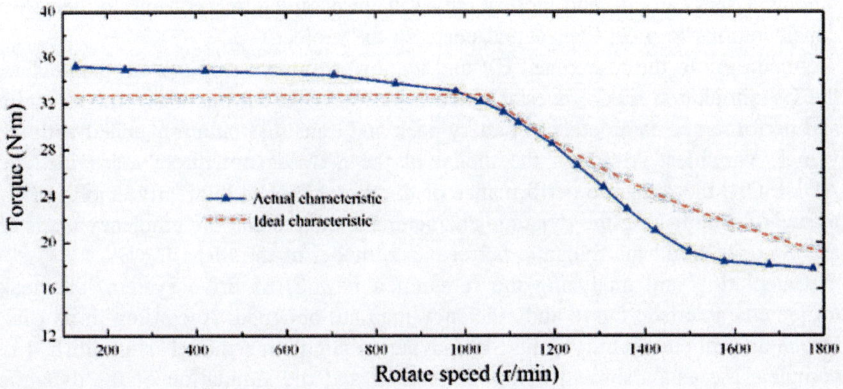

Fig. 6.3 Torque-speed profile of the SRM

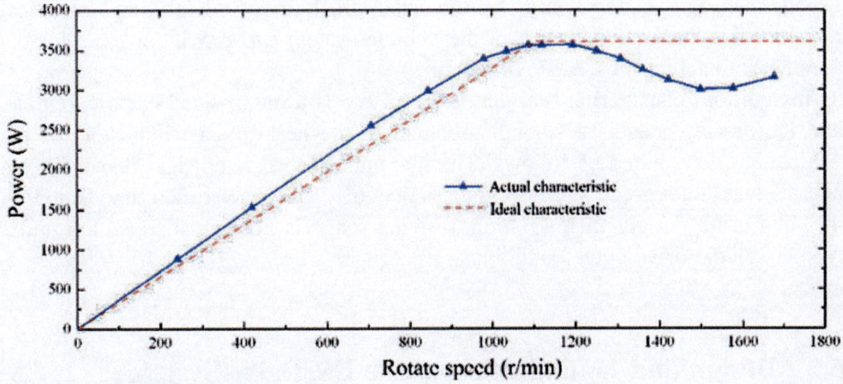

Fig. 6.4 Power-speed profile of the SRM

torque is also relatively large. It is very suitable for the requirements of EV starting conditions on the characteristics of the motor, such that the researched new type of SRM with two-phase excitation mode is very suitable as an EV traction motor.

6.4 Simulation Model of the EV

Advanced vehicle simulator (ADVISOR) is a simulation software developed by the National Renewable Energy Laboratory under MATLAB and Simulink software environments for various performance evaluations of vehicles, which is embedded with pure EV, traditional vehicles, hybrid vehicles, and other types of vehicle models. Users can edit and modify the software parameters according to their own considerations to meet their actual needs in the project.

According to the researched EV and its power transmission system parameters, the EV simulation model is established in ADVISOR. On this basis, the structure and performance parameters of battery pack and transmission are matched with the vehicle variables. Although the model of the SRM is not directly provided by ADVISOR, the dynamic performance of the motor can be indirectly employed by means of establishing the dynamic characteristic curves and the efficiency maps of the motor to build the dynamic performance model of the SRM for EV.

Calculating and analyzing the researched new SRM drive system, the peak torque characteristic curve and efficiency map are obtained. According to its peak torque and efficiency curve, the SRM dynamic simulation model is established to complete the establishment of the EV model and the simulation of the dynamic characteristics of the EV.

In addition, according to the EV layout and the vehicle design parameters, as shown in Table 6.2, the models for the key components of the vehicle are established, including battery pack, transmission, final drive, wheels, and electrical components. The establishment of the vehicle system can provide a basis for the parameter matching of the EV drivetrain.

In addition, considering that the designed EV is a micro-sized electric vehicle, and the driving road is a smooth urban road, the performance indicators of the vehicle are determined as follows. The maximum speed is not less than 55 km/h and the maximum grade ability is not less than 6%. The acceleration time from 0 to 40 km/h is not longer than 15 s, and driving range is not less than 80 km under typical vehicle drive cycle conditions.

6.5 Design and Optimization of the EV Drive System

With the development of the modern automobile industry, especially the gradual progress of new energy automobile technology, the performance parameter matching method focusing on the traditional vehicles, which consider dynamic performance as the key goals, can no longer meet the extra requirements of saving energy and improving the economy. As an important part of the EV drive system, the battery pack is the power source carrier for the drive motor. It plays a very important role in satisfying vehicle power and economy. Meanwhile, the reasonable selection of drivetrain parameters will greatly improve the EV driving performance

Table 6.2 The key parameters of the vehicle

Vehicle mass (kg)	Wind resistance coefficient	Windward area (m^2)	Rolling resistance coefficient	Tire radius (m)
550	0.3	1.7	0.01	0.282

under various working conditions. Therefore, the reasonable parameters matching among the battery pack, the drive motor, and the drivetrain can greatly improve the economy performance of the EV, and also meet the requirements on high power. As a result, the driving range of the EV could be greatly improved.

Although many literatures have studied the parameter matching of the EV drivetrain, most of the parameter matching is carried out based on meeting the vehicle power performance indicators. Meanwhile, the specific parameter selection and optimization for the battery and the transmission of the EV is not fully researched. Therefore, on the premise that the basic parameters of the drive motor and the vehicle are determined, the appropriate battery parameters and transmission parameters are taken as targets in this chapter. By matching the basic parameters of the motor and the vehicle, the purpose of improving the dynamic performance of the vehicle and improving the economic indicators of the EV can be achieved.

6.5.1 Power Battery Parameters Design and Optimization

The power source of the pure EV comes from battery. Therefore, the selection of battery structural parameters and performance parameters will directly affect the dynamic performance and the driving range of the pure EV. The choice of battery is of great significance for improving the performance of the pure EV.

Due to the special working environment and working conditions of EVs, their batteries are different from ordinary industrial batteries. High specific energy, high specific power, fast charging, small size are all performance requirements of EVs for their selection of batteries. As a kind of alkaline battery, the nickel-metal hydride storage battery is another important choice for EV batteries currently, no matter in terms of battery performance or cost, although the lithium battery is the most commonly used battery in EV. It should be noted that an optimization method is going to be proposed in this chapter, which could be suitable for all kinds of batteries.

According to the driving range of the EV, the design and calculation of battery capacity parameters can be carried out. When an EV is driving at the highest constant speed on a straight road, the load power of the battery can be calculated as

$$P_{bat} = \frac{1}{\eta_T \eta_{con} \eta_{bat}} \left(\frac{Mgf}{3600} V_{\max} + \frac{C_D A}{76140} V_{\max}^3 \right) \tag{6.11}$$

where P_{bat} is load power of the battery, η_{con} represents the efficiency of the motor controller, and η_{bat} refers to the discharge efficiency of the battery pack.

According to the requirements of the EV driving range, the output energy of the battery pack within the specified requirements is calculated as

$$E = 1000P_{bat}\frac{S}{V_{\max}} \tag{6.12}$$

where E is the output energy of the battery pack when the design driving range is reached, and S means the driving distance.

Considering that the discharge capacity of the battery pack has a certain relationship with the temperature of the battery and the discharge current, the discharge capacity of the battery pack is calculated as

$$C_{\text{bat}} = \frac{E}{U \times \xi(C)} \tag{6.13}$$

where C_{bat} is discharge capacity of the battery pack, U means battery cell voltage, and $\xi(C)$ represents the influence factor of temperature and discharge current during battery pack discharge.

After the required discharge capacity of the battery pack is obtained, the required number of batteries for the battery pack is expressed as

$$n = \frac{C_{bat}}{C_{sig}} \tag{6.14}$$

where C_{sig} is the rated capacity of the battery cell.

It can be seen from Eqs. (6.11–6.14) that the determination of battery cell capacity is the basis for calculating the number of batteries in the battery pack, so the selection of the rated capacity of battery cells is very critical. In order to study the influence of the rated capacity of battery cell on the power and driving range of the EV, a comprehensive evaluation method of battery cell capacity based on EV driving cycle conditions is proposed. According to the performance indicators of the EV power and driving range, the battery cell capacity and the number of batteries in the battery pack can be selected.

The driving cycle conditions for evaluating and testing the vehicle dynamic performance and economic performance include Economic Commission for Europe and Extra Urban Driving Cycle (ECE + EUDC), Urban Dynamometer Driving Schedule (UDDS), Highway Fuel Economy Test (HWFET), and Japan 1015 cycle condition. Since the main expected driving conditions of the designed micro EV in this chapter are urban road conditions, such that the average speed and maximum speed of the vehicle are relatively low, which meets the characteristics of the ECE driving cycle. Therefore, the ECE driving condition is selected when the designed prototype is simulated in the driving cycle condition.

Based on the preliminary calculation and selection of the EV drivetrain ratio of 3.48, comprehensively considering the power performance and the driving range of the EV, the effects of different battery cell capacities and different battery pack capacities of nickel-metal hydride batteries on the key indicators of the power performance and the driving range are investigated under ECE driving cycle condition. Among them, four different cell capacities of 28 Ah, 45 Ah, 60 Ah, and 80 Ah are selected for the battery pack. In addition, the acceleration time is the required time for the EV to accelerate from 0 to 40 km/h, and the maximum gradient refers to the gradient that the vehicle can climb with a constant speed of 10 km/h. Figure 6.5 shows the vehicle performance indexes obtained through simulation, where the vehicle maximum speed, acceleration time, maximum gradient, and driving range are presented, respectively.

The following observations can be drawn from the calculation results shown in Fig. 6.5. First of all, battery cells with various capacities have different effects on the power performance of the EV. Due to the increasing battery capacity, the total mass is also increased correspondingly. On the other hand, the increased mass will have certain impact on the power performance of the vehicle. Therefore, it can be observed that the two variables, battery capacity and mass, have coupling effects on the vehicle power performance. In addition, the capacity of the battery cell has little effect on the driving range performance of electric vehicles, while the battery pack capacity has greater influence on the driving range. The greater the capacity of the battery pack can be, the longer the driving range is.

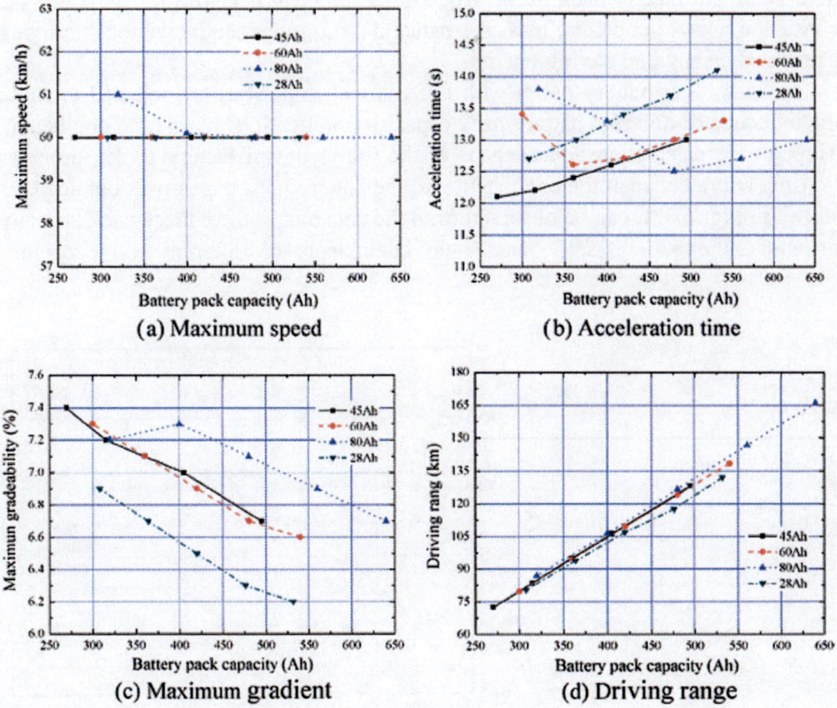

Fig. 6.5 Effects of battery capability on vehicle various performance

The change of the battery *SOC* value shows the energy utilization efficiency of EVs, which plays a vital role in the driving range of vehicle, such that the *SOC* value is also an important parameter index that needs to be considered. Figure 6.6 shows the relationship between battery pack capacity and battery *SOC* value, where the *SOC* is obtained after the EV drive a range with 23.8 km under the ECE urban driving cycle condition.

It can be seen from Fig. 6.6 that the *SOC* value of the battery pack gradually increases with the increase of the battery pack capacity, which is basically the same as the changing trend shown by the driving range. Because, the battery pack has more power remaining, the EV can reach longer distance. The *SOC* value only reflects the remaining power of the battery pack. In order to study the relationship between the two conflicting variables, the battery pack mass and the battery pack capacity, the *SOC* mass ratio coefficient is introduced by using the two variables, which is shown as

$$K_{SM} = \frac{SOC}{M_{bat}} \tag{6.15}$$

where M_{bat} is the total mass of the battery pack.

As expressed in Eq. (6.15), the *SOC* mass ratio coefficient K_{SM} can reflects the *SOC* value per battery pack mass. When this coefficient is larger, it means that the consumed power per battery mass is smaller and it is more conducive for improving battery efficiency and saving energy.

Similarly, four battery packs with different cell capacities are selected to study the influence of different battery pack capacities on the *SOC* mass ratio coefficient. Through the calculation and analysis of the four types of battery packs, the relationship curve between the coefficient and the battery pack capacity is obtained, as shown in Fig. 6.7. It can be observed from the data that as the battery pack capacity increases, the battery *SOC* mass ratio coefficients of different cell capacities

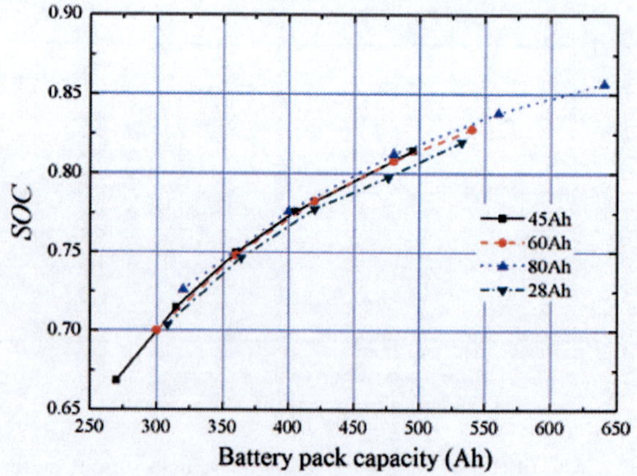

Fig. 6.6 Effects of battery capability on *SOC*

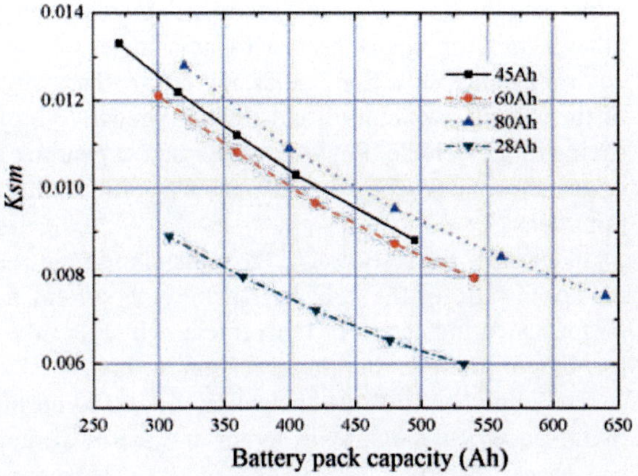

Fig. 6.7 Effects of battery capability on K_{SM}

gradually decrease, which can indicate that the larger battery pack capacity is not conducive to improve battery utilization efficiency and saving energy. In other words, the smaller the mass of the battery pack is, the better the battery utilization efficiency can be. As a result, there can be an obvious conflict between this result and the conclusion obtained from the previous analysis that the larger the capacity of the battery pack is, the greater the driving range of EVs could be, which indicates that the two goals of driving range and battery utilization efficiency need to be compromised. Therefore, on the premise of meeting the design driving range of EVs, the capacity of the battery pack should be reduced as much as possible, that is, the mass of the battery pack should be reduced.

Based on the requirements of the vehicle power performance and driving range performance, combined with the calculation results of the vehicle various performance indexes under different battery pack capacities given in Fig. 6.5, the battery cell capacity and battery pack capacity that meet the design requirements of the vehicle can be obtained. According to the obtained battery capacity that meets the requirements, the results can be compared with those of the *SOC* mass ratio coefficient shown in Fig. 6.7. In order to achieve the best battery utilization rate, it is obviously seen after comparison that the battery capacity with the largest coefficient value can be the best solution. Taking the designed EV and SRM in this chapter as an example, it can be clearly obtained that the optimal battery pack capacity is 315 Ah, and the battery cell model is the nickel-hydrogen storage battery with the capacity of 45 Ah.

6.5.2 Gearbox Parameters Design and Optimization

At present, the driving areas of EVs are mainly concentrated in cities or suburbs, such that the driving resistance of these vehicles varies greatly during running. Therefore, it is necessary to install a final drive and a transmission between the motor and the driving wheels in order to increase the torque range of the EV. On the other hand, for micro EVs with advantages of lightweight, energy-saving, and environmental protection, increasing the gearbox will inevitably increase the weight of the vehicle, which will have certain negative impact on the power and energy saving of the vehicle. Regarding the issue of gearbox selection, not only the number of gears for the gearbox, but also the feasibility for cancelling the gearbox should be both considered.

In terms of the selection of the transmission ratio, the traditional vehicles choose the speed ratio of the vehicle transmission system mainly according to the power performance indicators of the vehicle, which are the vehicle maximum speed, the maximum gradient, and the acceleration time.

The upper limit of the transmission ratio is normally determined by the maximum speed of the engine or motor and the maximum speed of the vehicle, which can be expressed as Eqs. (6.16) and (6.17), where $\prod_i$ means total transmission ratio.

$$\prod i \leq \frac{0.377 n_{\max} R}{V_{\max}} \tag{6.16}$$

$$\prod i = i_0 \cdot i_g \tag{6.17}$$

The lower limit of the transmission ratio is calculated according to the maximum gradient index designed for the vehicle, and the calculation process must consider the restriction requirements of road adhesion coefficient.

When the transmission ratio is calculated according to the maximum gradient, the influence of the air resistance in the moment of resistance on the vehicle is ignored due to the low vehicle speed. And the transmission ratio is calculated as

$$\prod i \geq \frac{F_{i\max} R}{\eta_t T_{\max}} \tag{6.18}$$

where $F_{i\max}$ is the resistance of the vehicle driving at the maximum gradient, and $T_{\max}$ means the maximum output torque of engine or motor.

Due to the influence of the road limit adhesion ability, the calculated transmission ratio is not proportional to the maximum gradient. For front-wheel driving vehicles, the calculation equation of the maximum gradient corresponding to the road limit adhesion ability is shown as

$$\tan \alpha_{\max} = b/(h_g + L/\phi) \tag{6.19}$$

where $\alpha_{\max}$ is the maximum slope angle, h_g means the height of mass center of the vehicle, and L represents the wheelbase of the vehicle.

Combining the vehicle design parameters and motor parameters, according to Eqs. (6.16–6.19), it can be obtained that the range of the transmission speed ratio should be between 3.15 and 4.05 in this chapter.

Since the calculated theoretical range of transmission ratio is relatively small and the vehicle mass is very important for the energy saving and environmental protection of micro EVs, a transmission with two gears or less is considered as the selection target. This also reflects the unique advantages of the motor as an EV traction power from another aspect, that is, the motor not only has superior speed and torque characteristics, but also greatly reduces the weight of the EV.

To fully improve the transmission efficiency of the EV, increase the driving range, improve the power performance of the vehicle, and increase the service life of the traction motor and the transmission system, a strategy for selecting and optimizing EV transmission gears and gear ratios based on the ECE urban cycle condition is proposed. Through optimization, the optimal number of gears and the optimal transmission ratio can be obtained to achieve the goal of improving the power performance of the vehicle and increasing the driving range. It needs to be pointed out that the transmission ratio mentioned below refers to the product of the transmission ratio and the main reducer transmission ratio. For the convenience of description, the transmission ratio is used in the following descriptions.

6.5.2.1 Selection of Transmission Ratio for Direct Gear

For the traction motor used in the micro EVs, its special torque and speed characteristics make the speed ratio range of the transmission system very small, such that a small transmission ratio range makes it possible for EVs to use direct gear drive, which can remove the gearbox and use the main reducer to achieve direct gear drive, resulting in greatly reducing the mass of the vehicle and improving the energy efficiency. On the other hand, at the same time, the issues of effectively improving the power performance of the vehicle and increasing the driving range also need to be considered. Therefore, the transmission ratio range is subdivided, and the best transmission ratio for direct gear drive is selected though the simulation under the ECE urban cycle condition.

Combining the selected battery model, the established SRM drive system model, and the vehicle model, the vehicle dynamic simulations with various transmission ratio values is carried out to investigate the indexes of the vehicle dynamic performance and the driving range, where the selected transmission ratio range is 3.15 to 4.05 with a step length of 0.1. After simulations, the test data are shown in Table 6.3.

According to the calculation data shown in Table 6.3, it can be seen that the transmission ratio has great impact on the maximum speed, the maximum gradient, the acceleration time, and the driving range under the condition of direct driving. In order to observe more intuitively, it is expressed in the form of a curve, as shown in Fig. 6.8.

Table 6.3 The vehicle performance in different transmission ratios

Transmission ratio	Maximum speed (km/h)	Maximum gradient (%)	Acceleration time (s)	Driving range (km)
3.15	66.3	6.4	13.2	83.89
3.25	64.3	6.6	12.9	83.8
3.35	62.4	6.9	12.6	83.71
3.45	60.5	7.1	12.3	83.56
3.55	58.9	7.4	12.1	83.28
3.65	57.2	7.6	11.9	82.96
3.75	55.6	7.8	11.7	82.63
3.85	54.2	8.1	11.6	82.3
3.95	52.8	8.3	11.5	81.91
4.05	51.5	8.5	11.4	81.53

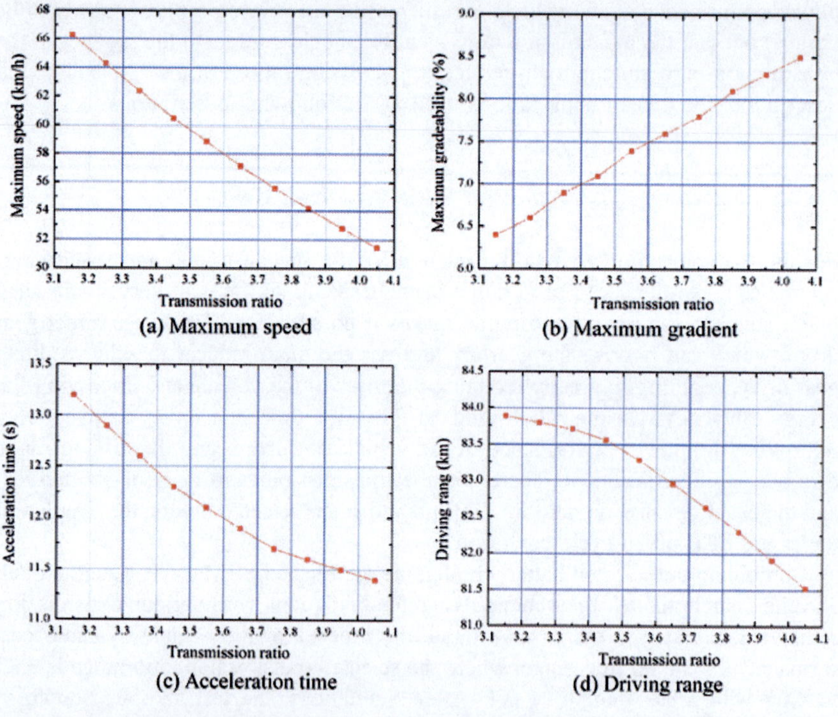

Fig. 6.8 The vehicle performance in different transmission ratios

It can be seen from Fig. 6.8 that under the direct driving condition, the maximum speed, the acceleration time, and the driving range decrease with the increase of the transmission ratio, while the maximum gradient shows a gradual increase. In addition, the changing trends of the acceleration time and the driving range are both nonlinear. Therefore, the optimal values of the maximum speed, the maximum gradient, the acceleration time, and the driving range can correspond to different transmission ratios. In other words, the four performance indicators cannot be maximized simultaneously. Therefore, in order to balance the relationships among the four performance indicators and maximize the performance of the vehicle, the comprehensive index coefficient of the vehicle performance is defined based on the weight coefficient, which is expressed as

$$K(\prod i) = \frac{k_1 \cdot V_{\max}}{(V_{\max})_{\max}} + \frac{k_2 \cdot i_{\max}}{(i_{\max})_{\max}} + \frac{k_3 \cdot S_{DR}}{(S_{DR})_{\max}} + \frac{k_4 \cdot (1/T_A)}{(1/T_A)_{\max}} \tag{6.20}$$

where k_1 is the weight coefficient of the maximum speed, k_2 means weight coefficient of maximum gradient, k_3 represents the weight coefficient of the driving range, k_4 refers to weight coefficient of the acceleration time, $i_{\max}$ denotes the maximum climbing degree of EV, S_{DR} is EV driving range, and T_A is acceleration time of EV.

On the premise of meeting the power performance of the vehicle, it is expected that the driving range of the EV is as large as possible, and the power performance of the vehicle can fully meet the driving requirements of each extreme working conditions of the EV, such that the weight coefficient of the driving range should be larger than other weight coefficient. The weight coefficient of driving range accounts for 55%, while the weight coefficient of maximum speed, the weight coefficient of maximum gradient, and the weight coefficient of acceleration time respectively account for 15%. Based on determining the weight coefficients of each index and the objective function, the comprehensive index coefficients of the vehicle performance under different transmission ratios is calculated using the vehicle parameters, which are shown by the red circle points in Fig. 6.9.

In order to obtain the transmission ratio corresponding to the largest index coefficient of vehicle performance based on the calculation results of 10 sets of discrete data shown in Fig. 6.9, a fitting curve is performed by polynomial fitting method, which is shown by the blue curve in Fig. 6.9. The cubic polynomial is expressed as

$$P(K) = 0.0005x^3 - 0.0274x^2 + 0.1877x + 0.6072 \tag{6.21}$$

Taking the design indexes of EV as the constraint condition, it can be obtained that the transmission ratio corresponding to the largest index coefficient of vehicle performance is 3.80, by solving the maximum value of the function in the transmission ratio range. Therefore, the optimal value of the transmission ratio is 3.80 in the direct gear drive.

To verify the results of various performance indexes of EVs before and after the transmission ratio optimization under the direct drive, the vehicle dynamic

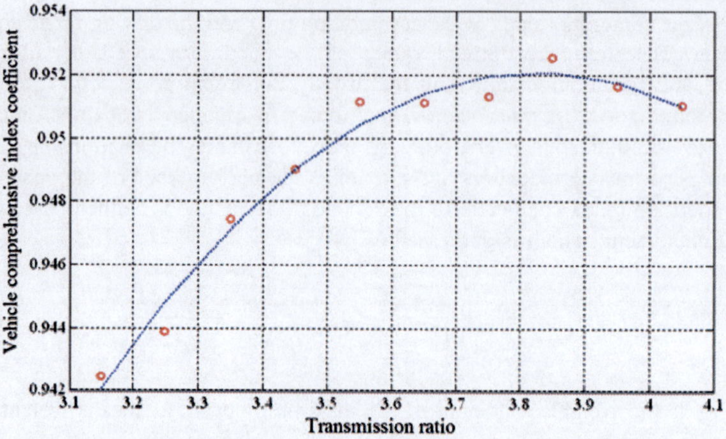

Fig. 6.9 The comprehensive index coefficients under different transmission ratios

performance simulations under the ECE urban cycle condition are performed based on the optimized transmission ratio. The simulation comparison results are obtained, as shown in Table 6.4.

From the comparison data in Table 6.4, it can be seen that although the maximum speed index decreases by 8.33% after optimization, the maximum gradient and the acceleration time indexes increase. The maximum gradient increase by 11.11%, and the comprehensive index coefficient increased by 0.37%, all of which indicate that the optimization has achieved the goal of improving the comprehensive performance of the vehicle.

6.5.2.2 Selection of Transmission Ratio for Two-Speed Drive Gearbox

Although the increase of the gearbox in EV can directly result in increasing the vehicle mass, the change in the transmission ratio can help the vehicle to satisfy various complex working conditions, improve the transmission efficiency, and increase the power performance and driving range of the vehicle.

Table 6.4 Comparison results before and after optimization in direct drive

Index	Before optimization	Optimized	Variation (%)
Maximum speed (km/h)	60	55	−8.33
Maximum gradient (%)	7.2	8	11.11
Acceleration time (s)	12.2	11.6	5.17
Driving range (km)	83.45	82.46	−1.19
Comprehensive index coefficient	0.9501	0.9536	0.37

Orthogonal experiment design method relies on the constructed orthogonal table to carry out combined experiments under the premise of maintaining the balance of the level of various factors, and find the optimal combination of factors and levels by means of relatively small numbers of experiments. In addition, the primary and secondary relationships of each factor and their interaction can also be determined through mathematical statistics. The orthogonal experiment design method has been widely used in practice, and the effectiveness is very obvious.

For the orthogonal optimization analysis of the two-speed gearbox, there are three main test factors, which include gear I, gear II, and the interaction between them marked by I × II. In terms of the selection of gear I and gear II levels, four levels are selected for investigation, as shown in Table 6.5. Therefore, the orthogonal optimization test of the two-speed gearbox is an orthogonal design problem with three-factor and four-level experiment.

In order to construct an orthogonal design table suitable for this experiment, the orthogonal Latin square method is used to construct a four-level orthogonal table L16 (4^5). Due to the balance of the orthogonal table, the first three columns are selected for orthogonal experiment design analysis. The table is shown in Table 6.6.

Considering the indexes for the orthogonal design, four indicators, i.e., the maximum speed, the maximum gradient, the acceleration time, and driving range, are also selected to be investigated. In addition, the optimization method is similar to that for the direct drive mode, that is, the transmission ratio is optimized by means of vehicle dynamic performance simulations under ECE urban cycle condition. According to the combination sequence of the orthogonal table L16(4^5), the performance indicators of the EV are simulated and analyzed under the ECE urban cycle condition to obtain the test index values under different combinations of gear I and gear II, as shown in Table 6.6. The influence of each gear and its transmission ratio on the power performance and driving range performance of the vehicle can be analyzed by means of the obtained results of the indexes. Meanwhile, the transmission ratio of the two-speed gearbox can also be optimized and obtained to achieve the purpose of improving the power performance of the EV and increasing the driving range by using the proposed evaluation index of the comprehensive index coefficient of the vehicle performance shown in Eq. (6.21).

Firstly, the maximum vehicle speed is used as the inspection index to analyze the influence of different gears and transmission ratio combinations on this index. According to the difference analysis method of the orthogonal experiment design, the maximum speed index is calculated and listed as shown in Table 6.7. According to the difference analysis, it can be known that the gear II has the greatest influence on the maximum speed, while the gear I and the interaction I × II between the both

Table 6.5 Variables and levels

Factors	levels			
	1	2	3	4
Gear I	4.05	3.95	3.85	3.75
Gear II	3.45	3.35	3.25	3.15

Table 6.6 $L_{16}(4^5)$ and parameters arrangement

Test number	I	II	I × II	Maximum speed (km/h)	Maximum gradient (%)	Acceleration time (s)	Driving range (km)
1	1	1	1	60.6	7.9	12.3	76.18
2	1	2	2	62.4	7.9	12.4	76.09
3	1	3	3	64.3	7.9	12.4	75.92
4	1	4	4	66.4	7.9	12.6	75.75
5	2	1	2	60.6	7.7	12.5	76.11
6	2	2	1	62.4	7.7	12.5	76.01
7	2	3	4	64.3	7.7	12.6	75.85
8	2	4	3	66.4	7.7	12.7	75.70
9	3	1	3	60.6	7.5	12.6	76.05
10	3	2	4	62.4	7.5	12.7	75.94
11	3	3	1	64.3	7.5	12.8	75.79
12	3	4	2	66.4	7.5	12.9	75.63
13	4	1	4	60.6	7.3	12.8	75.97
14	4	2	3	62.4	7.3	12.9	75.87
15	4	3	2	64.3	7.3	13.0	75.71
16	4	4	1	66.4	7.3	13.1	75.56

Table 6.7 Difference analysis for maximum speed

Factor		I	II	I × II
Average maximum speed	V_1 (km/h)	63.425	60.6	63.425
	V_2 (km/h)	63.425	62.4	63.425
	V_3 (km/h)	63.425	64.3	63.425
	V_4 (km/h)	63.425	66.4	63.425
Difference		0	5.8	0
Optimal scheme		–	II_4	–

have no influence on the maximum speed, which is also consistent with the calculation process of the vehicle maximum speed. The maximum speed index is the largest at II_4, and its data is 66.4 km/h.

Regarding the two indexes of maximum gradient and acceleration time, it can be seen from the data in Table 6.6 that both indexes are affected by gear and transmission ratio. For the acceleration time, the difference analysis shows that when the number of tests is 1, the acceleration time is the smallest, which is 12.3 s. For the maximum gradient, the maximum gradient of 7.9 appears at I_1, while the gear II and the interaction between the both have little effect on the maximum gradient index.

As two important vehicle dynamic performance indicators, the maximum speed and the maximum gradient can be calculated by means of their calculation equations, which are mainly determined by the high and low transmission ratios. However, the acceleration time index of EVs is very closely related to the gear and its combination. Meanwhile, there is no specific calculation equation for the driving range under cycle conditions, and the influence factors of this index are also uncertain. Therefore, it is necessary to conduct a detailed analysis on the influence factors of these two performance indicators.

The range analysis is then carried out on the two indicators of the acceleration time and the driving range, and the results are shown in Tables 6.8 and 6.9. It can be seen that the influence law on the two indicators is basically the same. The optimal combination of the two indicators is I_1, II_1, $(I \times II)_3$, and the corresponding acceleration time and maximum driving range are 12.3 s and 76.18 km, respectively. The gear I has greater influence on the acceleration time, while the gear II has greater impact on the driving range. The interaction of the two gears (I × II) has certain impact on both indicators, but the impact is small.

From the analysis, it can be observed that the optimal situation for the maximum speed index is II_4, the optimal situation for the maximum gradient index is I_1, while the optimal situations for the acceleration time and driving range are I_1 and II_1, such that the synchronization optimization for vehicle power performance and driving range cannot be realized. It is necessary to use the comprehensive index coefficients of the EV performance to balance the relationships among the various targets. The values of the weight coefficients for each index are the same as those of the direct gear drive mode.

The performance indexes are calculated under each test to obtain the comprehensive index coefficient of the EV performance, which is expressed in the form of curve, as shown in Fig. 6.10. It can be seen that when the test number is 4, the comprehensive index coefficient of the EV performance is the largest, which is the combination of I_1 and II_4. Meanwhile, the obtained index coefficients under each I have the same change trend, while the comprehensive index coefficient of the EV performance under each I changes significantly.

Similarly, the difference analysis method of the orthogonal experiment design is used to analyze the influences on the comprehensive index coefficient of the EV performance, and the calculated data are shown in Table 6.10. It can be seen through analysis that the optimal combination for the vehicle comprehensive index

Table 6.8 Difference analysis for accelerating time

Factor		I	II	I × II
Average acceleration time	T_1 (s)	12.425	12.55	12.675
	T_2 (s)	12.575	12.625	12.70
	T_3 (s)	12.75	12.70	12.65
	T_4 (s)	12.95	12.825	12.675
Difference		0.525	0.275	0.05
Optimal scheme		I_1	II_1	$(I \times II)_3$

Table 6.9 Difference analysis for driving range

Factor		I	II	I × II
Average driving range	S_1 (km)	75.985	76.053	75.885
	S_2 (km)	75.918	75.978	75.885
	S_3 (km)	75.853	75.818	75.885
	S_4 (km)	75.778	75.660	75.878
Difference (km)		0.207	0.925	0.007
Optimal scheme		I_1	II_1	$(I \times II)_{123}$

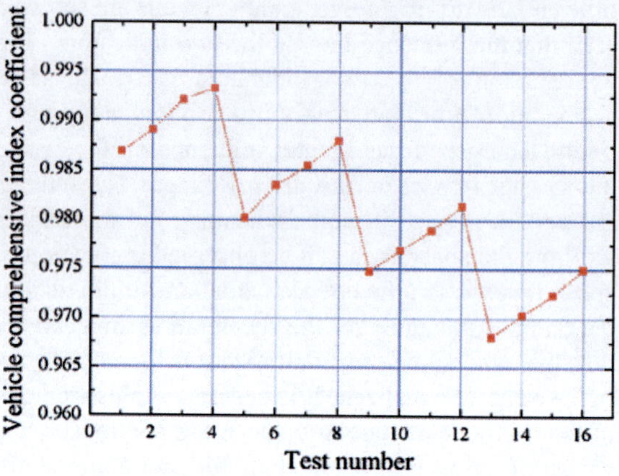

Fig. 6.10 The vehicle comprehensive index coefficients in different test number

Table 6.10 Difference analysis for vehicle comprehensive index coefficients

Factor		I	II	I × II
Average comprehensive index coefficient	K_1	0.9904	0.9775	0.9811
	K_2	0.9843	0.9800	0.9808
	K_3	0.9780	0.9823	0.9813
	K_4	0.9715	0.9844	0.9810
Difference		0.0189	0.0069	0.0005
Optimal scheme		I_1	II_4	$(I \times II)_1$

coefficients is I_1, II_4, and $(I \times II)_1$, and the corresponding optimal coefficient under this combination is 0.9933. The order of influence on the index coefficient is I, II, and (I × II), among which, the influence of I is the largest. The results can correspond to the analysis results according to Fig. 6.10.

Through orthogonal optimization design analysis described above, it can be known that taking the optimal comprehensive index coefficient of the EV performance as the objective function, the optimal combination of gear and transmission ratio is that the transmission ratio of the gear I is 4.05, and transmission ratio of the gear II is 3.15.

6.6 Simulation Results Analysis Under ECE Driving Cycle

Based on the matching analysis of battery parameters and transmission parameters, the matched EV under various cycle conditions are simulated and analyzed through the established electric vehicle performance simulation model. The optimal scheme is selected through the comparison and analysis of the results from various driving conditions. During this process, the main performance of the EV and the new excitation mode SRM is also simulated and analyzed.

6.6.1 Comparison of the Vehicle Performance Under Various Conditions

Since the EV is designed as a micro vehicle, the operating speed and operating environment of the vehicle are very similar to the ECE urban cycle condition. But, as a micro EV that can be used for multiple purposes, its actual operating environment and operating conditions cannot be consistent with a specific cycle operating condition. Therefore, it is very meaningful to conduct comprehensive performance analysis under multiple operating cycle conditions.

As mentioned in Sect. 6.5.1, the current driving cycle conditions for evaluating vehicle power performance and economic performance include European urban cycle condition (ECE), European suburban and urban cycle condition (ECE + EUDC), and American urban cycle condition (UDDS), and Japanese 1015 cycle condition (1015). In this section, the performance simulation analysis of the designed EV under four cycles is carried out, and the vehicle performance characteristics under the two design schemes of the gearbox are studied.

Since the maximum speed of the designed EV is relatively low, the maximum speed of each working condition, except ECE cycle condition, is higher than the designed vehicle speed. Therefore, the four test cycle conditions for the EV are modified simply to satisfy the dynamic performance of the vehicle researched in this chapter. The maximum speed of each cycle condition is set as the designed maximum speed of the researched vehicle. In other word, the following speeds of the vehicle under each driving cycle condition are calculated in corresponding proportions such that each operating condition can meet the operating speed of the designed EV. Then, the performance of the EV is simulated and analyzed under the modified conditions of each cycle.

Among the four indicators, the three performance indicators, the maximum speed, the maximum gradient, and the acceleration time has little relationship with vehicle driving cycle condition. As a result, the indicators can be computed separately and the calculated data are listed in Table 6.11. It can be seen from the comparison of the various power performance indicators in the table that compared to the two-speed gearbox drive, the three indicators of the maximum speed, the maximum gradient, and the acceleration time of the vehicle with direct drive are

Table 6.11 Comparison of vehicle dynamic performance

Transmission form	Mass (kg)	Maximum speed (km/h)	Maximum gradient (%)	Acceleration time (s)
Direct drive	50	55.0	11.6	8.0
Two-speed drive	85	66.4	12.6	7.9

lower. This is because the two-speed gearbox drive can make good use of the gear combination, increasing the maximum speed, the maximum gradient, and reducing the acceleration time.

For the vehicle driving range index, the simulation results under four driving cycle conditions are obtained and listed in Table 6.12. It can be seen from the comparison of the four driving cycle conditions that compared to direct drive, the driving range of the vehicle with two-speed gearbox drive in each cycle condition is smaller, and the difference ratio ranges from –4.35% to –9.73%. After further analysis for the reason of this difference, it can be seen that the increase in the two-speed gearbox drive leads to increase the vehicle mass, and also reduce the transmission efficiency of the drive train, which affecting the economy and driving range of the vehicle.

It can be seen through analysis that the two-speed gearbox drive mode has significant effect on improving the power performance of EVs, while the direct drive mode is significantly better than the two-speed gearbox drive in terms of driving range performance. Therefore, when selecting the EV transmission mode, it should be selected according to the specific operating requirements and working environments of the EV. For the EVs that focus on power performance and are also easy to charge, a two-speed gearbox drive or even a higher-speed gearbox can be considered. However, for the EVs that focus on economic performance without high requirements on power performance, the direct drive mode should be more suitable.

Through a comprehensive analysis of reducing the vehicle mass, improving the efficiency of the whole vehicle, and improving the energy utilization rate, the direct drive is selected in this chapter as the form of the transmission system for the designed micro EV. It is of great significance to improve the economic performance of the EVs and increasing the driving range of EVs.

Table 6.12 Comparison of vehicle driving range

Working condition	Direct drive (km)	Two-speed gearbox drive (km)	Difference percentage (%)
ECE	82.36	75.75	−4.92
ECE + EUDC	106.72	96.34	−4.35
UDDS	100.57	90.78	−9.73
1015	91.80	83.33	−9.23

6.6.2 Simulation Results in Optimal Design Scheme

Based on the battery selection scheme under driving cycle conditions and the transmission ratio optimized by the orthogonal optimization design method, as described in Sect. 6.5, the optimal matching scheme for the proposed EV is obtained through the matching results and comparative analysis. According to the optimal scheme, the established simulation model of EV is simulated and analyzed under the ECE urban cycle condition. In terms of vehicle indicators, two indicators of the vehicle speed and the *SOC* of battery pack are analyzed. For the two-phase excitation mode SRM is mainly analyzed from its working current and output torque. The four indicators obtained by simulation are shown in the form of curves, as shown in Fig. 6.11.

It can be seen from Fig. 6.11 that the matching scheme for the EV power system can well meet the driving requirements of the ECE urban cycle condition. The speed following response is good and greatly stable, and the change of *SOC* is also relatively stable. Considering the performance of the motor, the peak current of the SRM is about 62 A, and the peak output torque is about 50 N m, which meets the performance limits of the SRM, and there is still certain allowance. These can show

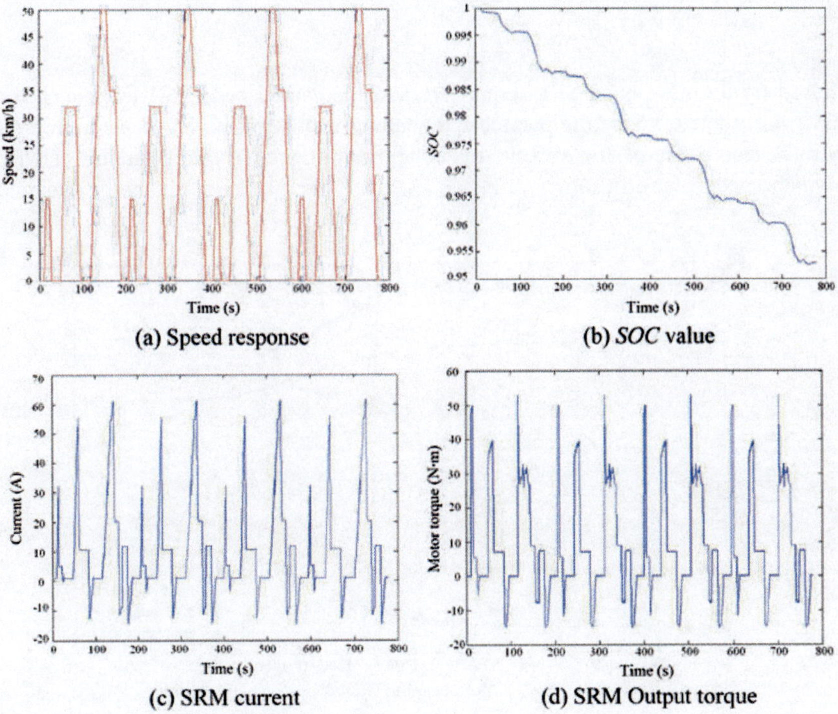

(a) Speed response (b) *SOC* value (c) SRM current (d) SRM Output torque

Fig. 6.11 Simulation results in ECE cycle

that the motor can well meet the driving performance requirements of the EV, and its superior torque and speed characteristics have great advantages in improving the power performance of the vehicle, increasing the driving range, and improving the efficiency of the drivetrain.

To study the dynamic operating characteristics of the SRM during the actual vehicle operating cycle condition, the real-time operating point of the SRM during the drive cycle is analyzed, as shown in Fig. 6.12, which is distribution diagram of the real-time working speed and torque points of the SRM during operation.

According to the distribution diagram of the actual working points of the motor, it can be seen that the working range of the motor is mostly concentrated in the rated torque range, which enables the SRM to meet the various complex driving requirements of the EV under the ECE urban cycle condition, and make the vehicle have a good speed following performance. It is of great significance to improve the power performance of the vehicle, increase the efficiency of the battery, and increase the driving range. In addition, the research on the distribution trend of the actual operating points provides a certain theoretical basis for the control strategy of the SRM and the vehicle.

6.7 Experiments for Matching Performance of the Vehicle with SRM

Based on the developed SRM controller and the established SRM test bench for EVs, in order to verify the parameter matching results of the EV drive system, a virtual test platform for vehicle dynamic performance is built on the electric dynamometer platform shown in Fig. 6.13. Using the platform, the motor dynamic

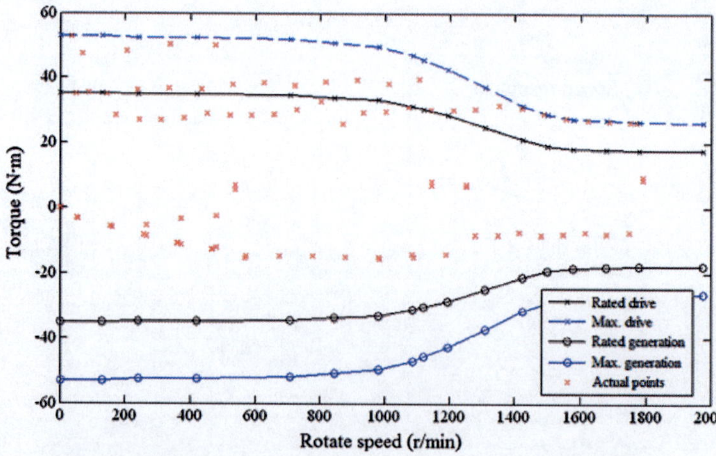

Fig. 6.12 Operation performance of the SRM in vehicle

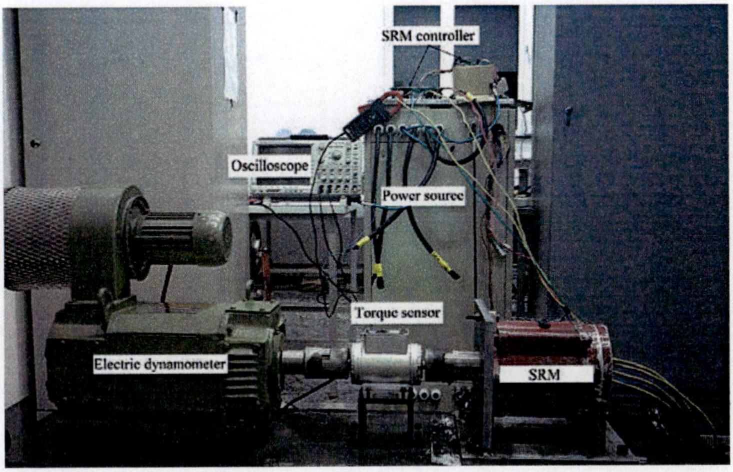

Fig. 6.13 The experiment platform

performance after the EV drive system is matched with the battery pack, the drive motor, and transmission system is tested and verified by means of the semi-physical simulation test.

6.7.1 Dynamic Performance Experiment in Slope Condition

Since the speed of the vehicle is low enough during the low-speed climbing process, the influence of air resistance can be ignored. Meanwhile, the acceleration resistance is also small during the climbing process with a constant speed, such that it can also be ignored. Therefore, when the vehicle is climbing at a constant low speed, the driving resistance of the vehicle is the sum of the slope resistance and the rolling resistance. Since both resistances are only related to the mass of the vehicle, the both can be regarded as a constant value.

The test intends to verify the ability of climbing a 6% slope at a speed of 10 km/h for the designed vehicle. During the ramp resistance application process, to describe the actual vehicle climbing process, the ramp resistance is uniformly applied to the SRM within 1 s. Combining the design parameter of the EV drivetrain, the test results of the SRM is obtained and presented in Figs. 6.14 and 6.15, through the semi-physical simulation platform, where change curves of the output torque and the current of the motor are shown, respectively.

From the analysis for the parameter change curves, it can be known that the designed EV drive system can effectively meet the requirements of climbing performance indicators, and the output torque is stable. The motor bus current value is

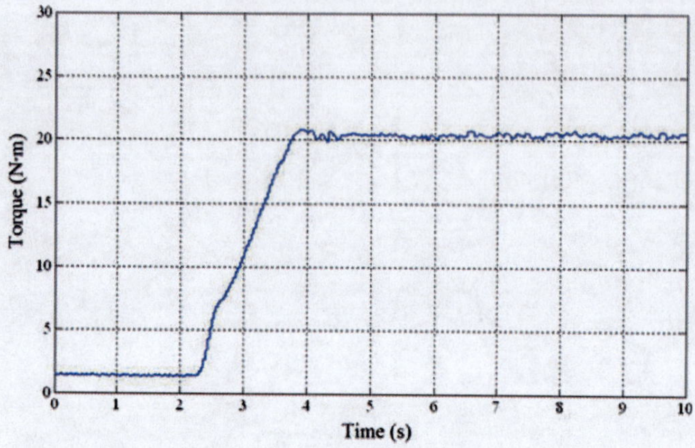

Fig. 6.14 Output torque of the SRM in slope condition

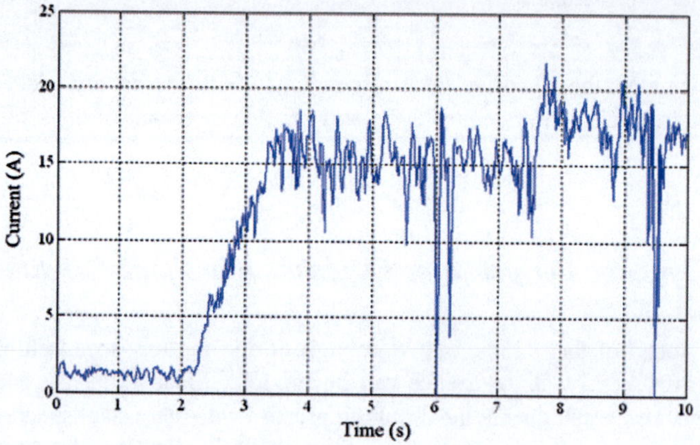

Fig. 6.15 Current of the SRM in slope condition

about 15 A. Through index parameter analysis, it is concluded that the parameter design and optimization for the SRM and EV drive system can well meet the requirements of the climbing performance.

6.7.2 Dynamic Performance Experiment in Accelerating Condition

Acceleration performance is also a very important parameter in the power performance indicators of EVs. In the process of vehicle acceleration, the driving resistance caused by air, wheel rolling, and vehicle acceleration also affects the overall force characteristics of the vehicle, such that the load torque is a dynamic and non-linear parameter.

The test intends to verify the acceleration ability of the designed EV and the SRM when vehicle accelerates from 10 to 50 km/h. The load torque forced to the EV is computed according to Eqs. (6.1–6.5). On this basis, the output torque and current of the SRM are dynamically tested and the test data are shown in Figs. 6.16 and 6.17, which are the dynamic curves of the output torque and the bus current under this working condition.

It can be analyzed from the output torque and current change curve that the load torque of the EV during acceleration has a highly time-varying nonlinear characteristic, and the SRM and the EV drive system can well meet the torque dynamic demands during the acceleration of the vehicle. The motor runs at a stable current of 40 A, which is within the range of motor design parameters. Therefore, it can be observed that the design parameters of the motor and the EV drivetrain can well meet the acceleration performance requirements of the EV.

6.7.3 Dynamic Performance Experiment in ECE Cycle Condition

In the process of parameter matching and design of EV drive system, an optimization method for drivetrain parameters based on ECE urban cycle condition is proposed in this chapter. To verify the performance of the SRM and the design parameters of the EV drive system, the reference speed of the drive motor are obtained by calculation using the required vehicle drive speed under this working condition in Fig. 6.11a. Meanwhile, the test program of semi-physical simulation

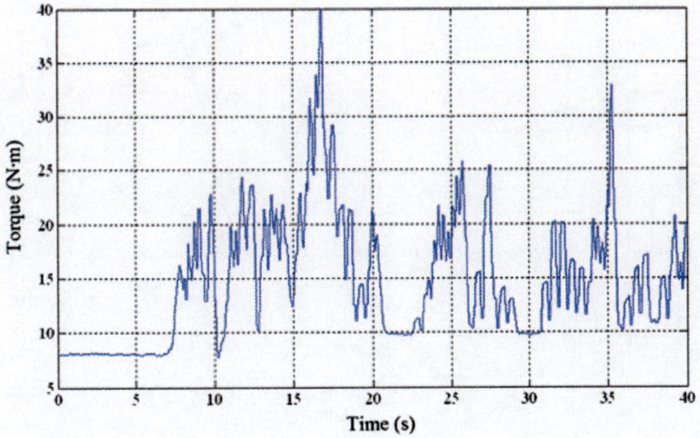

Fig. 6.16 Output torque of the SRM in accelerating condition

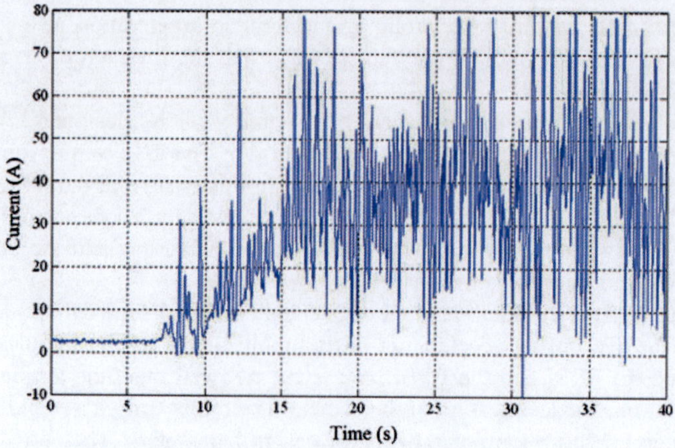

Fig. 6.17 Current of the SRM in accelerating condition

under ECE urban cycle working condition is compiled, and on this basis, the matching parameter of the designed EV drive system is verified. The test results of the motor speed, the output torque, and the bus current are obtained, and their change curves are shown in Figs. 6.18, 6.19, and 6.20, respectively.

The design parameters of the drive system are verified by using the written test program for the ECE urban cycle condition. And the following speed curve of the SRM is shown in Fig. 6.18. It can be seen from the figure that the SRM drive system can well follow the vehicle speed requirements and has a better dynamic response. It should be noted that the motor in test is set an idling speed with 200 r/min, because the electric dynamometer can't stop completely when the drive cycle is operating. In practice, the EV driven by electric motor do not need idling condition because of the perfect starting performance of the motor.

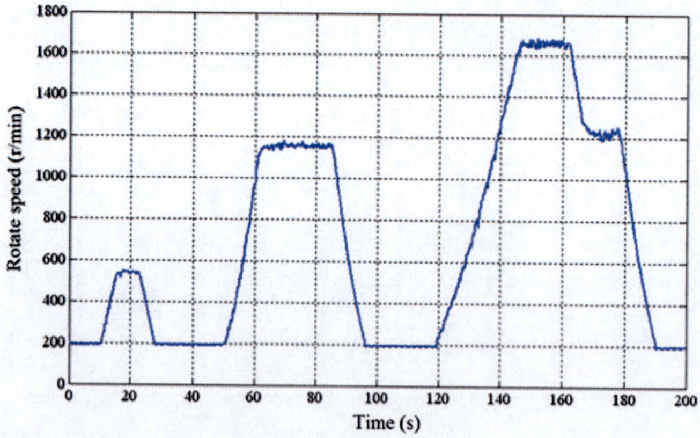

Fig. 6.18 Speed response of the SRM in ECE cycle condition

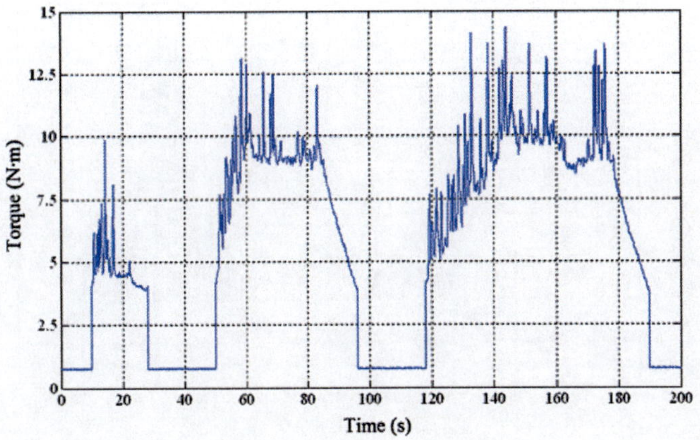

Fig. 6.19 Output torque of the SRM in ECE cycle condition

Figures 6.19 and 6.20 show the motor output torque characteristics and the current change curves under ECE urban cycle drive condition. It can be seen that under vehicle dynamic load condition, the motor output torque characteristics and the current characteristics remain normal. At the same time, the motor speed is stable and the dynamic response is good. Therefore, it is believed that the designed motor and the design parameters of the EV can well meet the requirements of drive characteristics for the EV under the ECE urban cycle condition.

6.7.4 SOC *Calculating and Analysis in ECE Cycle Driving Condition*

As expressed in Sect. 6.5.1, the *SOC* value of the battery pack is of great significance for evaluating the energy efficiency of EVs. In order to simulate the electric power consumption and *SOC* value during the ECE urban cycle drive condition of the EV, the *SOC* change trend curve during the test is computed and established through formula derivation and calculation.

The battery-related data, especially the *SOC* value, cannot be directly obtained from the platform, because the battery pack for the EV is simulated by the direct current power source in the platform. Therefore, these data need to be solved mathematically through equal calculations. Utilizing the current change curve with time obtained during the test, which is shown in Fig. 6.20, the capacitance consumed during the test is calculated by the numerical integration method. In addition, the use of the electrical appliances needs to be considered in the calculation process because it will affect the *SOC* value of the battery pack. Therefore, the capacitance consumed by the vehicle during the test is expressed as

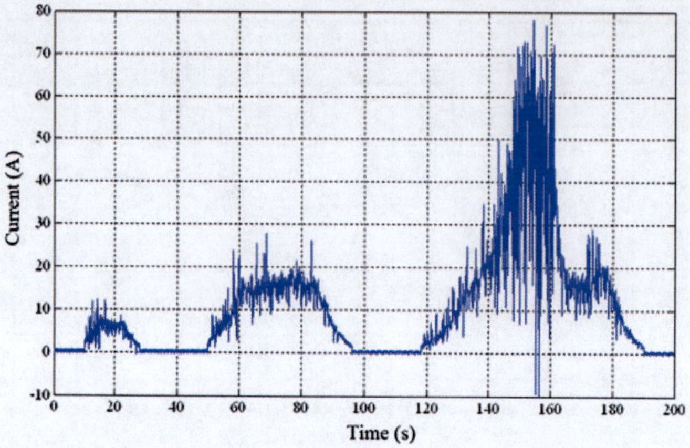

Fig. 6.20 Current of the SRM in ECE cycle condition

$$C_{\text{exp}} = \int_0^t I(t)dt + \frac{P_e t}{U_c} \tag{6.22}$$

where C_{exp} is the capacitance consumed by the vehicle during the test, $I(t)$ means the relationship between current and time in the test, P_e denotes total power of vehicle electrical appliances, and U_c represents electrical voltage.

According to the selected capacity of the electric vehicle battery pack in 6.5.1, the *SOC* dynamic change value during the test can be obtained as

$$SOC = \frac{C_{\text{bat}} - C_{\text{exp}}}{C_{bat}} \tag{6.23}$$

Through the experiment, the data of the motor bus current with time is obtained. Then, the *SOC* value of the battery pack during the operation of the EV is computed by means of the Eqs. (6.22) and (6.23). In order to observe more intuitively, the *SOC* change data are expressed in the form of curve, as shown in Fig. 6.21. Among them, four ECE urban cycle conditions are considered in the calculation process.

It can be seen from Fig. 6.21 that the *SOC* value of the battery pack is everlastingly decreasing and the decreasing value is the same as the theoretical analysis data, although the experimental calculation value is slightly higher. This is because the factors such as the losses of the battery internal resistance and the battery discharge efficiency are not considered in the calculation process. On the other hand, the experimental calculation value changes more smoothly, because the voltage changes for the battery pack are more complicated than those for the DC power source.

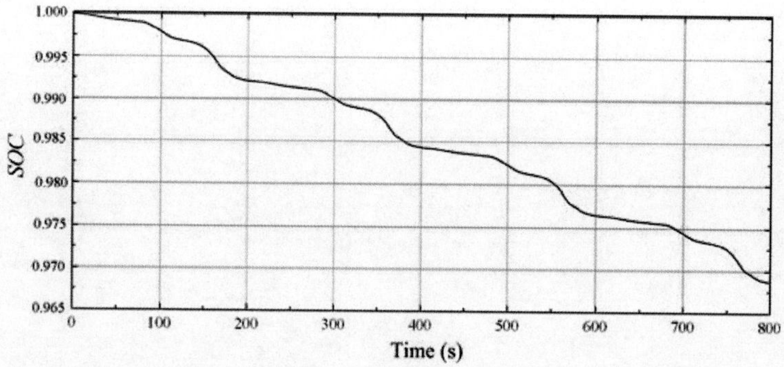

Fig. 6.21 *SOC* of the battery in ECE cycle condition

References

1. Chau KT, Chan CC, Chunhua L. Overview of permanent magnet brushless drives for electic and hybrid electric vehicles. IEEE Transactions on Industry Electronics, 2008, 55(6): 2246–2257.
2. Emadi A, Young Joo L, Rajashekara K. Power electronics and motor drives in electric, hybrid electric and plug-in hybrid electric vehicles. IEEE Transactions on Industry Electronics, 2008, 55(6): 2237–2244.
3. Amini MH, Karabasoglu O. Optimal operation of interdependent power systems and electrified transportation networks. Energies 2018,11: 196.
4. Hudson AL, Gandevia SC, Butler JE. A principle of neuromechanical matching for motor unit recruitment in human movement. Exercise and Sport Sciences Reviews, 2019, 47(3): 157–168.
5. Jiang X, Hu J, Jia M, et al. Parameter matching and instantaneous power allocation for the hybrid energy storage system of pure electric vehicles. Energies, 2018, 11(8): 1933–1951.
6. Hall E, Ramamaurthy MSS, Balda JC. Optimum speed ratio of induction motor drivers for electric vehicle propulsion. Proceedings of the 2001 Applied Power Electrical Conference, California, 2001: 371–377.
7. Hall E. Electric propulsion with emphasis on switched reluctance motor and the motor optimum speed ratio. George Washington University, 2002: 20–120.
8. Willamson SS, Emadi A. Comprehensive efficiency modeling of electric traction motor drives for hybrid electric vehicle propulsion applications. IEEE Transactions on Vehicular Technology, 2007, 56: 1561–1571.
9. Peng T. Parameter matching of parallel hybrid electric vehicle powertrain. Chinese Journal of Mechanical Engineering, 2003, 39(2): 69–73.
10. Chen Q, Shu H, Chen B, et al. Parameter matching and control strategy of parallel power train system for CNG-electric hybrid urban bus. International Journal of Electric and Hybrid Vehicles, 2012, 4(3): 248–259.
11. Liu H, Lei Y, Fu Y, et al. Parameter matching and optimization for power system of range-extended electric vehicle based on requirements. Proceedings of the Institution of Mechanical Engineers Part D: Journal of Automobile Engineering, 2020, 234(14): 1–13.
12. Chan CC. The state of the art of electric and hybrid vehicles. Proceedings of the IEEE, 2002 (2): 247–275.

Chapter 7
Torque Coordination Control of Distributed Drive Electric Vehicle with SRM

7.1 Introduction

In order to improve the driving stability of the vehicle, traditional vehicles often use anti-lock braking systems [1], direct yaw moment control systems [2], and acceleration slip regulation systems [3] to control the vehicle. These control systems mainly rely on the vehicle braking system to change the driving torque of the vehicle. For distributed drive electric vehicle (DDEV), the driving torque directly acts on the driving wheels, and the output torque of each wheel can be changed in real time [4, 5]. For improving the driving stability of the vehicle, distributed drive can make the output torque of the vehicle more accurate and respond more quickly. However, the torque response difference of the coaxial motors and the different adhesion capabilities of the coaxial wheels for DDEV lead to the poor vehicle stability under straight line driving conditions. To improve the DDEV straight line driving stability, the problem of the difference of phase angle (DPA) between coaxial motors and complex roads are researched in this chapter.

At present, the research of driving stability control for the DDEV mainly focuses on yaw stability control and acceleration slip regulation. Between them, the yaw stability control has a good control effect for improving the driving stability of the DDEV [6, 7]. The yaw rate and the sideslip angle are two important indicators to evaluate the yaw stability of the vehicle, and the driving stability of the vehicle can be improved by controlling them. Some literatures [8, 9] proposed a direct yaw moment control method based on the sideslip angle in order to improve the stability of the yaw motion when turning. Meanwhile, some papers [10] proposed a comprehensive control strategy based on the yaw rate and the sideslip angle for yaw stability of the vehicle. In addition, acceleration slip regulation can control the wheel slip rate within a certain range, such that the maximum adhesion performance between the wheel and the ground can be maintained [11, 12]. Many literatures [13–15] proposed a torque distribution method for electric vehicles driven by four-wheel hub motors based on drive stiffness and slip rate in order to improve the

Y. Zhu, *The Key Technologies for Powertrain System of Intelligent Vehicles Based on Switched Reluctance Motors*, Recent Advancements in Connected Autonomous Vehicle Technologies 1, https://doi.org/10.1007/978-981-16-4851-9_7

stability of the vehicle. Although some of the proposed control methods improve the straight-line driving stability of the vehicle to a certain extent, these methods do not consider the essential reasons for the unstable driving of the vehicle, such as the influence caused by the motor and the road. It is of great significance to improve the driving stability of the vehicle to avoid the occurrence of vehicle instability from the root. Therefore, in order to improve the straight-line driving stability of DDEV, the optimization controls of the DPA between coaxial motors and complex roads are carried out through some control methods in the chapter. The DPA controller and the straight driving stability controller are respectively designed based on the proposed control methods of the DPA and complex roads, and the effectiveness of the proposed controller is verified though co-simulation.

7.2 DDEV Model

7.2.1 *Vehicle Dynamics Model*

Assuming that the effect of the suspension and the steering system on vehicle handling performance are ignored, the motion model of the vehicle can be described by a linear two-degree of freedom (2-DOF) model, which is shown in Fig. 7.1. Because the 2-DOF vehicle model ignores the role of the suspension, it is believed that the vehicle only performs longitudinal, lateral and yaw motion parallel to the ground. The longitudinal speed u of the vehicle along the X-axis direction remains unchanged, such that only movement along the Y-axis and Z-axis needs to be considered.

The resultant force along the Y-axis and the moment around the center of mass for 2-DOF model are showed as

$$\begin{aligned} \sum F_Y &= F_{Y1}\cos\delta_f + F_{Y2} \\ \sum M_Z &= aF_{Y1}\cos\delta_f - bF_{Y2} \end{aligned} \tag{7.1}$$

where F_{Y1} and F_{Y2} are lateral forces of the front and rear axles, a and b respectively represent the distances from the center to the front and rear axles, and δ_f refers to the

Fig. 7.1 2-DOF vehicle model

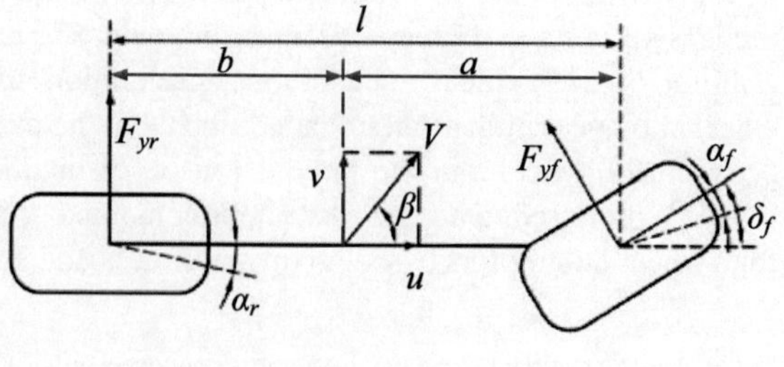

steering angle of front wheel. It should be noted that cos $\delta_f \approx 1$, because δ_f is too small. Therefore, Eq. (7.1) can be written as follows

$$\begin{aligned} \sum F_Y &= k_1\alpha_1 + k_2\alpha_2 \\ \sum M_Z &= ak_1\alpha_1 - bk_2\alpha_2 \end{aligned} \tag{7.2}$$

where k_1 and k_2 are the tire slip stiffness of the front and rear wheels, α_1 and α_2 are the tire slip angle of the front and rear wheels, respectively.

The slip angle of the front and rear wheels of the vehicle can be approximated and expressed by

$$\begin{aligned} \alpha_1 &= \beta + \frac{a\omega_r}{u} - \delta_f \\ \alpha_2 &= \beta - \frac{b\omega_r}{u} \end{aligned} \tag{7.3}$$

where β is the vehicle sideslip angle, ω_r is the vehicle yaw rate, and u is the vehicle longitudinal velocity.

Therefore, motion differential equation of the 2-DOF vehicle is obtained from Eqs. (7.2) and (7.3), which is shown as

$$\begin{aligned} (k_1 + k_2)\beta + \frac{1}{u}(ak_1 - bk_2)\omega_r - k_1\delta_f &= m(\dot{v} + u\omega_r) \\ (ak_1 - bk_2)\beta + \frac{1}{u}(a^2k_1 + b^2k_2)\omega_r - ak_1\delta_f &= I_z\dot{\omega}_r \end{aligned} \tag{7.4}$$

where m is the mass of the vehicle, v is the vehicle lateral velocity, and I_z is the yaw moment of inertia. The differential equation contains parameters such as the vehicle sideslip angle and yaw rate, which can reflect the movement of the vehicle.

After obtaining the expression of the 2-DOF vehicle model, the steady-state steering characteristic of the vehicle characterize the steady-state response of the vehicle. When the vehicle is in a stable state, the yaw rate is a constant value, and the lateral acceleration and yaw angle are both 0. Therefore, the yaw rate in the steady state is expressed as follows

$$\omega_r = \frac{u}{(1 + Ku^2)L}\delta_f \tag{7.5}$$

where L represents the distances from the front axle to rear axle, and K is the Stability index. Among them, when $K = 0$, the yaw rate ω_r is proportional to the vehicle speed u, and the vehicle is a neutral steering. When $K < 0$, the vehicle is over steering. The over-steered vehicles tend to lose stability when reaching critical speed. When $K > 0$, the vehicle is understeer. The turning radius of the understeered vehicles increases as the vehicle speed increases.

7.2.2 Vehicle Modeling Based on CarSim

CarSim is a professional vehicle dynamics simulation software, which has good performance in analyzing vehicle dynamics, calculating vehicle performance characteristics, developing active controllers, and designing next-generation active safety systems. CarSim allows users to build complex scenarios and test event sequences, which can help companies to implement vehicle analysis and simulation. And this software directly provides users with vehicle configuration parameters, simulation status, simulation solver, and post-processing of simulation results. Its user interface is shown in Fig. 7.2.

A variety of models are available in CarSim, and the body structure parameters, aerodynamics, transmission system, braking system, steering system, suspension system, and tire model can be modified according to requirements. In addition, CarSim can set up vehicle simulation conditions, such as driving, braking, shifting, steering, and road conditions. The simulation solver can not only realize the solving operation of the software itself, but also realize the co-simulation operation with other software due to its standardized interface which can be linked to MATLAB/Simulink, NI LabVIEW, FMI/FMU and other software. The post-processing interface can clearly and intuitively display the simulation results for users in the form of animations and charts.

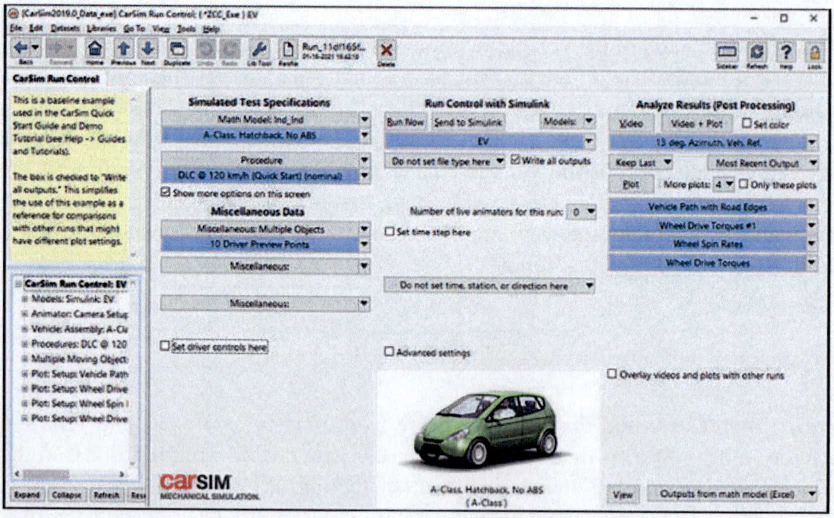

Fig. 7.2 User interface in CarSim

In this chapter, the vehicle structure parameters are set, the types of power and transmission system, braking and steering system, suspension system and tire model are selected, and the vehicle dynamics model is also established. The vehicle dynamics modeling process based on CarSim is shown as follows.

(1) Structure parameter configuration.

According to the system design requirements, the A-class vehicle is selected as the test vehicle, and its structural parameters are modified on the basis of the existing parameter configuration. Table 7.1 shows the main parameters of the researched DDEV in this chapter.

(2) Power and transmission system.

According to the characteristics of the power and transmission system of the DDEV, the power and transmission system of the vehicle dynamics model is set. The transmission system model of the DDEV with external power input is shown in Fig. 7.3. Compared to traditional fuel vehicles, DDEV does not require complicated fuel engine systems and power transmission systems, but uses motors that are installed near the wheels to directly drive the wheels. First, the vehicle drive mode is set as four-wheel drive mode. Then, the engine, clutch, transmission, transfer case, and differential are set to external input. In order to transfer the drive torque of the established external motor model to the software, the input port of CarSim is set to IMP_MY_OUT_D1_R, IMP_MY_OUT_D1_L, IMP_MY_OUT_D2_R, IMP_MY_OUT_D2_L. At the same time, in order to output the wheel speed calculated by the solver to the external motor, the output port of CarSim is set to AVy_R1, AVy_L1, AVy_R2, AVy_L1.

(3) Braking and steering system.

This article mainly studies the influence of driving torque on the driving stability of the vehicle, without considering the braking situation. As a result, the braking system parameters in CarSim are set according to the default parameters. Since the detailed information such as steering torque and deflection need not be considered in the test simulation process, a manual rack and pinion steering system is selected in this vehicle model.

Table 7.1 The main parameters of the researched DDEV

Parameters	Sign	Value	Unit
Vehicle mass	m	583	kg
Front wheelbase	a	1100	mm
Rear wheelbase	b	1250	mm
Centroid height	h	540	mm
Transmission ratio	i	3	-
Yaw moment of inertia	Iz	750	kg m^2
Wheel radius	R	250	mm

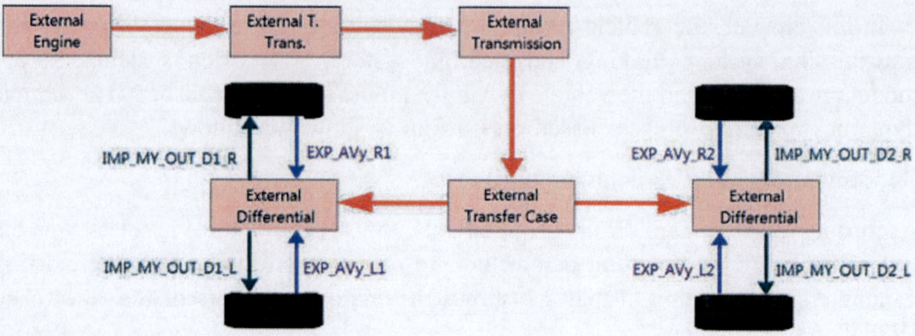

Fig. 7.3 Powertrain of the DDEV

(4) Suspension system.

Since the drive motor of the DDEV is generally installed in or near the drive wheel, the unsprung mass of the vehicle and the moment of inertia of the vehicle are greatly increased. The front and rear suspensions of the vehicle model are all independent suspensions. The unsprung masses of the front and rear axles are both 60 kg, and the moments of inertia of the four wheels are 0.6 kg m^2.

(5) Tire model.

Several tire models are available in CarSim, which can calculate the vertical force of the tire, the shear force on the ground, and the moment caused by the deformation of the tire. Figure 7.4 shows the longitudinal tire force, lateral tire force, and aligning moment under different vertical loads.

7.2.3 *Co-Simulation Model of DDEV*

Since CarSim has not yet developed an electric motor model, when analyzing the driving state of DDEV, it is necessary to establish a motor model and corresponding control strategies for co-simulation through MATLAB/Simulink. Figure 7.5 is a co-simulation model of DDEV driven by four switched reluctance motors.

The vehicle dynamics model and driver model are provided by CarSim, from which the speed of the four wheels and the throttle signal can be obtained. And through post-processing, the driving state of the vehicle can be observed, including the longitudinal speed, the yaw rate, the sideslip angle of the vehicle, and other parameters. CarSim standardized interface provides convenience for co-simulation with other software. The torque of the four motors is transmitted to CarSim through a standardized interface, and CarSim calculates the speed of the vehicle and the wheel speed of each wheel according to the solver.

Through the simulation test of acceleration performance and straight-line driving stability performance, the validity of the co-simulation model of DDEV driven by switched reluctance motor is verified. First, the acceleration performance of the

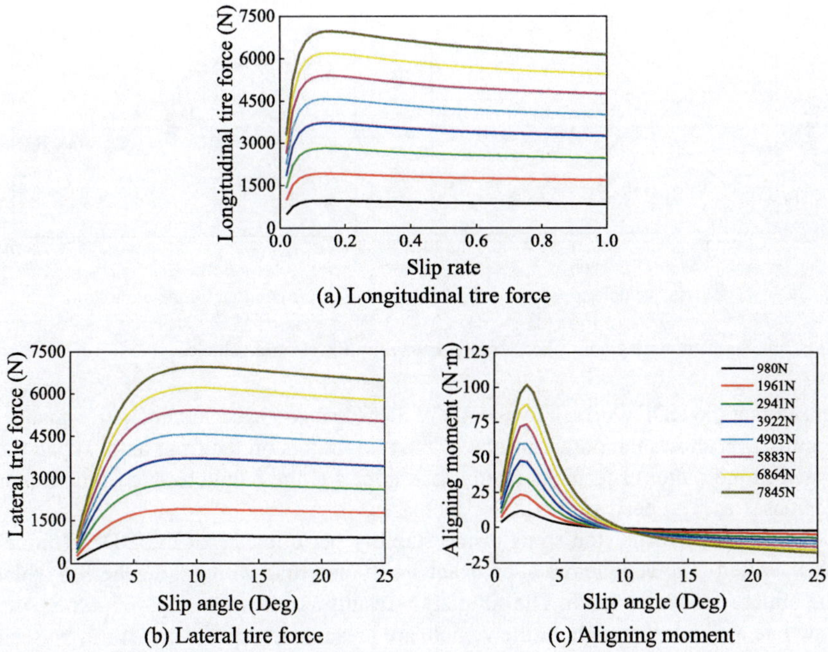

Fig. 7.4 Tire model used in this chapter

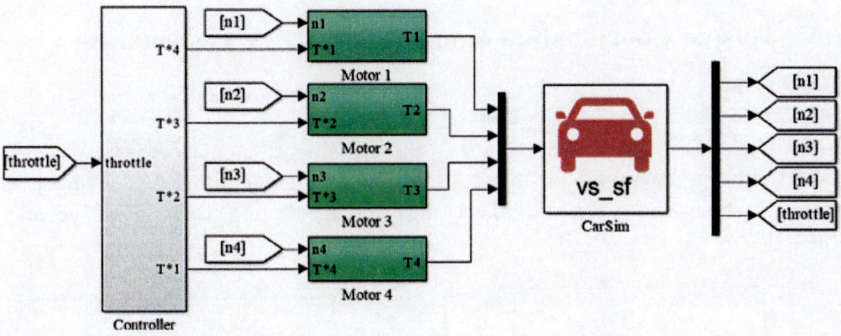

Fig. 7.5 DDEV system model

vehicle model is verified. The vehicle accelerates on a road with good adhesion when the initial speed is 0. The simulation results of the acceleration performance are shown in Fig. 7.6, where vehicle speed and output torque of motors are presented, respectively.

Figure 7.6a shows the speed curve of the electric vehicle model. The DDEV accelerates from 0 km/s, and the vehicle speed reaches 68 km/s after 10 s. This

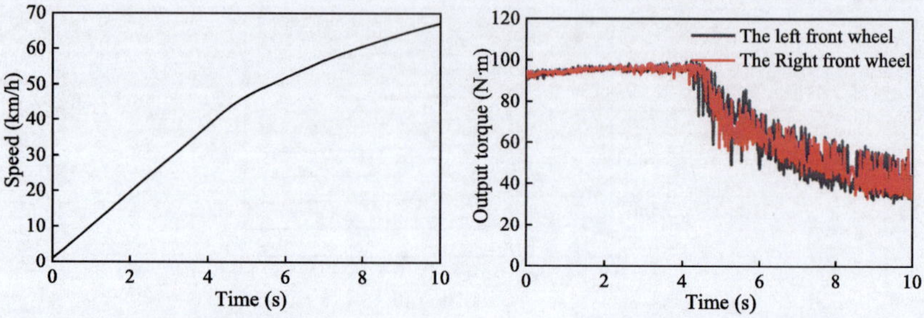

(a) Vehicle speed (b) Output torque of motors

Fig. 7.6 Simulation diagram of acceleration capability for electric vehicle

shows that the DDEV driven by the 4 kW SRM has good acceleration performance. Figure 7.6b shows the output torque of the two motors on the front axle. When the speed of the motor exceeds its rated speed after 4 s, the output torque of the motor decreases as the speed of the motor increases.

On the other hand, the straight-line stability performance of the DDEV model was verified. The vehicle run at constant speed on a road with good adhesion when the initial speed is 50 km/h. The simulation results are shown in Fig. 7.7, where the yaw rate and sideslip angle of the vehicle are presented, respectively. It can be seen that the DDEV co-simulation model can maintain good straight-line driving stability on a good road and without interference from the external environment.

7.3 Torque Coordination Control for DDEV Considering the DPA

When the DDEV is driven by the SRMs, the difference of phase angle (DPA) between coaxial motors may cause the torque output of the coaxial wheels to conflict, which can adversely affect the straight driving stability of vehicle.

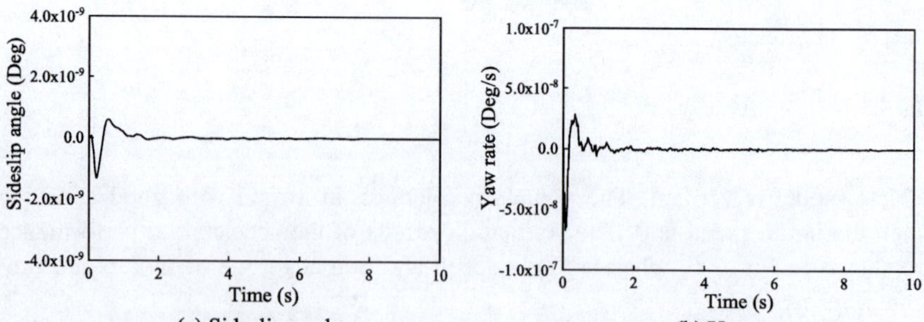

(a) Sideslip angle (b) Yaw rate

Fig. 7.7 Simulation diagram of straight driving stability for electric vehicles

Therefore, in order to improve the straight driving stability of vehicle, the influence of the DPA on the driving stability should be further studied.

7.3.1 Analysis of DPA for DDEV

For the DDEV driven by four SRMs, DPA is often generated between coaxial motors when the vehicle is under driving condition. For example, when the vehicle returns to a straight line after turning, the phase angle of the coaxial SRM cannot be kept consistent due to the different rotation speeds of the left and right wheels. In addition, when a vehicle is driving on a complex road, DPA between the coaxial SRMs is generated due to the difference in the slip conditions of the wheels on both sides. In short, there could be a difference in the movement of the wheels on both sides for DDEV, which may cause DPA.

It can be known that the electromagnetic torque is related to position angle and current. Therefore, when the position angles of two coaxial SRMs are different, the output torques of each motor can be diverse even though the currents for both SRMs are completely the same, which may cause the stability problems for DDEV. When the same DC current is applied to the working phases of the left SRM (LSRM) and right SRM (RSRM), the magnetic flux line distribution of the two motors with different phase angles is shown in Fig. 7.8. Because the rotor position angles of LRSM and RSRM are different, the degree of distortion of the magnetic flux lines of force is different, which ultimately leads to different electromagnetic torques of the two motors. Therefore, the DPA between the coaxial motors causes the output torque of the coaxial wheels to be out of sync.

On the other hand, a significant disadvantage of SRM is that the torque ripple is large. Larger torque ripple can aggravate the torque difference of the coaxial motor,

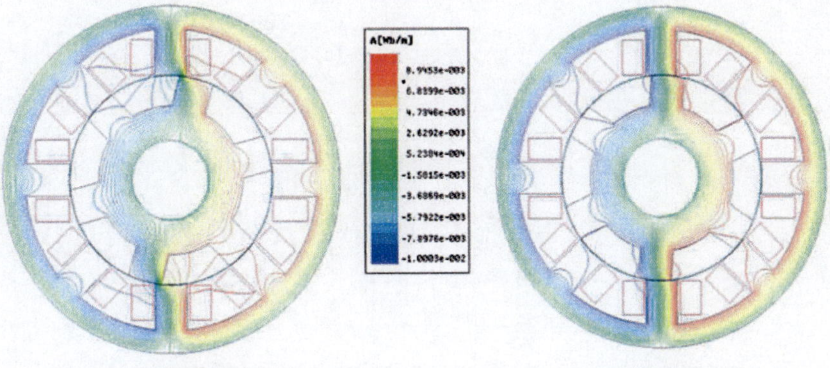

(a) LSRM (b) RSRM

Fig. 7.8 Coaxial switched reluctance motor for DDEV

increasing the instability factor of the vehicle, especially under some special conditions when large torque output is needed for the vehicle. Figure 7.9a presents one kind of typical output torque of two coaxial SRMs, where the torque for one of the motors is at the peak position when the other is at the valley. It can be clearly seen that there are large torque differences between the LSRM and RSRM installed on both sides of one axle, which are shown in Fig. 7.9b.

DDEV is independently driven by four motors, and the output torque of each wheel does not affect each other. However, due to the lack of rigid connections between the coaxial wheels, it is difficult to ensure that the coaxial wheels output the same torque when the vehicle is running under complex road conditions. Next, the influence of the unsynchronized driving torque of the left and right wheels on the vehicle stability is analyzed.

Assuming that the steering wheel angle remains a constant of zero, the tire characteristics of each driving wheel are consistent, and the deformation of the components of the steering system is ignored. The yaw motion equation is expressed as

$$I_z\omega_r = \frac{B}{2}(F_{x1} - F_{x2} + F_{x3} - F_{x4}) \tag{7.6}$$

Fig. 7.9 Output torque and torque difference of the motors

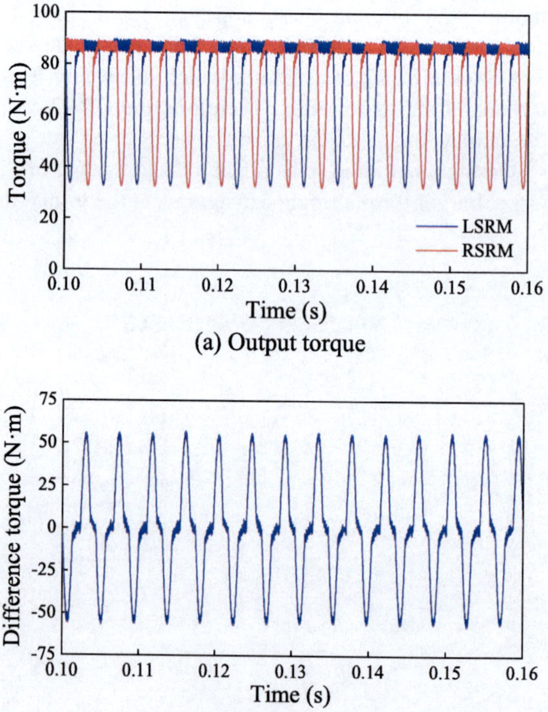

(a) Output torque

(b) Torque difference

where F_{x1}, F_{x2}, F_{x3}, and F_{x4} are the driving force of the left front wheel, right front wheel, left rear wheel, and right rear wheel, as shown in Fig. 7.10, and B refers to the wheelbase of the vehicle.

According to Eq. (7.6) and Fig. 7.10, it can be seen that the motion state of a vehicle is closely related to its driving force. If the longitudinal forces of the coaxial wheels are the same, the yaw rate ω_r of the vehicle is 0, and the vehicle remains in a stable straight-line driving state. However, if the longitudinal forces of the coaxial wheels are different, this balance will be lost, causing the vehicle to yaw or even run to one side. The longitudinal sum force could cause a counterclockwise yaw motion for the vehicle at the center of mass, if the driving forces from both sides are different.

It can be further obtained through analysis that the torque differences caused by DPA from both SRMs can influence the motion state and the driving stability of the vehicle during straight line driving. Therefore, to improve the straight-line driving stability of DDEV, it is necessary to detect and control the DPA between coaxial SRM.

7.3.2 DPA Control System Design

The DPA between coaxial motors has an adverse effect on the straight-line stability of DDEV. Therefore, in order to improve the driving performance of the vehicle, a control system that can eliminate DPA is designed. The DPA control system includes a DPA calculator that makes the system respond more quickly, a control selector in four-wheel drive mode, and a DPA controller that takes into account the yaw rate of vehicle. Figure 7.11 presents the block diagram of driving system and control system of the DDEV, including vehicle dynamics system, SRM drive system and DPA control system.

7.3.2.1 DPA Calculator

The main function of the DPA calculator is to determine the phase relationship of the coaxial switched reluctance motor, which is represented by DPA. Generally, the

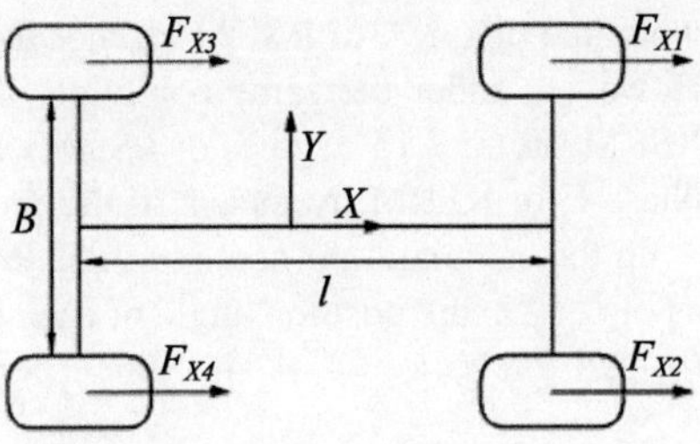

Fig. 7.10 Vehicle driving force in four-wheel drive

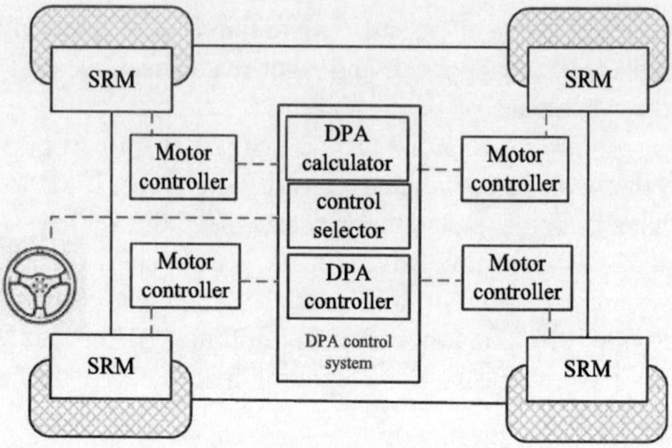

Fig. 7.11 Driving system and DPA control system of DDEV

rotor position of each motor can be obtained by the sensor, such that the DPA between coaxial motors can be calculated directly. If the phase angles of SRMs are simply subtracted, the range of the DPA would be obtained as

$$\Delta\theta = \pm\frac{360}{2 \times q \times N_r} \tag{7.7}$$

where $\Delta\theta$ is the DPA of two motors, + means that the phase angle of the left motor is ahead of that of the right one, − means that the phase angle of the left motor is behind that of the right one, q is the phase number of the motor, and N_r is the pole number of the rotor.

In this paper, the range of the DPA for two four-phase 8/6 SRMs is between −15 degrees and 15 degrees through the simple calculation method as Eq. (7.7). However, the calculated range may not be the optimal for expressing DPA due to the symmetry of the SRM structure. That is to say, the equivalence of excitation process for the four-phase windings of A, B, C, and D. It should be noted that the DPA of the two motors can be regarded as zero when the position angle of the LSRM working phase is equal to the position angle of the RSRM working phase even though the two working phases are different. After further study, a new DPA calculation method is proposed and demonstrated, as shown in Fig. 7.12. The control range of DRP is narrowed down to between −7.5 and 7.5 due to the symmetrical structure of SRM, which makes DPA control response faster. Figure 7.13 shows two coaxial SRMs of the DDEV, which have different phase angles. According to the simple calculation method, the DPA should be 10 degrees when the phase A of RSRM catches up with that of LSRM as shown in Fig. 7.13. However, under the same condition, the DPA turns to 5 degrees (20 degrees of RSRM subtract 15 degrees of LSRM) when the phase B of LSRM catches up with phase C of RSRM by means of the proposed DPA calculation method.

In the calculation process of the DPA between the four-phase 8/6 SRMs, shown as Fig. 7.12, the position angle of each phase at the left and right of motor represent θ_{Li} and θ_{Ri} (i = 1, 2, 3, 4), which period is 60 degrees. The difference of angle

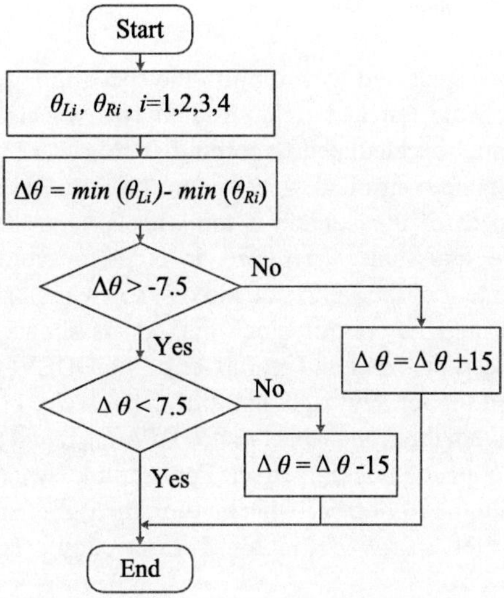

Fig. 7.12 DPA calculation process

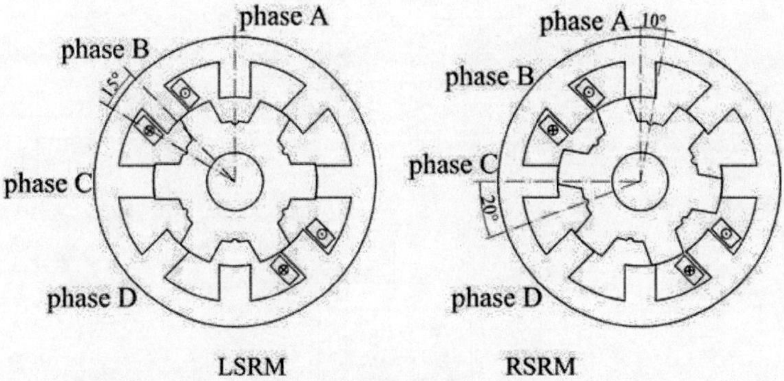

Fig. 7.13 Coaxial SRMs on both sides of DDEV

$\Delta\theta$ = min (θ_{Li}) − min (θ_{Li}) is DPA of the both motors. If $|\Delta\theta\ |$ < 7.5 and $\Delta\theta$ > 0, the relative position of the left motor rotor is forward, and the rotor of right motor catches up with that of left motor. If $|\Delta\theta\ |$ < 7.5 and $\Delta\theta$ < 0, the relative position of the right motor rotor is forward, and the rotor of left motor catches up with that of right motor. If $\Delta\theta$ > 7.5, the rotor of left motor catches up with that of right motor of $\Delta\theta$ − 15. If $\Delta\theta$ < 7.5, the rotor of right motor catches up with that of left motor of $\Delta\theta$ + 15.

7.3.2.2 Control Selector

The controller motor is selected according to the speed difference and the DPA of the coaxial SRMs on the left and right sides. If two wheels are driven independently, the DRP can be calculated according to Fig. 7.12. But for four-wheel independent drive, if the controlled wheel is on the same side, it would cause the extra lateral movement of the vehicle if the wheels controlled to slow down or speed up are on the same side. Therefore, in order to avoid the adverse effects caused by the adjustment of the same side motors, a new motor coordination control method is designed, and its control block diagram is shown in Fig. 7.14. When there are PPAs between the front and rear axles of the DDEV, the selected motor is always distributed on the left and right sides of the vehicle. In the process of DPA control, it is ensured that the front-wheel motor DPA θ_{front} and the rear-wheel motor DPA θ_{rear} have the opposite tracking effect. For example, when the left front-wheel motor catches up with the right front-wheel motor, the right rear-wheel motor needs to catch up with the left rear-wheel motor. The selection process of the regulated

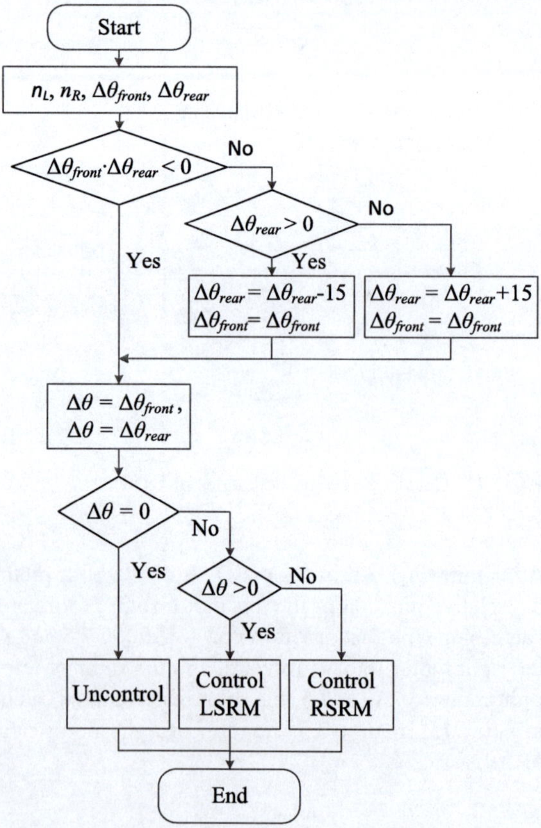

Fig. 7.14 Strategy for selecting the controlled motor

motor in the four-wheel drive mode actually produces DPA with the opposite sign. Furthermore, select the regulated motor according to the motor speed and DPA. When the DPA $\Delta\theta$ is greater than zero, the LSRM is regulated. Otherwise, the RSRM is regulated when the DPA $\Delta\theta$ is less than zero.

7.3.2.3 DPA Control System

The process of eliminating DPA is the chasing process of the motor rotors on both sides, which will inevitably increase the speed difference of the wheels on both sides of the vehicle, resulting in poorer driving stability of the vehicle. In order to eliminate the DPA among SRMs and keep stability of the vehicle during the chasing process, a DPA control strategy with feedback control of vehicle yaw rate is proposed in this paper. Figure 7.15 shows the DPA control system and the SRM control system, where the DPA control system includes a DPA calculator, a control selector, and a DPA controller with yaw rate feedback. In addition, DPA controllers include DPA PID controllers, DPA PWM controllers and yaw rate feedback controllers.

The working process of the DPA controller to eliminate DPA is further illustrated through a block diagram of a speed-down SRM drive system for DDEV, shown in Fig. 7.15. The DRP is regarded as the control target. The speed of the motor is adjusted by the DPA controller to realize the phase synchronization of the motor. In order to reduce the influence of the DPA control process on the vehicle straight line driving stability, the vehicle yaw rate is taken as the subordinate control target. In addition, the SRM is controlled by torque control strategy, where the reference torque and estimated torque are converted into the target voltage of the motor through PID controllers. Therefore, the speed regulation of the SRM is achieved by torque controller and the proposed DPA control strategy.

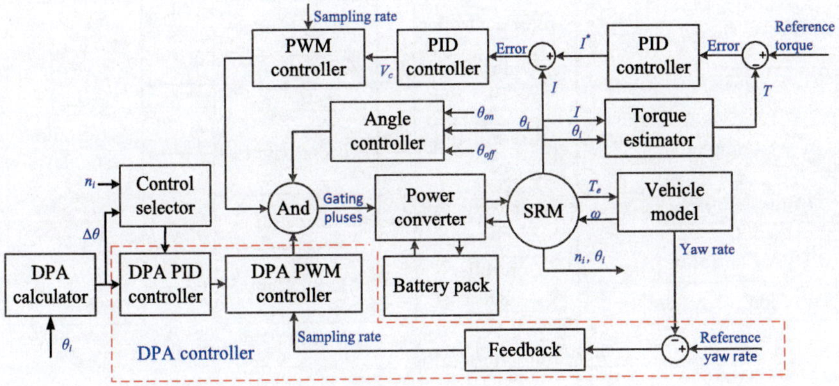

Fig. 7.15 DPA and SRM control system for DDEV

7.3.3 *Simulation and Results Analysis*

Through co-simulation of CarSim and MATLAB, the effectiveness of the DPA control system for DDVE is verified in this chapter. The vehicle dynamic model of DDEV and the driver model are established through CarSim, and the SRM model, motor controller, and the proposed DPA control system are constructed through MATLAB, as shown in Fig. 7.16. The DPA control system obtains the adjustment signal of each motor according to the phase angle and speed of each motor, and adjusts the output torque of the SRM through the motor controller, and finally realizes the DPA control. Table 7.1 lists the main parameters of the vehicle and the SRM.

Generally, the driving modes of DDEV are divided into four-wheel drive and two-wheel drive. The DPA can directly influence the driving performance of the two driving modes. Therefore, both driving modes are considered and employed in this paper to verify the effectiveness of the proposed DPA control strategy.

7.3.3.1 Results Analysis Under Two-Wheel Drive Mode

According to the DPA control system, the drive conditions for two-wheel drive DDEV with and without DPA control are simulated, where the initial speed is 50 km/h, the steering wheel angle is 0 degrees, and the initial DPA is 7 degrees. The simulation results are shown in Fig. 7.17, where vehicle speed, DPA, yaw rate, speed difference between LSRM and RSRM, and torques with DPA control are presented, respectively. Furthermore, to verify the effectiveness of the DPA control considering the yaw rate as a subordinate control target, the proposed DPA control strategy is divided into two conditions, DPA control with and without yaw rate feedback (YRF).

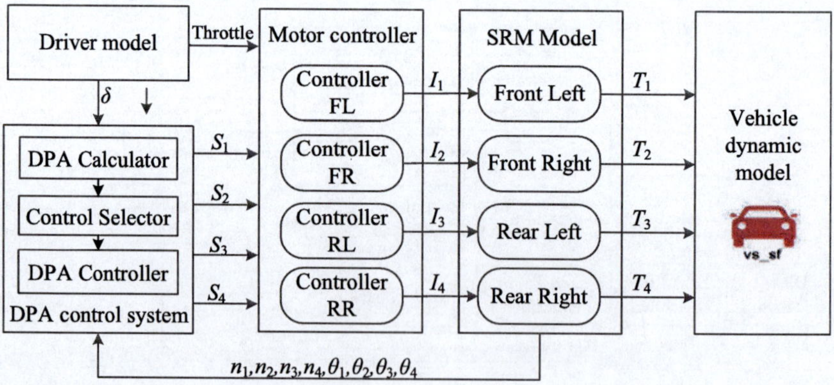

Fig. 7.16 Co-simulation model

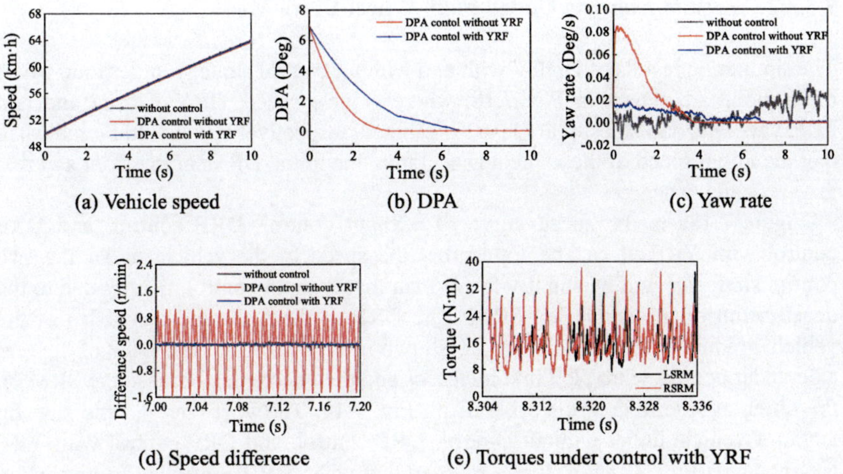

Fig. 7.17 Simulation results for two-wheel drive mode

Figure 7.17a, b shows the variation curves of vehicle speed and DPA for DDEV in two-wheel drive mode, respectively. It can be seen that the DPA control effectiveness is obvious as the DRP is controlled to be steady after 6.7 s, and DRP gradually approached zero. Figure 7.17c, d presents yaw rate of vehicle and the speed difference between LSRM and RSRM under condition without control, DPA control with YRF, and DPA control without YRF. The maximum yaw rate of vehicle without DPA control is 0.03 deg./s, and the frequency of yaw motion is relatively large, which seriously affects the straight line driving performance of the vehicle. However, the yaw rate of vehicle with DPA control is lower than 0.001 deg./s when the vehicle is controlled to be steady. Therefore, the DPA control strategy plays a significant role on improving the yaw motion caused by DPA. In addition, it can be seen that the maximum yaw rate with and without YRF is 0.082 deg./s and 0.02 deg./s during the DPA control process. Compared to that without feedback control, the DPA control with YRF generates a smaller yaw motion during the control process. Figure 7.17d further illustrates the advantage of the DPA control strategy through the speed difference of LSRM and RSRM. Compared to the condition without control, the speed difference with DPA control is reduced by 90%. Figure 7.17e shows the output torque curves of LSRM and RSRM under DPA control between 8.305 s and 8.338 s. Obviously, the output torques generated by LSRM and RSRM keep concordant after the DPA is controlled to be steady. Therefore, the proposed DPA control strategy can have an excellent effect on improving the driving stability of DDEV under two-wheel drive mode.

7.3.3.2 Results Analysis Under Four-Wheel Drive Mode

The simulation results of DDEV with and without control strategy under four-wheel drive mode are shown in Fig. 7.18, where vehicle speed, DPA of front and rear axle, yaw rate, and sideslip angle are presented, respectively. It should be indicated that the initial speed of the vehicle is 40 km/h, the initial DPAs of the front and rear axle is 7 and 5 degrees.

Figure 7.18a is the speed curve of without control, DRP control, and DRP control with YRF. It can be found that the speed of the vehicle under the two control strategies will be slightly lower than that without control, mainly due to the deceleration of the controlled SRM. The DRP of the front and rear axles of the vehicle gradually approached zero, as shown in Fig. 7.18b, c. The DRP of the rear axle under control with YRF first increases and then decreases because the SRM on the different side is controlled based on Fig. 7.14. The yaw rate and the sideslip angle of vehicle under without control, DRP control, and DRP control with YRF are shown as Fig. 7.18d, e. It can be seen that when DRP controlled to be steady, the yaw rate and sideslip angle under DRP control and DRP control with YRF are far less than these under without control. The yaw rate and sideslip angle of vehicle under without control are 0.039 deg./s, 0.003 deg., respectively. However, the values are both less than 0.001 deg./s and 0.0001 deg. under two other control strategies. Compared to DRP control, the DRP control with YRF during the control process is more stable, and the yaw rate and sideslip angle of vehicle are significantly reduced.

In order to further evaluate the performance of DRP control system, various road condition and vehicle speed are selected and analyzed. The time to stabilize, the maximum yaw rate in the control process, and the maximum yaw rate to stabilize

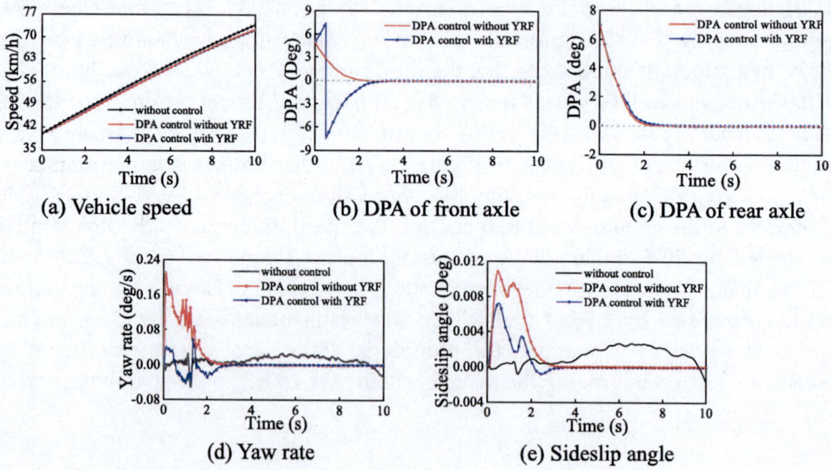

Fig. 7.18 Simulation results for four-wheel drive mode

are obtained and shown in Table 7.2. It can be seen that the proposed control system can play a very good control role under various road condition and vehicle speeds. When DRP controlled to be steady, the yaw rate under DRP control and DRP control with YRF is much smaller than that without control. In addition, compared to DRP control, the yaw rate under DRP control with YRF is lower, which is better for keeping the stability of the vehicle in DRP control process, though the time to stabilize can increase.

Table 7.2 Control results under various driving conditions

Vehicle speed	Road adhesion	Control strategy	Control completion time	Maximum yaw rate during the control process	Maximum yaw rate after stabilization
10 km/h	$\mu = 0.2$	Without control	–	–	0.04 deg./s
		Control without YRF	3.6 s	0.153 deg./s	<0.0001
		Control with YRF	4.5 s	0.062 deg./s	<0.0001
	$\mu = 0.8$	Without control	–	–	0.038 deg./s
		Control without YRF	3.1 s	0.131 deg./s	<0.0001
		Control with YRF	3.8 s	0.036 deg./s	<0.0001
50 km/h	$\mu = 0.2$	Without control	–	–	0.053 deg./s
		Control without YRF	2.9 s	0.147 deg./s	<0.0001
		Control with YRF	3.3 s	0.055 deg./s	<0.0001
	$\mu = 0.8$	Without control	–	–	0.039 deg./s
		Control without YRF	2.5 s	0.135 deg./s	<0.0001
		Control with YRF	2.8 s	0.051 deg./s	<0.0001

7.4 Torque Coordinated Control of DDEV Under Road Conditions

7.4.1 *Straight Driving Vehicle Stability Analysis on Complex Roads*

The stability of a car refers to the ability of the vehicle to maintain the same state of motion or quickly adjust to the original driving state when the vehicle is subjected to external disturbances during normal driving. Specifically, it is the ability of the vehicle to resist sideslip or tipping during driving. Excellent vehicle stability has an important influence on the driving speed, safety performance and passing performance of the vehicle. The objective evaluation method of vehicle stability is to measure physical quantities related to stability through instruments, such as the yaw rate, the lateral acceleration, the sideslip angle and the steering force. Since this article mainly studies vehicle drive control, the yaw rate and slip rate are selected as control variables to achieve the purpose of judging the stability of the vehicle.

The driving wheel slips or even run to one side due to changes in the road adhesion conditions. In response to this problem, the straight-line driving stability of DDEV is analyzed, and dynamical model of this vehicle system can be expressed as follows

$$M(\dot{u} - v\omega_r) = F_{x1} + F_{x2} + F_{x3} + F_{x4} \tag{7.8}$$

$$M(\dot{v} - u\omega_r) = F_{y1} + F_{y2} + F_{y3} + F_{y4} \tag{7.9}$$

$$I_z\dot{\omega}_r = \frac{B}{2}(F_{x1} - F_{x2} + F_{x3} - F_{x4}) + a(F_{y1} + F_{y2}) - b(F_{y3} + F_{y4}) \tag{7.10}$$

where F_{x1}, F_{x2}, F_{x3}, and F_{x4} represent the Longitudinal force of the left front wheel (LFW), right front wheel (RFW), left rear wheel (LRW), and right rear wheel (RRW), and F_{y1}, F_{y2}, and F_{y3} and F_{y4} represent the lateral force of the LFW, RFW, LRW, and RRW, as shown in Fig. 7.19.

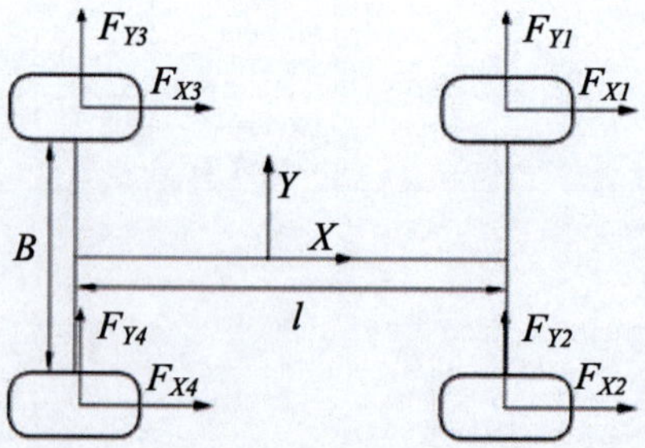

Fig. 7.19 The driving wheel force diagram on straight driving

The yaw rate is an important indicator to measure the stability of the vehicle, reflecting the vehicle yaw movement. When the vehicle is driving in a straight line, the yaw rate of the vehicle is maintained at a constant of zero to improve the driving stability performance of vehicle. When the yaw rate exceeds the stability threshold, it will increase the risk of vehicle sideslip. Therefore, limiting the yaw rate to a smaller range can increase the straight-line driving stability of vehicle.

It is obvious that the maximum adhesion force F_φ that the ground can provide is related to the ground adhesion coefficient φ. When the ground tangential reaction force F_x caused by the torque acting on the driving wheels is greater than the maximum adhesion force F_φ that the ground can provide, the driving wheels will slip. Therefore, in order to prevent the driving wheel from slipping, the ground tangential reaction force should not be greater than the adhesion force that the ground can provide, namely

$$\frac{T_x - T_f}{R} = F_x \leq F_\varphi = F_z\varphi \tag{7.11}$$

where T_x is the output torque of the driving wheels, T_f is the resistance torque to overcome the road resistance, and F_z is the normal force of the driving wheels.

In general, the slip rate s represents the sliding state of driving wheel, which is the proportion of the sliding component of the driving wheel during the movement, and it can be expressed as

$$s = \frac{\omega R - u}{\omega R} \tag{7.12}$$

where ω represent the wheel rotate speed. When the slip rate s is 0, the wheel speed is equal to the vehicle speed, and the wheel is purely rolling at this time. When the slip ratio s is 100%, the wheel speed is much higher than the vehicle speed, and the wheel is purely sliding at this time. Purely sliding wheels not only affect the stable state of the vehicle, but also accelerate tire wear and reduce tire life.

According to Eq. (7.11), when the vertical load of the driving wheel remains constant, the maximum adhesion force is proportional to the adhesion coefficient. The relationship between the slip rate and the adhesion coefficient is shown in Fig. 7.20. It can be seen that as the slip rate s increases, the longitudinal adhesion coefficient φ first increases and then decreases. When the slip ratio s is between 15 and 20%, the longitudinal adhesion coefficient φ is the largest. The lateral adhesion coefficient φ_l decreases as the slip rate s increases. When the slip rate s is 100%, the lateral force coefficient φ_l is the smallest, and the vehicle are prone to sideslip under the influence of lateral force. The maximum adhesion force is closely related to the sliding state of the tire and the ground. Therefore, the driving stability of vehicle is improved through restricting the sliding state between the driving wheels and the ground.

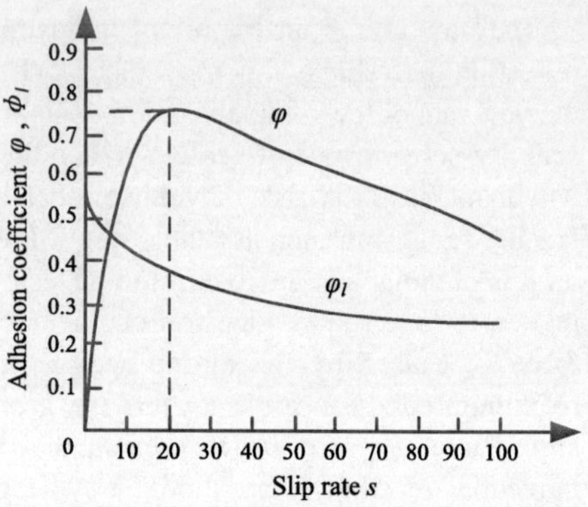

Fig. 7.20 Relation between slip rate and adhesion coefficient

When the vehicle is running on a road with good road adhesion, the wheels do not slip excessively, and the actual longitudinal driving force of the wheels depends on the output torque of the in-wheel motor. When the road adhesion coefficient is small, the maximum longitudinal driving force that the road can provide at this time cannot meet the dynamic requirements of the vehicle, and the wheels will slip excessively.

Therefore, when the vehicle is driving on a road with a high adhesion coefficient, the driving force provided by the motor is generally within the driving force range provided by the road. At this time, the wheels do not slip excessively, and the vehicle has good acceleration performance and straight driving stability.

When the vehicle is driving on a road with a low adhesion coefficient, the driving force provided by the motor is greater than that provided by the road. At this time, the four wheels of the vehicle will quickly slip excessively, which will cause severe wear of the vehicle tires. While reducing the driving capacity of the vehicle, it weakens the vehicle ability to resist lateral interference. Once the vehicle is slightly disturbed in the lateral direction, the vehicle will lose driving stability and may cause dangers such as tail-flicking.

When a vehicle is driving on a road with different road adhesion coefficients on both sides, the driving force of the wheels on the high adhesion road is less than the maximum driving force provided by the road, while the actual driving force of the wheels on the low adhesion road is greater than the maximum driving force provided by the road. At this time, the wheels on the low-adhesion road will produce serious slippage. If there is a large difference in the driving force of the vehicles on both sides, and the driving force is not controlled, the vehicle will run to one side and lose the ability to drive straight. In addition, the wheels on low-adhesion roads have been in an excessive slip state, and the driving stability of the vehicle will be greatly reduced.

7.4.2 Straight Driving Vehicle Stability Control System Design

7.4.2.1 Yaw Moment Controller

Sliding mode control is a nonlinear control system, which is mainly manifested in the discontinuity of control action. This sliding mode control has fast response speed and good robustness, and it will not become insensitive due to parameter changes or external disturbances. At the same time, the main disadvantage of this control is the serious chattering problem. The yaw motion controller is designed based on the principle of sliding mode control, and the design process is showed as follows.

According to Eq. (7.10), the differential equation of yaw rate is obtained as

$$\dot{\omega}_r = \frac{a(F_{y1} + F_{y2}) - b(F_{y3} + F_{y4}) + M_z + M_\omega}{I_z} \tag{7.13}$$

where M_ω is the additional yaw moment provided by the four driving motors, and M_z is the disturbance moment caused by unstable road conditions.

First, the sliding mode function is determined by

$$S = (\omega_r - \omega_d) + k_1 \left(\int \omega_r dt - \int \omega_d dt \right) \tag{7.14}$$

where k_1 represents the weighting coefficient of the yaw angle error, and ω_d is the desired yaw rate of the vehicle. Since only the straight-line vehicle stability is considered, the desired yaw rate of the vehicle is a constant of zero. The designed sliding mode function can be expressed as

$$S = \omega_r + k_1 \int \omega_r d \tag{7.15}$$

The differential equation of sliding mode function can be obtained as

$$\dot{S} = \dot{\omega}_r + k_1 \omega_r \tag{7.16}$$

According to Eqs. (7.13) and (7.16), the differential equation of sliding mode function can be modified as

$$\dot{S} = \frac{M_\omega + M_z}{I_z} + \frac{a(F_{y1} + F_{y2})}{I_z} - \frac{b(F_{y3} + F_{y4})}{I_z} + k_1 \omega_r \tag{7.17}$$

Second, the sliding mode control law is designed according to the exponential approach

$$\dot{S} = -k_2\text{sgn}(S) - k_3S \tag{7.18}$$

where k_2 represents the speed of the system approaching the sliding surface, k_3 represents the convergence speed after reaching the sliding surface, k_2 and k_3 are both greater than 0. In order to weaken the control system chattering caused by the discontinuous function sgn(S), the saturation function sat(S) is applied

$$sat(S) = \begin{cases} \frac{S}{k_4} & |S| < k_4 \\ \text{sgn}(S) & |S| > k_4 \end{cases} \tag{7.19}$$

where k_4 represents the thickness of the boundary layer of the sliding surface.

Finally, the control system stability is verified based on the Lyapunov stability. Lyapunov function is defined as

$$V = \frac{1}{2}S^2 \tag{7.20}$$

$$\dot{V} = S\dot{S} = S[-k_2 sat(S) - k_3 S] \tag{7.21}$$

when $sat(S) = \frac{S}{k_4}$, $\dot{V} = S\dot{S} = -\left(\frac{k_2}{k_4} + k_3\right)S^2 \leq 0$;

when $sat(S) = \text{sgn}(S)$, $\dot{V} = S\dot{S} = -\left[\frac{k_2}{|S|} + k_3\right]S^2 \leq 0$.

According to the Lyapunov principle, the designed sliding mode control satisfies

$$V(0)=0; \quad \forall x \neq 0, \quad V(S) > 0 \quad \text{and} \quad \dot{V}(S) < 0 \tag{7.22}$$

Therefore, the control system meets the stability control requirements, and the control system gradually stabilizes. According to Eqs. (7.16) and (7.17), the additional yaw moment can be obtained

$$\begin{aligned} M_\omega = &- a(F_{y1} + F_{y2}) + b(F_{y3} + F_{y4}) - M_z \\ &- I_z(k_2 sat(S) + k_3 S + k_1 \omega_r) \end{aligned} \tag{7.23}$$

7.4.2.2 Acceleration Slip Regulation Controller

Fuzzy control is an intelligent control method which can imitate human reasoning and decision-making process. It performs nonlinear control on the controlled object based on fuzzy set theory, fuzzy language and fuzzy logic inference. The dynamic characteristics of the controlled object are included in the input and output fuzzy sets and fuzzy rules of fuzzy control, such that it can control the system well without considering the mathematical model of the controlled object. However, the fuzzy control is too dependent on expert knowledge and experience, and often shows a strong case pertinence, such that the fuzzy control rules are relatively limited. If fuzzy control is combined with other control methods, a better control effect can be achieved. Therefore, in order to enhance the sensitivity of the control system, a fuzzy PID control method is proposed based on the advantages of fuzzy control.

Fuzzy PID control combines fuzzy control and PID control, absorbs the advantages and makes up for the deficiencies of the two control methods, and improves the adaptability of the control system. In the fuzzy control, the error e and the error change rate ec are taken as the input, and the fuzzy control rules are formulated according to the PID control parameter adjustment experience. Then, the parameters of the PID controller are adjusted adaptively through fuzzy reasoning. Compared with the traditional PID control, the fuzzy PID control can adjust the PID control parameters in real time according to the human control strategy and experience, so that the controlled object has a good dynamic and static stable state.

Based on fuzzy PI control, an acceleration slip regulation controller of DDEV is designed, as shown in Fig. 7.21. In the process of vehicle running, When the car is running, the slip condition of the wheels is judged by the slip rate the slip rate is used to judge the slip state of the wheel. Therefore, the wheel slip rate error e and the slip rate error change rate ec are taken as the control objectives of the acceleration slip regulation controller, where slip the rate error $e = S_d - S_i$, the slip rate error change rate $ec = \dot{S}_d - \dot{S}_i$, and S_d, S_i represent the desired slip rate and actual slip rate of driving wheel respectively. For the acceleration slip regulation controller, fuzzy control is added on the basis of PI control. According to the designed fuzzy control rules, the parameters of the proportional link and the integral link are adjusted in real time to improve the adaptability of the acceleration slip regulation controller.

First, the wheel slip rate error e is transformed by the PI controller, and the driving anti-skid torque is obtained T_{slip}

$$T_{slip} = K_p e(\lambda) + K_I \int_0^t e(\lambda)dt \tag{7.24}$$

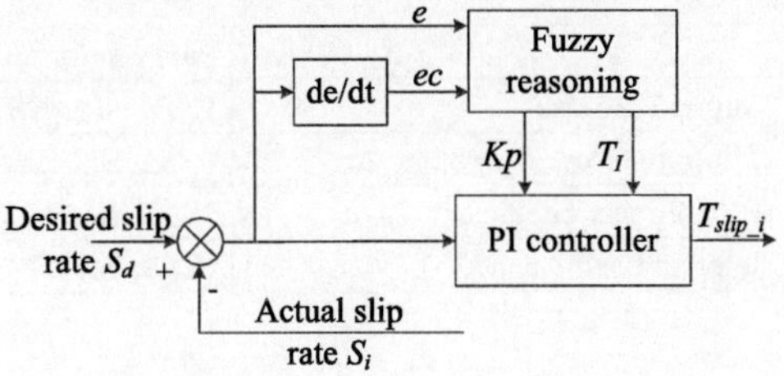

Fig. 7.21 The acceleration slip regulation controller

where K_p and K_I represent the proportionality coefficient and integration coefficient in PI control respectively.

Then, the wheel slip rate error e and the slip rate error change rate ec are controlled by fuzzy control, and the proportionality coefficient K_p and integration coefficient K_I of PI control are obtained. The design process of the acceleration slip regulation controller based on fuzzy PI control is shown as follows.

(1) Determine the discourse domain.

According to Fig. 4.20, when the wheel's slip rate is between 15 and 20%, the longitudinal and lateral adhesion coefficients of the driving wheel are the largest. Therefore, the desired slip rate of the driving wheel S_d is set as 20%. When the vehicle is running on the road, the wheel slip rate error e varies from −0.8 to 0.2, and the slip rate error ec varies from −30 to 30. The quantization factors of slip rate error and slip rate error change rate are defined as 10 and 0.2 respectively. Therefore, the discourse domain of the slip rate error e and the slip rate error change rate ec are [−8, 2] and [−6, 6] respectively. The output variable of the fuzzy controller is the adjustment coefficient of the PI controller. The range of the proportionality coefficient and integration coefficient are set between 0 to 100 and between 1 to 10 respectively, and their quantization factors are 0.1 and 1 respectively. Therefore, the discourse domain of proportionality coefficient and integration coefficient is [0, 20] and [0, 10] respectively.

(2) Determine the membership function.

The discourse domain of wheel slip rate error is divided into five grades, which are negative large, negative medium, negative small, zero, and positive small respectively, and the discourse domain set of slip rate error can be expressed as {negative large, negative medium, negative small, zero, positive small} = {Nd, NM, NS, ZE, PS}. Similarly, the wheel slip rate error change rate, proportional coefficient and integral coefficient are divided into five, four and four grades respectively. These discourse domains and fuzzy sets are shown in Table 7.3.

The input parameters of the fuzzy controller are exact value, which needs to be transformed into fuzzy value through fuzzification. In addition, triangles are selected as the membership function of the fuzzy controller. Figure 7.22 shows the membership functions of slip rate error e, slip rate error change rate ec, proportional coefficient K_P and integration coefficient K_I.

Table 7.3 Discourse domain and fuzzy sets

	Fuzzy sets	Discourse domain
Slip rate error e	{ND, NM, NS, ZE, PS}	[−8, 2]
Slip rate error change rate ec	{NM, NS, ZE, PS, PM}	[−6, 6]
Proportional coefficient K_P	{Z, S, M, B}	[0, 20]
Integral coefficient K_I	{Z, S, M, B}	[0, 10]

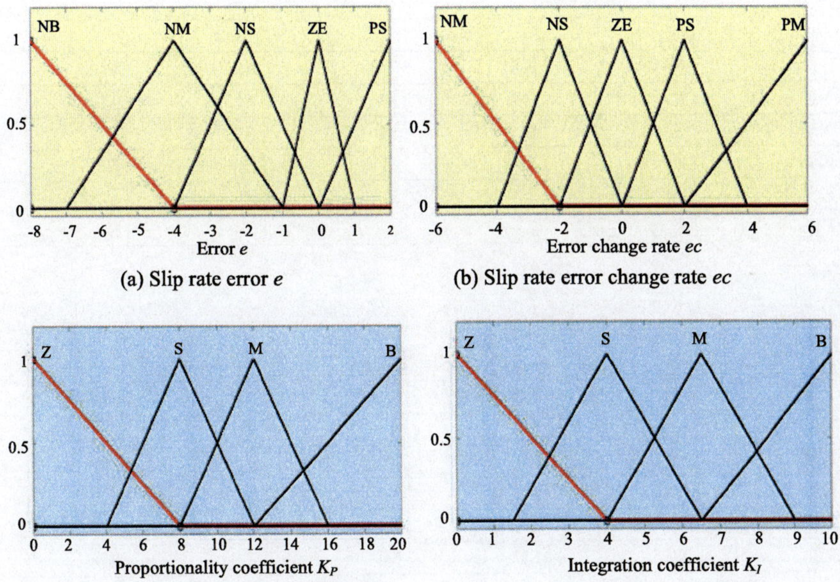

(a) Slip rate error e (b) Slip rate error change rate ec

(c) Proportional coefficient K_P (d) Integral coefficient K_I

Fig. 7.22 Membership functions of input variables

(3) Establish fuzzy rules.

The control rules of fuzzy controller are determined by analyzing the influence of PI control proportion link and integral link on the driving anti-skid torque T_{slip}. When the wheel slip rate is much higher than the desired slip rate, the driving wheel slip is more serious. At this time, the proportional link should respond quickly to reduce the output torque of the driving motor. When the driving wheel slip rate is close to the desired slip rate, the adhesion effect of the wheel is better, and the effect of the proportional link should be weakened. On the other hand, in order to eliminate the static deviation of the control system, the control effect of the integral link is enhanced. Therefore, the fuzzy control rules of the proportionality coefficient K_P and integration coefficient K_I are shown in Tables 7.4 and 7.5.

(4) Defuzzification.

Through fuzzy reasoning on the fuzzy set of the input, the fuzzy set of the output is obtained. This fuzzy set cannot adjust the coefficient of PI control system directly, so it needs to be solved to get the precise output value. At present, the commonly used defuzzification methods in fuzzy control include area gravity center method, maximum mean value method and maximum membership degree method. The designed fuzzy controller adopts area gravity center method in this chapter. The defuzzification results of the proportionality coefficient K_P and integration coefficient K_I are shown in Fig. 7.23.

Table 7.4 K_P fuzzy control rules

e	*ec*				
	NM	NS	ZO	PS	PM
NB	B	B	M	M	M
NM	B	B	M	S	Z
NS	B	M	S	Z	S
ZO	S	Z	Z	Z	S
PS	S	Z	S	S	S

Table 7.5 K_I fuzzy control rules

e	*ec*				
	NM	NS	ZO	PS	PM
NB	B	M	S	Z	S
NM	M	S	Z	S	M
NS	S	Z	S	M	B
ZO	Z	S	M	M	B
PS	S	M	M	B	B

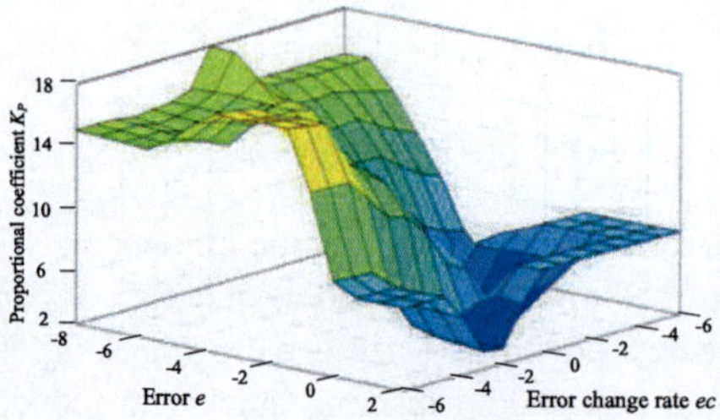

(a) Proportionality coefficient *Kp*

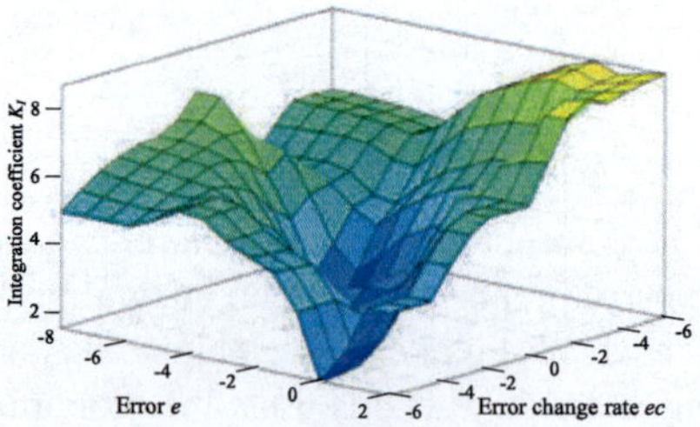

(b) Integration coefficient K_I

Fig. 7.23 Output of the fuzzy controller

7.4.3 Torque Coordinated Control

According to the designed straight driving vehicle stability control system, the additional yaw moment M_{ω} and the driving anti-skid moment T_{slip} are obtained. In order to distribute the torques calculated by the upper controller to the four in-wheel motors in an appropriate manner, the vehicle straight driving stability control is realized. The rule-based torque coordinated control method is adopted to adjust the driving force of the left and right wheels in real time under the premise of ensuring that the wheels do not excessively slip. To improve the driving stability of the vehicle on complex roads, the torque of the four wheels is distributed based on the driving state of the vehicle, and the specific method is as follows.

(1) Desired torque distribution.

When a vehicle drives on the road, the desired torque that can allow the vehicle to run normally is evenly distributed to the four wheels.

(2) Additional yaw moment distribution.

In order to restrain the yaw motion generated by external factors for the vehicle, an additional yaw moment is applied to the vehicle. This additional yaw moment is distributed to each wheel, and the additional torque of each wheel is

$$[T_{yaw\_1} \quad T_{yaw\_2} \quad T_{yaw\_3} \quad T_{yaw\_4}] = \frac{R}{2B} M_{\omega}[1 \quad -1 \quad 1 \quad -1] \tag{7.25}$$

where T_{yaw_1}, T_{yaw_2}, T_{yaw_3}, and T_{yaw_4} is the additional torque of the LFW, RFW, LRW, and RRW, respectively.

(3) Driving anti-skid torque distribution

In order to improve the driving stability of the vehicle, the additional driving anti-skid torque of the coaxial wheels should remain the same. The additional driving torque of the front and rear axle wheels is

$$\begin{aligned} T_{slip\_f} &= \max(T_{slip\_1}, T_{slip\_2}) \\ T_{slip\_r} &= \max(T_{slip\_3}, T_{slip\_4}) \end{aligned} \tag{7.26}$$

where T_{slip_f} and T_{slip_r} are the additional driving anti-skid torque of the front and rear axle wheels, T_{slip_1}, T_{slip_2}, T_{slip_3}, and T_{slip_4}, is the additional driving anti-skid torque of the LFW, RFW, LRW, and RRW, respectively.

When a vehicle is driving on a complex road, the vehicle motion state is judged by the yaw rate and the wheels slip rate. First, set the vehicle yaw rate control threshold. When the actual yaw rate is greater than the threshold, the yaw motion controller is activated to calculate the additional yaw moment against the vehicle deviation. Second, set the desired slip rate of the driving wheels. When the actual slip rate is greater than the desired slip rate, the acceleration slip regulation controller is activated to calculate the driving anti-skid control torque of each wheel. Based on the above analysis, the driving torque of each wheel is obtained as

$$\begin{aligned} T_{x1} &= \frac{T_{des}}{4} + T_{slip\_f} + T_{yaw\_1} \\ T_{x2} &= \frac{T_{des}}{4} + T_{slip\_f} + T_{yaw\_2} \\ T_{x3} &= \frac{T_{des}}{4} + T_{slip\_r} + T_{yaw\_3} \\ T_{x4} &= \frac{T_{des}}{4} + T_{slip\_r} + T_{yaw\_4} \end{aligned} \tag{7.27}$$

where T_{x1}, T_{x2}, T_{x3}, and T_{x4} is the driving torque of the LFW, RFW, LRW, and RRW, respectively.

7.4.4 *Simulation Results Analysis*

In order to verify the effectiveness of the proposed straight driving vehicles stability control strategy on complex roads, a co-simulation system based on the DDEV model, the straight driving vehicle stability control system and the torque coordinated control method is carried out on the step road and the split road, and shown in Fig. 7.24. The yaw comment controller and the acceleration slip regulation controller respectively calculate the additional yaw moment and the driving anti-skid torque to maintain the driving stability of the vehicle, and the driving torque is distributed to the four wheels through the torque coordination controller.

7.4.4.1 Result Analysis of the Step Road Condition

The driving conditions of DDEV on the step road are simulated under with and without straight driving stability control, where the initial speed is 0, and the road adhesion coefficient is shown in Fig. 7.25. The simulation results are shown in Fig. 7.26, where include the vehicle speed, four wheels slip rate, the wheel rotate speed, the driving torque, yaw rate, and sideslip angle.

Figure 7.26a present the variation curves of vehicle speed under with and without straight driving stability control. The vehicle speed under the control modes reaches 48 km/h after 10 s, and due to the change of the road adhesion conditions,

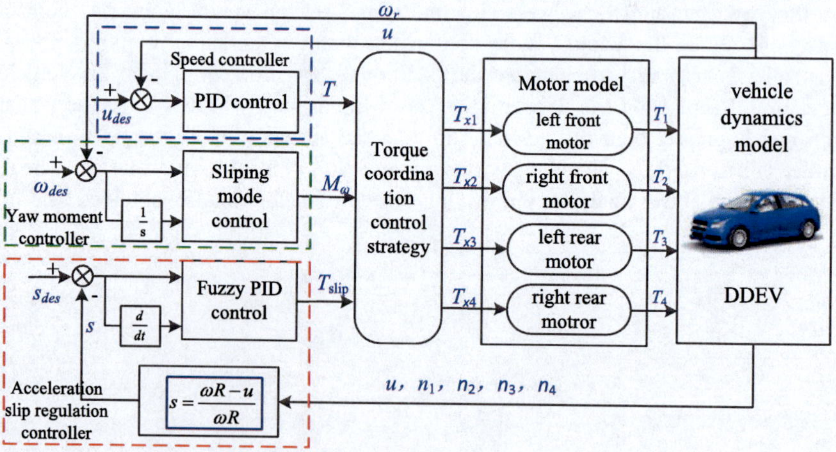

Fig. 7.24 The co-simulation system of straight driving vehicle stability control

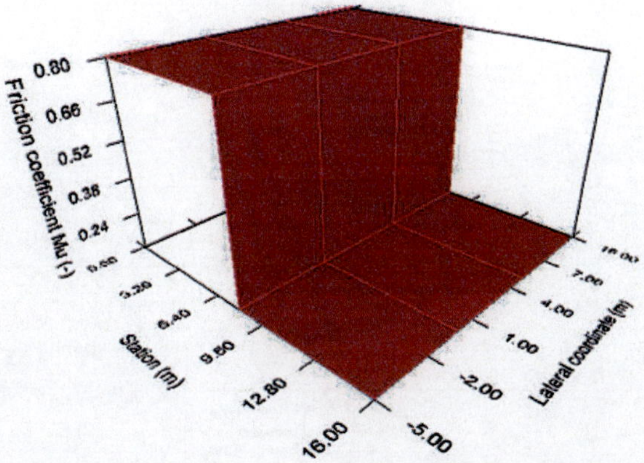

Fig. 7.25 The step road

the vehicle acceleration ability is significantly reduced after 2.3 s. Four wheels slip rates under the two control modes are shown in Fig. 7.26b, c. When the vehicle enters the low adhesion road, the wheels quickly slip. For without straight driving stability control, the wheel slip rate quickly reaches 0.9, and the slip rate gradually decreases as the vehicle speed increases. According to Fig. 4.2, the vehicle is in a dangerous situation, because the lateral adhesion ability of the vehicle is very small at this time, and a small lateral force can cause the vehicle to side slip. However, for with control conditions, the wheel slip rate first increases and then rapidly decreases, and the wheel slip rate of front and rear axle remain near the optimal slip rate of 0.2 after 4.5 s. Therefore, the proposed straight driving stability control strategy plays an important role in preventing wheel slip and making full use of the adhesion force of the road. Figure 7.26d, e, f, g are four wheels rotate speed and driving torque under two control strategies, respectively. For without control, the wheels rotate speed of front and rear axle increase rapidly, reaching the maximum speed of 4340 r/min at 4.6 s and 5.2 s, respectively, and the output torque of wheels decreases as the vehicle speed increases. For with straight driving stability control, the driving torque of the front and rear wheels is rapidly reduced to 28 N m and 26 N m when the wheels are slipping, so that the wheels maintain a good adhesion ability to the ground. The wheels rotate speed of front and rear axle increases slowly, and reached 546 r/min at 10 s. Figure 7.26h, i show the vehicle yaw rate and the sideslip angle, respectively. It can be seen that the yaw rate fluctuates when the front and rear axle wheels of the vehicle enter the low adhesion road, but the vehicle can maintain stable driving on the other road. Therefore, the proposed straight driving stability control strategy plays a very good role in improving the driving stability of vehicles under the step road.

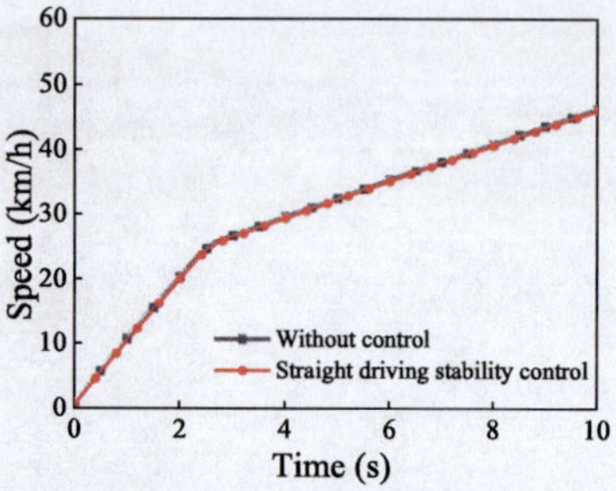

(a) The vehicle speeds

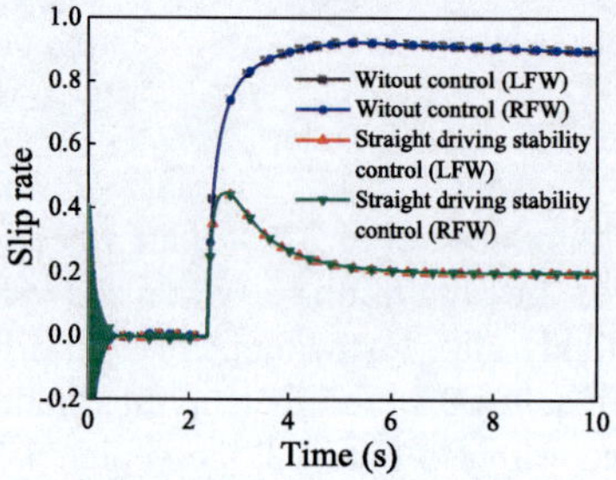

(b) The wheel slip rate of front axle

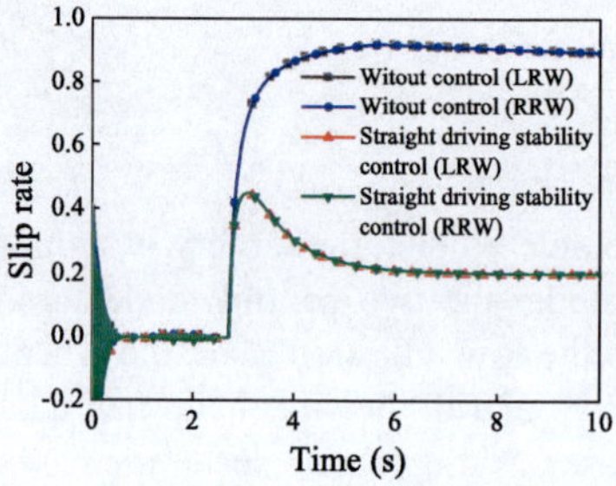

(c) The wheel slip rate of rear axle

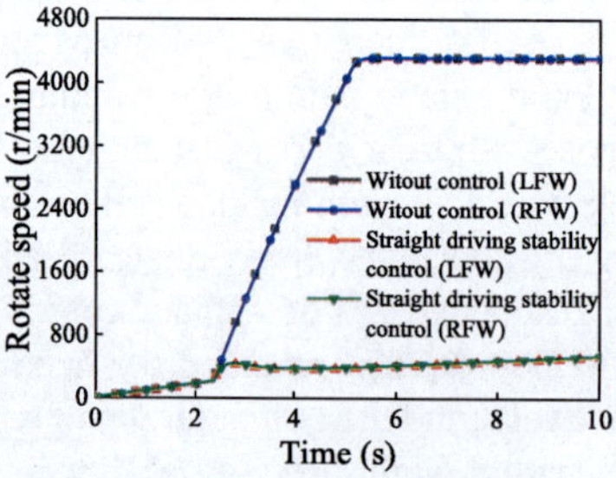

(d) The wheel rotate speed of front axle

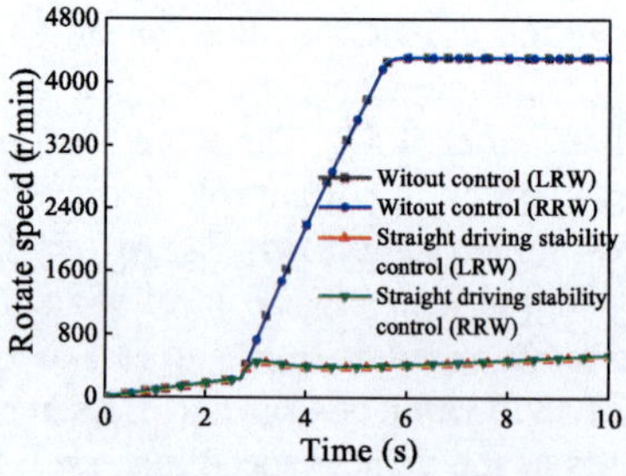

(e) The wheel rotate speed of rear axle

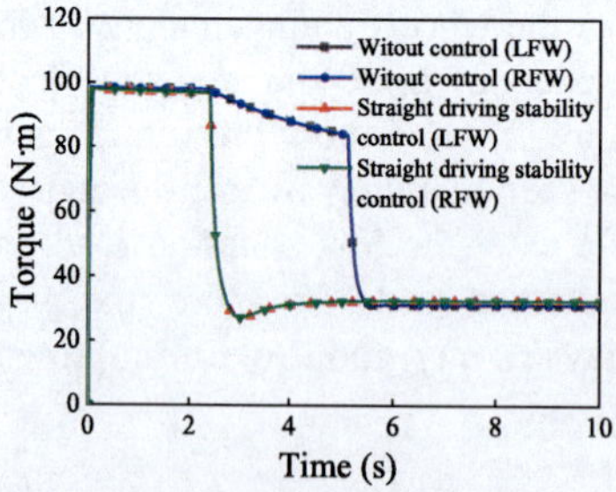

(f) The wheel driving torque of front axle

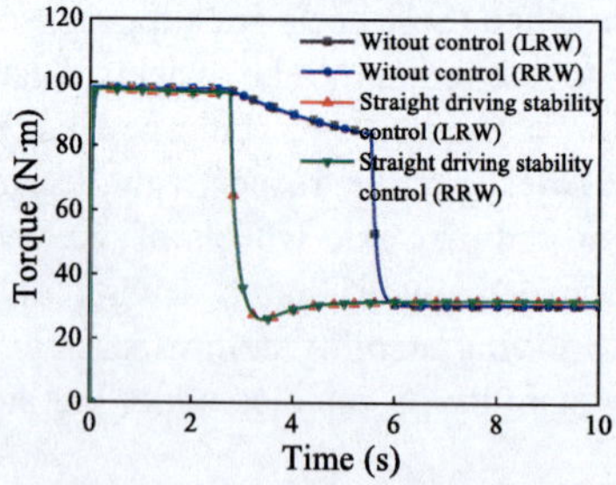

(g) The wheel driving torque of rear axle

Fig. 7.26 Simulation results under the step road condition

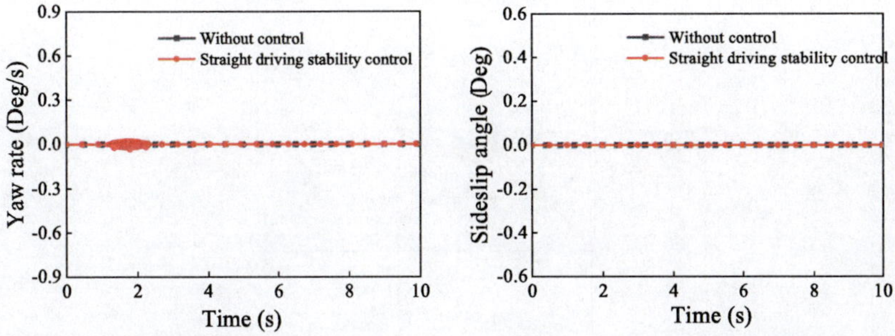

(h) Yaw rate (i) Sideslip angle

Fig. 7.26 (continued)

7.4.4.2 Result Analysis of the Split Road Condition

The simulation of DDEV on the split road are performed, where the initial speed of the vehicle is 0, the left road adhesion coefficient is 0.1, and the right road adhesion coefficient is 0.6, shown in Fig. 7.27. The simulation results under with and without straight driving stability control shown in Fig. 7.28, where include the vehicle speed, the wheel slip rate of front and rear axle, the wheel driving torque of front and rear axle, yaw rate, and sideslip angle.

The vehicle speed curves under with and without straight driving stability control are presented in Fig. 7.28a. For without control, the vehicle has better acceleration performance, and the vehicle speed can reach 85 km/h after 10 s. For with straight driving stability control, the vehicle speed is 64 km/h after 10 s. Figure 7.28b, c show the wheel slip rate of the front and rear axle under the two control conditions. For without control, when the left wheel of the vehicle enters the

Fig. 7.27 The split road

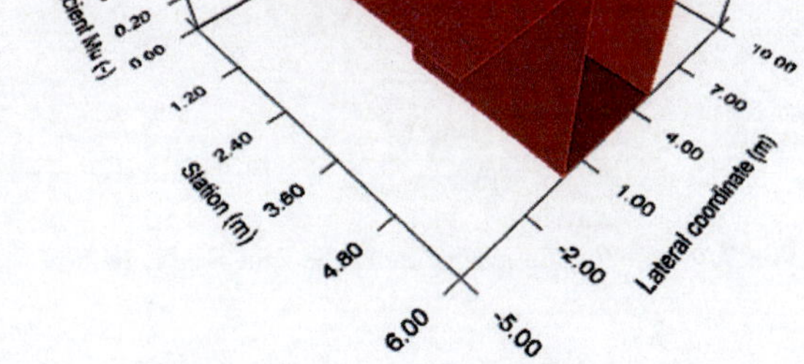

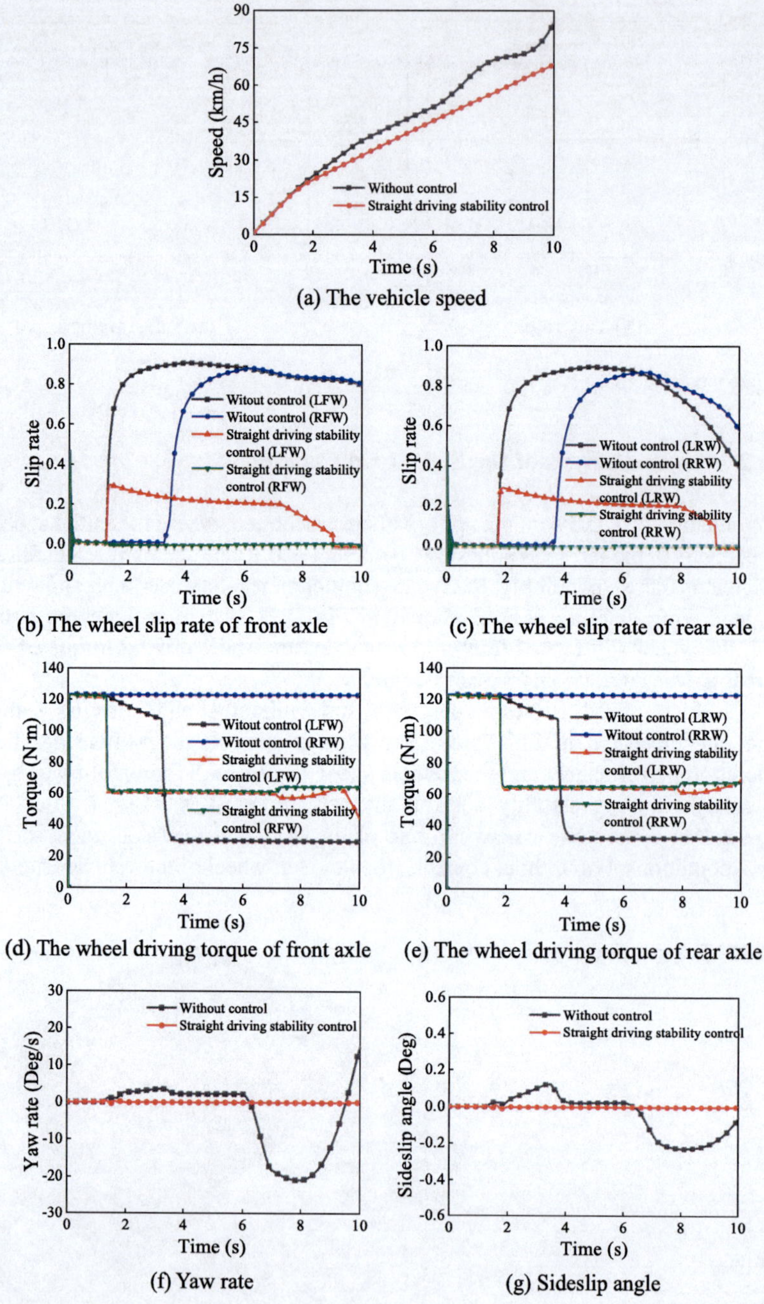

Fig. 7.28 Simulation results under the split road condition

low adhesion road, the wheel slip rate increases, reaching 0.91 and 0.89, respectively. Before 3.6 s, the right wheel slip rate of the vehicle is almost unchanged because the left wheel is on a high adhesion road. After 3.6 s, the vehicle runs to one side due to the large yaw motion, which caused the right wheel to enter the low adhesion road. For with control, the left wheel slip rate increases rapidly when the left wheel enters the low adhesion road. As the wheel slip phenomenon and the yaw motion are controlled by straight driving stability control system, the left wheel slip rate is reduced to near the optimal slip rate of 0.2. Figure 7.28d, e show the wheel driving torque of the front and rear axle under two control strategies. For without control, the driving torque of the FLW and FRW is reduced to 29 N m at 3.7 s and 4 s, respectively. For with control, the coaxial wheels driving torque is reduced due to the increase in the slip rate, and the wheels driving torque of the front and rear axle are reduced to 57 N m and 63 N m, respectively. Figure 7.28f, g present the vehicle yaw rate and the sideslip angle under the two control strategies, respectively. For without control, the yaw rate and the sideslip angle are both greater than zero, and the vehicle tends to turn to the left. After 3.4 s, the wheels on both sides enter on the low adhesion road, which leads to a sharp decrease in vehicle stability. For with control, the vehicle maximum yaw rate and maximum sideslip angle are 0.03 deg./s and 0.002 deg., respectively, and the vehicle has a good straight driving stable performance. In addition, compere to without control, the straight driving stability of the DDEV with control is superior, although the dynamic property of the DDEV is reduced.

References

1. Murata S. Innovation by in-wheel-motor drive unit. Vehicle System Dynamics, 2012, 50(6): 807.830.
2. Shibahata Y., Shimada K., Tomari T. Improvement of vehicle maneuverability by direct yaw moment control. Vehicle System Dynamics, 1993, 22(5-6): 465–481.
3. Yin D, Omae M., Shimizu H, et al. Active stability control strategy based on maximum transmissible torque estimation. Journal of Automotive Safety & Energy, 2011, 2(1): 34–38.
4. Liu H, Chen X, Wang X. Overview and prospects on distributed drive electric vehicles and its energy saving strategy. Przeglad Elektrotechniczny, 2012, 88(7): 122–125.
5. Alipour H, Sabahi M, Bannae Sharifian MB. Lateral stabilization of a four wheel independent drive electric vehicle on slippery roads. Mechatronics, 2015: 275–285.
6. Shino M, Nagai M. Yaw-moment control of electric vehicle for improving handling and stability. JSAE Review, 2001, 22(4): 473–480.
7. Motoyama S, Isoda K, Mimuro T. Quantitive evaluation for yaw control ability. AVEC, 1998: 443–447.
8. Kim HS, Hyun YJ, Nam KH. Disturbance observer-based sideslip angle control for improving cornering characteristics of In-Wheel Motor electric vehicles. International Journal of Automotive Technology, 2018, 19(6): 1071–1080.
9. Yu Z, Bo L, Lu X, et al. Direct yaw moment control for distributed drive electric vehicle handling performance improvement. Chinese Journal of Mechanical Engineering, 2016, 29 (3): 486–497.

10. Wang W, Zhang W, Zhao Y. Integrated stability control strategy of In-Wheel Motor driven electric bus. International Journal of Automotive Technology, 2020, 21(4): 919–929.
11. Chen AQ, Kang BS, Chen H, et al. Acceleration slip regulation of distributed driving electric vehicle based on road identification. IEEE Access, 2020, 99: 165–172.
12. Zhou S J, Luo YT, Huang X D, et al. Slip rate identification and traction control of 4WD electric vehicle. Journal of South China University of Technology (Natural Science Edition), 2008, 5: 132–139.
13. Maeda K, Fujimoto H, Hori Y. Four-wheel driving-force distribution method based on driving stiffness and slip ratio estimation for electric vehicle with in-wheel motors. Vehicle Power & Propulsion Conference, 2012.
14. Yin D, Oh S, Hori Y. A novel traction control for EV based on maximum transmissible torque estimation. IEEE Transactions on Industrial Electronics, 2009, 56(6): 2086–2094.
15. Nam K. Lateral stability control of In-Wheel-Motor-Driven electric vehicles based on sideslip angle estimation using lateral tire force sensors. IEEE Transactions on Vehicular Technology, 2012, 61(5): 1972–1985.

Chapter 8
Comprehensive Control of in Wheel SRM-Suspension System

8.1 Introduction

Recently, the in-wheel switched reluctance motor (IWSRM) has become one of the competitive in-wheel motors (IWMs) because of its advantages of low cost, high torque density, integrated configuration, direct drivetrain system, and efficient power transfer [1, 2]. However, the use of IWMs including IWSRM, can undoubtedly increase the vehicle unsprung mass, causing vertical vibration and ride comfort problems of electric vehicles (EVs) [3]. Moreover, IWSRM also has some disadvantages of high torque ripple and unbalanced radial force due to a great nonlinear stepping magnetic field [4, 5], which can directly result in the vehicle vibration problem of EVs. In addition, besides the driving condition, combined with the characteristics of switched reluctance generator (SRG) complex generating principle, the braking energy recovery and vertical vibration caused by IWSRMs under the regenerative braking condition should also be considered together [6, 7].

In the past several years, there are many researches about all kinds of methods that have been done to reduce the vibration of IWSRM, which includes the structure optimization [8, 9], the power converter design [10, 11], and the control strategies [12, 13]. Generally, according to the propulsion system for EVs, there are two driving modes, centralized drive and in-wheel motor drive [14]. Nevertheless, differing from centralized drive system with SRM, IWM drive is coupled with the suspension of EVs. Thus, to achieve the gold of the vibration control of EVs, the IWM and suspension should be considered as a whole. The unbalanced radial force generated by motor and the unsprung mass from IWSRM-suspension system can directly result in the vehicle vibration and influence the ride comfort of EVs. For increasing the dynamic performance of the IWM-suspension system, the study of the coupling performance between the IWM and the suspension have been done in [15, 16]. Meanwhile, based on different kinds of conditions, the influence theory of the balanced radial force on the vertical vibration performance of EVs was also described. In the meantime, to improve the whole vehicle dynamic performance, a

Y. Zhu, *The Key Technologies for Powertrain System of Intelligent Vehicles Based on Switched Reluctance Motors*, Recent Advancements in Connected Autonomous Vehicle Technologies 1, https://doi.org/10.1007/978-981-16-4851-9_8

lot of active control strategies for suspension have been presented in [17, 18]. On the other hand, for increasing the dynamic performance of the whole system, a variety of dynamic vibration absorbing structures were designed and developed in [19, 20]. According to these studies mentioned above, it can be obviously found that previous studies showed the dynamic performance of the IWSRM-suspension system have obvious improvements via the principle analysis, active suspension control, and the use of the dynamic vibration absorbing structure. But in fact, besides these methods mentioned above, the control strategies of motor also play a significant role for the vehicle vibration suppression [21, 22].

When SRM begins to work, to quickly generate the maximum torque, current chopping control (CCC) method is generally adopted to control the current of each phase, while the method could result in much vibration of EVs. Differing from CCC method, with phase current steadily increase, pulse width modulation (PWM) could decrease the vibration of EVs. Based on these reasons, a modified PWM control method for IWSRM in starting stage is presented to reduce the vertical vibration of vehicle. On the one hand, to further suppress the vibration of vehicle, angle position control (APC) is applied as a supplementary method for CCC or PWM in this chapter. However, as far as authors know, the control method of APC has not been employed to decrease the vibration of EVs. According to the comparisons of the new developed controllers (CCC-F and PWM-F) and the traditional ones (CCC and PWM), the conclusions can be acquired that each of them performs differently with the imparity of conditions. As a result, based on driving modes, a switchable controller of IWSRM is designed in this chapter.

In addition, according to these studies mentioned above, they almost focused on driving condition of EVs, with little regard for the regenerative braking condition. However, the regenerative braking condition is one of the most universal conditions because of the feature of regenerative braking energy recovery. Thus, the condition should be considered when the vibration suppression problem of EVs is discussed. Under regenerative braking condition, focusing on the problem of vibration suppression for EVs with IWSRM, the effect of eccentricity of IWSRM on vibration of EVs is analyzed and a vertical vibration control strategy is proposed in this chapter. Meanwhile, for further reducing the vibration of EVs, the switch angles of IWSRM based on speed and braking strength are also optimized.

8.2 The Model of IWSRM

8.2.1 The Structure Model of IWSRM

Focusing on the structure type of IWSRM, there are two typical drive systems, which is respectively inner rotor IWSRM and outer rotor IWSRM. Based on the features of the two drive systems, it can be known that outer rotor IWSRM has a great advantage to be applied in EVs, on account of its simple structure and the integrated design of wheels, drive systems, braking systems, and suspension

Table 8.1 IWSRM arguments

Arguments	Value	Arguments	Value
Rotor diameter D_s	360 mm	Stator diameter D_a	290 mm
Rotor yoke height h_{cr}	35 mm	Stator yoke height h_{cs}	20 mm
Rotor pole β_r	7.5 deg.	Stator pole β_s	9 deg.
Air gap g	0.3 mm	Second air gap g_i	22.5 mm
Motor rated power P	4 kW	Motor rated Torque T	42 N m

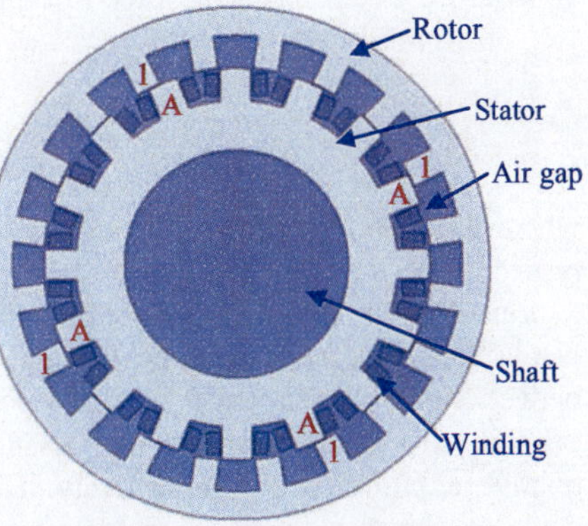

Fig. 8.1 IWSRM structure diagram

systems. Thus, in this chapter, a 16/20 four-phase outer rotor IWSRM is selected as the research object, whose structure arguments and structure diagram are presented in Table 8.1 and Fig. 8.1, respectively.

8.2.2 The Mathematic Model of IWSRM

8.2.2.1 Electromechanical Equations

Considering that the mutual inductance between the phases on electromagnetic characteristic is much slight, the mutual inductance between the phases can be neglected in the actual calculation process. In this chapter, the magnetic common energy (MCE) W'_m can be expressed by the integral of the phase flux $\psi(\theta, i)$ to the phase current i. Meanwhile, it should be noted that the phase flux is the function of the rotor position angle θ and the phase current i. Therefore, the equation can be described as

$$W_m = \int_0^i \psi(\theta, i)di \tag{8.1}$$

According to the principle of virtual displacement, the generalized electromagnetic force can be described by the partial derivative of the total MCE W'_m with respect to generalized displacement x. Thus, the generalized phase force F_k generated by the electromagnetic attraction can be expressed as

$$F_k = \frac{\partial W_m}{\partial x}\Big|_{i=const} = \int_0^i \frac{\partial \psi_k(\theta, i)}{\partial \theta} di \tag{8.2}$$

The total generalized electromagnetic force can be computed by the sum of generalized electromagnetic force of each phase. Therefore, it can be described as

$$F = \sum_{k=1}^{4} F_{\mathrm{k}} \tag{8.3}$$

where k refers to different excited phases, which could be 1, 2, 3, and 4, respectively.

Equations (8.1) to (8.3) are the fundamental equations of SRM. The generalized displacements also refer to the air gap of SRM I_g, the relative angular position between stator and rotor θ, and the core length of stator I_z, respectively. The generalized output forces are the driving torque T_e, the radial force F_r, and the relative axial force F_h, respectively. However, the relative axial force can be neglected, because its value is much slight and the core length of stator is relatively large. Thus, the driving torque and the relative axial force have much greater impact on the dynamic performance of EVs. Furthermore, the driving torque and the relative axial force can be calculated as

$$T_e = \frac{\partial W_m}{\partial \theta}\Big|_{i=const} = \int_0^i \frac{\partial \psi(\theta, i)}{\partial \theta} di \tag{8.4}$$

$$F_r = \frac{\partial W_m}{\partial I_g}\Big|_{i=const} = \int_0^i \frac{\partial \psi(\theta, i)}{\partial I_g} di \tag{8.5}$$

According to the Eqs. (8.4) and (8.5), it can be found that the key to the calculation of torque and force is to obtain the flux characteristics of switched reluctance motor. Meanwhile, it should be noted that the flux is the function of phase inductance and phase current, which can determine the electromagnetic equation and circuit equation.

8.2.2.2 The Equation of Electromagnetic Coupling

A widely used method to obtain the flux is by means of Fourier series and data fitting. The initial data can be obtained either through experimental measurements or finite element analysis. The magnetic flux or inductance, can be used to facilitate the flux calculations, which is employed in this chapter.

Assuming that the initial position is that the stator poles and rotor poles are not aligned, and the corresponding angle is 0 degree, the alignment position can be π/N_r. Meanwhile, it should be noted the phase inductance is the function of position angle and phase current. The corresponding rotor angular position and winding current can be written in Fourier series as Eq. (8.6), where N_r is the number of rotor poles and $L(\theta, i)$ is the inductance.

$$L(\theta,i) = L_0(i) + L_1(i)\cos(N_r\theta + \varphi_1) + \sum_{n=2,3,\cdots}^{\propto} L_n(i)\cos(nN_{r\theta} + \varphi_n)$$
$$= \sum_{n=0}^{\propto} L_n(i)\cos(nN_r\theta + \varphi_n) \quad (8.6)$$

The factor L_n can be acquired by the phase inductance L_a at the aligned position, the phase inductance L_u at the unaligned position, and the phase inductance L_m at the semi-positive position. Ignoring the higher-order harmonics, the accuracy obtained by the third-order Fourier series could meet the requirements, such that the relationships can be calculated as

$$L_0(i) = \frac{1}{2}\left[\frac{1}{2}(L_a + L_u) + L_m\right] \quad (8.7)$$

$$L_1(i) = \frac{1}{2}(L_a - L_u) \quad (8.8)$$

$$L_2(i) = \frac{1}{2}\left[\frac{1}{2}(L_a + L_u) - L_m\right] \quad (8.9)$$

Among them, the phase inductance L_u of the pole-to-groove misaligned position can be assumed as a constant because of the relatively large air gap. Meanwhile, when the pole-to-pole is aligned, L_a and L_m can be computed as

$$L_a(i)|_{\theta=\frac{\pi}{N_r}} = \sum_{n=0}^{N} a_n i^n \quad (8.10)$$

$$L_m(i)|_{\theta=\frac{\pi}{2N_r}} = \sum_{n=0}^{N} b_n i^n \quad (8.11)$$

where a_n and b_n is polynomial fitting coefficient. According to Eq. (8.6), the kth phase inductance can be written as

$$\begin{aligned} L_k(\theta, i_k) &= \sum_{n=0}^{\infty} L_n(i_k)\cos(nN_r\theta + \varphi_n) \\ &= \frac{1}{2}\left[\cos^2(N_r\theta) - \cos(N_r\theta)\right]\sum_{n=0}^{N} a_n i_k^n + \sin^2(N_r\theta)\sum_{n=0}^{N} b_n i_k^n \\ &\quad + \frac{1}{2}L_u\left[\cos^2(N_r\theta) + \cos(N_r\theta)\right] \end{aligned} \tag{8.12}$$

where i_k refers to the kth phase current. However, since the phase inductance is a partial differential equation of the flux to the winding current, the kth phase flux can be expressed as

$$\begin{aligned} \psi(\theta, i_k) &= \int_0^{i_k} L_k(\theta, i_k)di_k \\ &= \frac{1}{2}\left[\cos^2(N_r\theta) - \cos(N_r\theta)\right]\sum_{n=0}^{N} c_n i^n + \sin^2(N_r\theta)\sum_{n=0}^{N} d_n i^n + \frac{1}{2}L_u i_k\left[\cos^2(N_r\theta) + \cos(N_r\theta)\right] \end{aligned} \tag{8.13}$$

where the integral factors of polynomial fitting coefficient a_n and b_n can be computed by $c_n = a_{n-1}/n$ and $d_n = b_{n-1}/n$, respectively. Moreover, the values of c_0 and d_0 are both zero because of continuity. It can be concluded that the flux is also a Fourier series model of the rotor angle position and phase current. In this chapter, the original data are obtained by finite element analysis.

8.2.2.3 The Equation of Mechanics

According to the Eq. (8.13), the partial differential of the phase flux to position angle can be acquired as

$$\frac{\partial \psi(\theta, i_k)}{\partial \theta} = \sin(N_r\theta)\sum_{n=0}^{N} e_n i_k^n + \sin(2N_r\theta)\sum_{n=0}^{N} f_n i_k^n \tag{8.14}$$

where the values of e_n, f_1, e_0, e_1, and f_n are $(1/2)N_r c_n$, $(1/2)N_r(2d_1\text{-}c_1\text{-}L_u)$, zero, $(1/2)N_r(c_1\text{-}L_u)$, and zero, respectively.

Moreover, the output electromagnetic forces include two parts, one is radial force, the other is the tangential force which provides the driving power for SRM by the form of torque. Then, the kth phase output torque can be expressed as Eq. (8.15).

$$T_k = \int_0^{i_k} \frac{\partial \psi(\theta, i_k)}{\partial \theta} di_k = \sin(N_r\theta)\sum_{n=1}^{N} \frac{1}{n} e_{n-1} i_k^n + \sin(2N_r\theta)\sum_{n=1}^{N} \frac{1}{n} f_{n-1} i_k^n \tag{8.15}$$

The total output torque T_e is the sum of the output torques of each phase. Thus, it can be expressed as

$$T_e = \sum_{k=1}^{4} T_k \tag{8.16}$$

According to the characteristics of the phase inductance, it can be seen that the phase inductance is a nonlinear function of rotor angle position and phase current. The Fourier fitting results are consistent with the original data, which verifies the effectiveness of the modeling for SRM by the analytic Fourier fitting method. In general, the torque varies with the phase current at different rotor positions and the output torque of SRM is the maximum at the alignment position. But in fact, the SRM can output the desired, continuous, stationary torque with minimal vibration by controlling the phase current operating in the appropriate state and waveform.

Similarly, the vertical electromagnetic force between the rotor and the stator poles can be obtained. Since the air gap l_g is relatively unchanged, when the rotor and stator magnetic poles move, the edge flux and mutual inductance can be ignored, and the radial force of the *k*th phase can be expressed as

$$F_{rk} = -\frac{1}{2} i_k^2 \frac{L_k(\theta, i_k)}{l_g} \tag{8.17}$$

The SRM is a kind of motor that conforms to the principle of minimum reluctance, and the rotor tends to have a minimum reluctance position in the case of excitation. In order to generate sufficient reluctance torque, the radial electromagnetic force between the stator and rotor poles can be high correspondingly, due to the double salient pole structure.

The radial forces, as the main source of vibration and noise of IWSRM, are theoretically equal to each other in opposite directions, such that they balance each other. However, the air gap of IWSRM is usually asymmetric due to the mechanical tolerance and load, which could result in the situation that different magnetic radial forces are located between the opposite poles, and the actual unbalanced radial force is not zero. Moreover, a small eccentricity can lead to large unbalanced residual radial force due to the air gap is small. Therefore, the application of IWSRM could cause the results that the vibration caused by the unbalanced radial forces can be delivered to the vehicle body, further result in the problem of the ride comfort of EVs.

The relative eccentricity can be defined as

$$\varepsilon = \frac{h_e}{l_g} \tag{8.18}$$

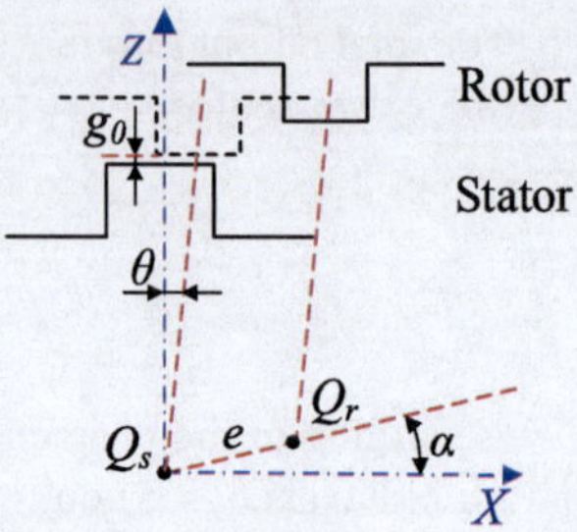

Fig. 8.2 The structure of IWSRM eccentricity

where as shown in Fig. 8.2, l_g is the radial air gap length without eccentricity, and h_e refers to the absolute eccentricity in the radial direction. In actual situations, the maximum relative eccentricity can generally reach 30–60%.

Then, the vertical attraction force between the rotor and stator poles can be described as

$$F_{rm} = -\frac{1}{2} i^2 \frac{L_k(\theta, i_k)}{l_g(m)} \tag{8.19}$$

The total vertical unbalanced radial force is the difference of the upward and downward vertical components of the radial forces imposed on the exciting magnetic poles of SRM. Thus, the total vertical unbalanced radial force of kth phase can be expressed as

$$F_{rk} = F_{rm} - F_{rn} = \frac{1}{2} i^2 \frac{L_k(\theta, i_k)}{l_g(n)} - \frac{1}{2} i^2 \frac{L_k(\theta + \pi, i_k)}{l_g(m)} \tag{8.20}$$

where m and n respectively refer to the poles of the upward and downward directions. According to the structure of the IWSRM researched in this chapter, it can be known that $|m–n|$ = 8.

For the system of vehicle, the vertical components of the radial forces should be paid attention to, because it can intensify eccentricity and further aggravate the vehicle vibration. Thus, the vertical radial force for the kth phase, as shown in Fig. 8.3, can be expressed as

$$F_{rk-Z} = F_{rk} \cos \theta \tag{8.21}$$

The total vertical unbalanced radial force is the sum of the vertical unbalanced radial force of each phase. As a result, the total vertical unbalanced radial force of IWSRM can be described as

$$F_{r-Z} = \sum_{k=1}^{4} F_{rk-Z} \tag{8.22}$$

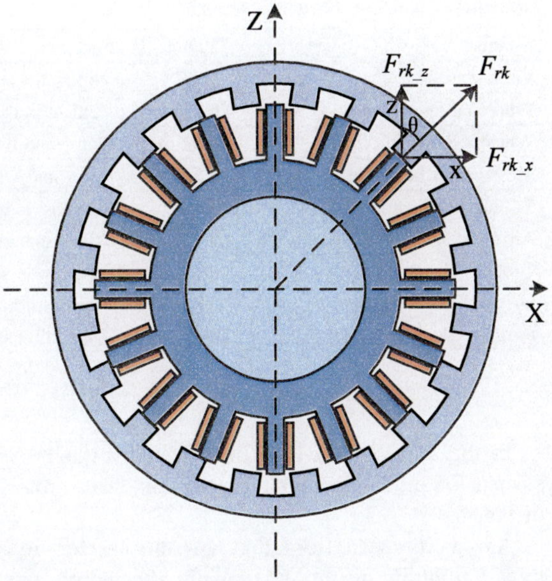

Fig. 8.3 The vertical radial force of IWSRM

According to Eq. (8.20), it can be seen that the radial force between the stator and rotor is quadratic function of phase current. The vertical radial forces of IWSRM is different between the opposite poles, and are inversely proportional to the air gap. Due to the eccentricity, an increase in the radial force from one direction may lead to a decrease in radial force from the opposite direction, further leading to the result that the magnitude of the unbalanced radial force increases. Therefore, with the increase of eccentricity, the radial force increases sharply. Thus, it can be concluded that the eccentricity ratio has a significant effect on the radial force. For a small air gap, eccentricity ratio tends to be relatively larger such that it can cause higher residual radial force. In addition, according to the analysis of the Eq. (8.12) and the Eq. (8.19), they can derivate that the inductance of each phase can be expressed in Fourier series form, consisting of fundamental harmonic components related to the motor speed. As a result, the residual radial force of the IWSRM transmitting to the vehicle also has multiple harmonic frequencies.

8.3 The Model of a Quarter IWSRM-Suspension System

To study the effect of IWSRM eccentricity on the ride comfort of EVs, it is necessary that the vehicle dynamic system model is built. In this chapter, the vehicle factors are listed in Table 8.2.

Table 8.2 Factors of the vehicle system

Factors	Value	Factors	Value
Vehicle mass m	1020 kg	Windward area A	1.4 m^2
Wheel radius r	0.25 m	Vehicle body mass m_b	200 kg
Wheelbase L	1.62 m	Stator and bearing mass m_{sb}	20 kg
Mass center height h_g	0.58 m	Rotor and wheel mass m_{rw}	35 kg
Rolling coefficient f	0.014	Suspension damping c_s ($\times 10^3$)	1.6 Ns/m
Air resistance coefficient C_D	0.4	Wheel damping c_t	350 Ns/m
Ground adhesion coefficient φ	0.7	Suspension stiffness k_s ($\times 10^4$)	1.8 Ns/m
Distance of rear axle to center b	0.86	Wheel stiffness k_t ($\times 10^5$)	1.9 Ns/m
Ramp angle α	0 deg.	Axle stiffness k_{sb} ($\times 10^6$)	5 Ns/m

In this chapter, the EVs is driven by the IWSRM, and a quarter IWSRM-suspension model, including suspension, stator, rotor, and wheel, is built, as shown in Fig. 8.4.

As can be seen from Fig. 8.4, the acceleration of EV body in the vertical direction during operation is mainly affected by two factors. On the one hand, the excitation generated by road surface is transmitted to the EV body through the suspension system, resulting in vertical acceleration. On the other hand, when the EV works, there could be an eccentricity between the stator and the rotor, due to the small air gap. As a result, the small eccentricity can cause large residual unbalanced radial force which then transmits to the vehicle body through the suspension and further aggravates vehicle vibration.

According to the quarter IWSRM-suspension system, the vibration equation of the quarter IWSRM-suspension system can be described as

$$\begin{aligned}
&m_b\ddot{x}_b + k_s(x_b - x_{sb}) + c_s(\dot{x}_b - \dot{x}_{sb}) = 0 \\
&m_{sb}\ddot{x}_{sb} + k_s(x_{sb} - x_b) + k_{sb}(x_{sb} - x_{rw}) + c_s(\dot{x}_{sb} - \dot{x}_b) + F_d = 0 \\
&m_{rw}\ddot{x}_{rw} + k_t(x_{rw} - x_z) + k_{sb}(x_{rw} - x_{sb}) + c_t(\dot{x}_{rw} - \dot{x}_z) - F_d = 0
\end{aligned} \tag{8.23}$$

where c_s and c_t respectively represent the damping of the suspension and the wheel; x_b, x_{sb}, and x_{rw} are respectively the vertical displacements of the body, the vertical displacements of stator-bearing, and the vertical displacements of rotor; F_d denotes the total vertical unbalanced radial force of the IWSRM generated by eccentricity; m_b, m_{sb}, and m_{rw} refer to the mass of the body, the mass of the stator-bearing, the mass of the rotor-wheel, respectively; k_s, k_{sb}, and k_t are the stiffness of the suspension, the stiffness of axle, and the stiffness of wheel, respectively; x_z represents the road surface excitation, which can be described through random white noise.

$$\dot{x}_z(t) = -2\pi n_1 u x_z(t) + 2\pi n_0 \sqrt{G_0 u} w(t) \tag{8.24}$$

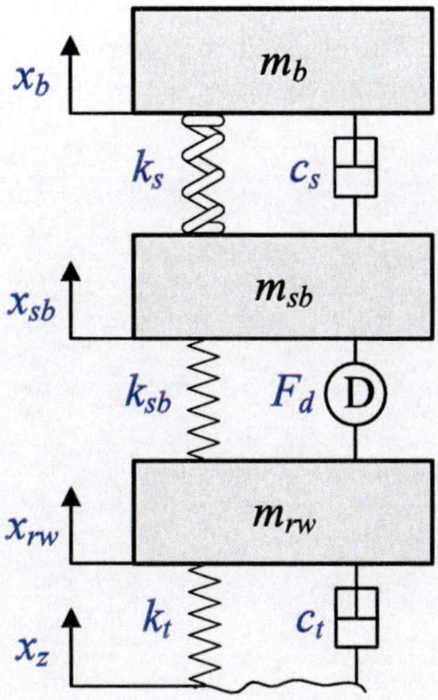

Fig. 8.4 The quarter IWSRM-suspension model

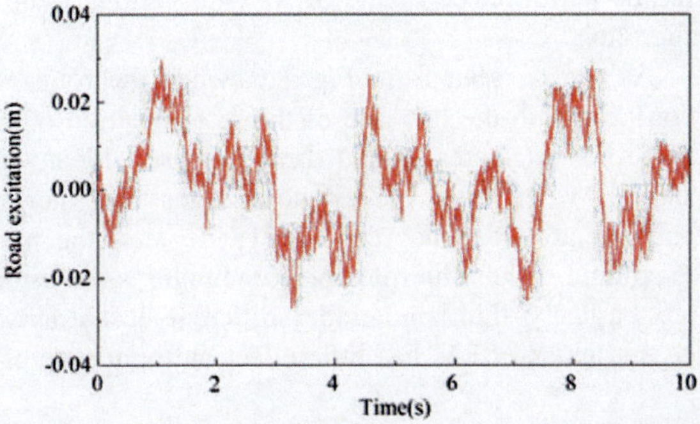

Fig. 8.5 The road surface excitation

where G_0 refers to road roughness coefficient; n_0 and n_1 refer to the reference space frequency and the lower cut-off spatial frequency of road roughness, respectively; $w(t)$ denotes the random white noise. Taking B grade road as an example, the surface excitation generated when the vehicle speed is 40 km/h is illustrated in Fig. 8.5.

8.4 The Vibration Control Under Driving Condition

8.4.1 The Effect of Eccentricity on the Torque of IWSRM

8.4.1.1 The Effect of Eccentricity Ratio

In this chapter, the torque performance of SRM with different eccentricity ratios is analyzed by finite element analysis (FEA). The rotor eccentricity direction angle α is set as 90 degrees and the constant phase current is set as 12 A. Under different

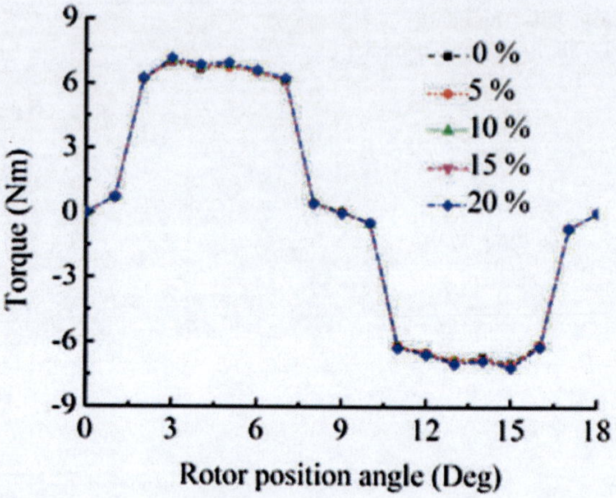

Fig. 8.6 The torque of IWSRM based on different eccentricity ratios

eccentricity ratios, FEA method is used to analyze the influence of eccentricity on torque performance of the SRM with various rotor position angles, as shown in Fig. 8.6.

As can be seen from Fig. 8.6, when the rotor eccentricity of the IWSRM is constant, with the increase of the rotor position of the IWSRM, the torque of the IWSRM first increases and then decreases. Meanwhile, the changing trend is the same in value when the torque changes to be negative. When the rotor position angle is around 3 degrees and 15 degrees, the motor torque value reaches the maximum. When the rotor position angle is the same, the torques of the IWSRM are basically the same under different eccentricity ratios. Therefore, the rotor eccentricity of SRM has little effect on the torque of the motor.

8.4.1.2 The Effect of Eccentricity Angle

When the rotor eccentricity ratio is set as 10%, the winding constant current is set as 12 A, under different rotor eccentric direction angle α, FEA is also used to analyze the influence of eccentricity on torque performance of SRM, as shown in Fig. 8.7. It can be seen from Fig. 8.7 that when the rotor eccentricity angle α of the IWSRM is constant, with the increase of the rotor position of the IWSRM, the torque of the IWSRM first increases and then decreases. After that, the changing trend is the same in value when the torque changes to be negative. When the rotor position angle is around 3 degrees and 15 degrees, the motor torque value reaches the maximum. When the rotor position angle is the same, the torque value of the IWSRM basically remains unchanged with the increase of the eccentric direction angle of the rotor α. Therefore, the eccentric direction angle of the SRM has almost no effect on the torque of the motor.

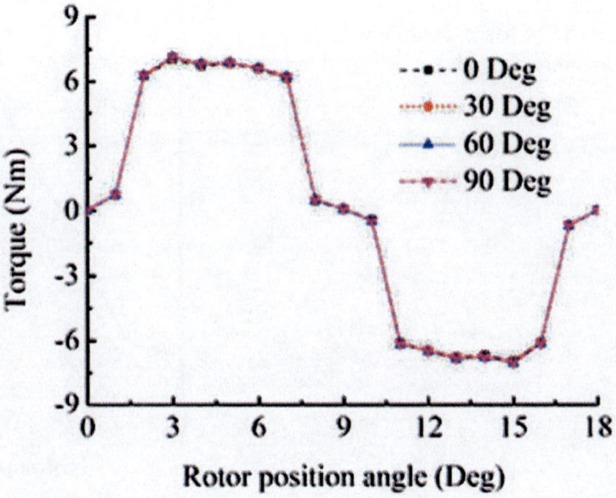

Fig. 8.7 The torque of IWSRM based on different eccentricity angles

8.4.2 The Effect of Eccentricity on the Inductance of IWSRM

8.4.2.1 The Effect of Eccentricity Ratio

The inductance performance of SRM with rotor angles under various eccentricity ratios is analyzed by FEM. The rotor eccentricity direction angle α is set as 90 degrees and the phase current is set as a constant of 12 A. Under different eccentricity ratios, FEA is used to analyze the influence of eccentricity ratio on inductance performance of SRM, as shown in Fig. 8.8. As can be seen from Fig. 8.8, when the rotor eccentricity ratio of the IWSRM is constant, the inductance of the IWSRM increases first and then decreases as the rotor position of the IWSRM increases. When the rotor position is around 9 degrees, the inductance of the motor reaches the maximum. When the rotor position is constant, with the increase of rotor eccentricity ratio of IWSRM, the inductance value basically remains unchanged. Therefore, the eccentricity ratio of IWSRM has almost no effect on the inductance of the motor.

8.4.2.2 The Effect of Eccentricity Angle

When the rotor eccentricity ratio is set as 10%, the winding constant current is set as 12 A, under different rotor eccentricity direction angle α, FEA is also used to analyze the influence of the eccentricity angle on inductance performance of SRM, as shown in Fig. 8.9. As can be seen from Fig. 8.9, when the rotor eccentricity angle of the IWSRM is constant, the inductance of the IWSRM increases first and then decreases as the rotor position of the IWSRM increases. When the rotor position is about 9 degrees, the inductance of the motor reaches the maximum.

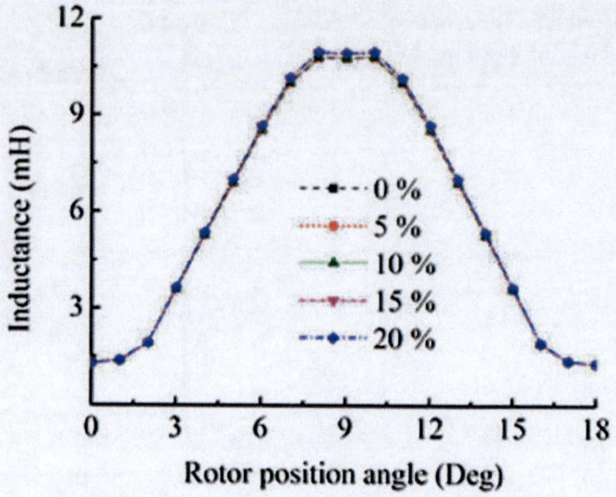

Fig. 8.8 The inductance of IWSRM based on different eccentricity ratios

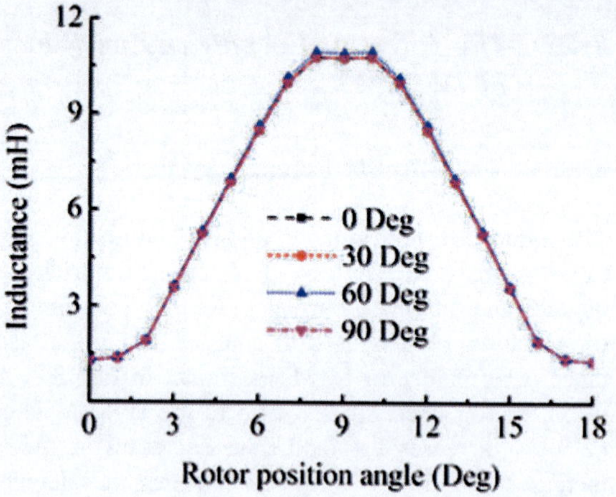

Fig. 8.9 The inductance of IWSRM based on different eccentricity angles

When the rotor position is constant, with the increase of rotor eccentricity angle of IWSRM, the inductance value basically remains constant. Therefore, the eccentricity angle of IWSRM has almost no effect on the inductance of the motor.

8.4.3 The Effect of Eccentricity on the Radial Force of IWSRM

8.4.3.1 The Effect of Eccentricity Ratio

In the process of the EV operation, the unbalanced radial force is caused by the eccentricity of IWSRM, then the produced unbalanced radial force is delivered to

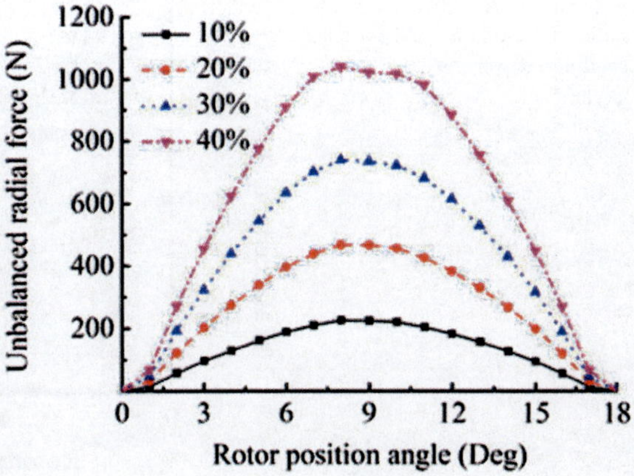

Fig. 8.10 The unbalanced radial force of IWSRM based on different eccentricity ratios

the body of EV, which can further result in the vertical acceleration of vehicle. Therefore, the unbalanced radial force characteristic curve varying with the rotor position angle of SRM is very important for the eccentric fault analysis of SRM. In order to accurately analyze the influence of the unbalanced radial electromagnetic force on the different eccentric fault states of the rotor, the unbalanced radial force performance is analyzed in this chapter by means of FEA under various eccentricity ratios. When the rotor eccentricity direction angle α is set as 90 degrees and the phase current is set as a constant of 12 A, the influence of eccentricity ratio on the unbalanced radial force varying with rotor position angles is analyzed and the results can be found in Fig. 8.10.

It can be seen from the Fig. 8.10, when the rotor eccentricity ratio of the IWSRM is constant, the unbalanced radial force of the IWSRM increases first and then decreases as the rotor position of the IWSRM increases. When the rotor position is about 9 degrees, the unbalanced radial force of the motor reaches the maximum. When the rotor position is constant, with the eccentricity ratio of IWSRM increases, the unbalanced radial force of the IWSRM increases gradually, and the difference of the unbalanced radial force of the motor increases first and then decreases gradually. When the rotor position is about 9 degrees, the difference of the unbalanced radial force of the motor reaches the maximum. Therefore, the greater the eccentricity ratio of SRM is, the higher the electromagnetic radial force can be, which could result in larger vibration of EVs.

8.4.3.2 The Effect of Eccentricity Angle

When the rotor eccentricity ratio is set as 10%, the phase constant current is set as 12 A, FEA is also used to analyze the influence of the eccentricity angle on the unbalanced radial force performance of SRM under different rotor eccentric direction angle α, as shown in Fig. 8.11. It can be seen from the Fig. 8.11, when the rotor eccentricity angle of the IWSRM is constant, the unbalanced radial force of

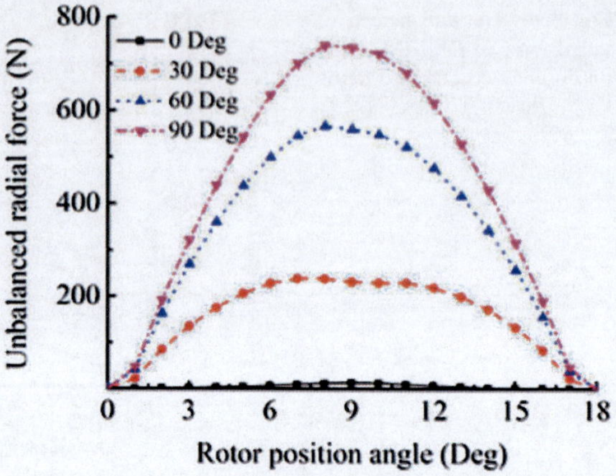

Fig. 8.11 The unbalanced radial force of IWSRM based on different eccentricity angles

the IWSRM increases first and then decreases as the rotor position of the IWSRM increases. When the rotor position is about 9 degrees, the unbalanced radial force of the motor reaches the maximum. When the rotor position is constant, with the eccentricity angle of IWSRM increases, the unbalanced radial force of the IWSRM increases gradually, and the difference of the unbalanced radial force of the motor increases first and then decreases gradually. When the rotor position is about 9 degrees, the difference of the unbalanced radial force of the motor also reaches the maximum. Therefore, the greater the rotor eccentricity angle of SRM is within the range of 0 degree to 90 degrees, the higher the electromagnetic radial force can be, and the greater the impact on the vibration of EVs should be.

8.4.4 The Effect of the Unbalanced Radial Force on the Vehicle Performance

According to the Eq. (8.23), to achieve the effect analysis of the unbalanced radial force on EVs dynamic performance, the Laplace transform is performed and expressed as

$$\begin{aligned} &(m_b S^2 + c_s S + k_s) X_b(S) - (c_s S + k_s) X_{sb}(S) = 0 \\ &(m_{sb} S^2 + c_s S + k_s + k_{sb}) X_{sb}(S) - (c_s S + k_s) X_b(S) = k_{sb} X_{rw}(S) - F_d \\ &(m_{rw} S^2 + c_t S + k_t + k_{sb}) X_{rw}(S) - (c_t S + k_t) X_z(S) = k_{sb} X_{sb}(S) + F_d \end{aligned} \tag{8.25}$$

According to the Eq. (8.25), the two transfer functions, the sprung mass displacement and the vehicle vertical acceleration to the unbalanced radial force $H_b(S)$ and $H_A(S)$, are obtained as

$$H_b(S) = \frac{(k_s + c_sS)(m_{rw}S^2 + c_tS + k_{sb} + k_t) - k_{sb}(k_s + c_sS)}{(k_s + c_sS)^2(m_{rw}S^2 + c_tS + k_{sb} + k_t) + k_{sb}^2(m_bS^2 + c_sS + k_s) - (m_bS^2 + c_sS + k_s)(m_{sb}S^2 + c_sS + k_s + k_{sb})(m_{rw}S^2 + c_tS + k_{sb} + k_t)}$$

$$H_A(S) = \frac{2((k_s + c_sS)(m_{rw}S^2 + c_tS + k_{sb} + k_t) - k_{sb}(k_s + c_sS))(k_{sb}^2S^2 - S^2(m_{sb}S^2 + c_sS + k_s + k_{sb})(m_{rw}S^2 + c_tS + k_{sb} + k_t))^2}{((k_s + c_sS)^2(m_{rw}S^2 + c_tS + k_{sb} + k_t) + k_{sb}^2(m_bS^2 + c_sS + k_s) - (m_bS^2 + c_sS + k_s)(m_{sb}S^2 + c_sS + k_s + k_{sb})(m_{rw}S^2 + c_tS + k_{sb} + k_t))^3} \tag{8.26}$$

Therefore, the response characteristics curve produced by the unbalanced radial force is described in Fig. 8.12. It can be seen from the figure that as the radial force frequency increases, the vertical acceleration response of sprung mass increases first and then decreases, until the vertical acceleration response of sprung mass reaches the maximum when the frequency is 2 Hz. Thus, there is a conclusion that the influence of the unbalanced radial force on the vehicle vertical vibration is obvious at low frequencies. Especially, it should be noted that it is more obvious when the value of the radial force frequency is from 2 to 60 Hz. In the control system of SRM, as one of the typical DC motors, it needs a special power converter to periodically excite each phase of SRM with the help of the control unit. As a result, one of the characteristics of SRM is that it works in stages, resulting in periodic excitation frequencies. For the direct driving system with outer rotor SRM, the excitation frequency of IWSRM is calculated as follows

$$f_\varepsilon = \frac{\omega}{2\pi}N_r = \frac{v}{2\pi r}N_r \tag{8.27}$$

where ω refers to the speed of the IWSRM, r means the radius of the outer rotor of the IWSRM, and N_r denotes the number of poles of the outer rotor of the IWSRM. The relationship curve of excitation frequency and vehicle speed can be obtained by Eq. (8.27), as shown in Fig. 8.13, where it can be seen that as the vehicle speed

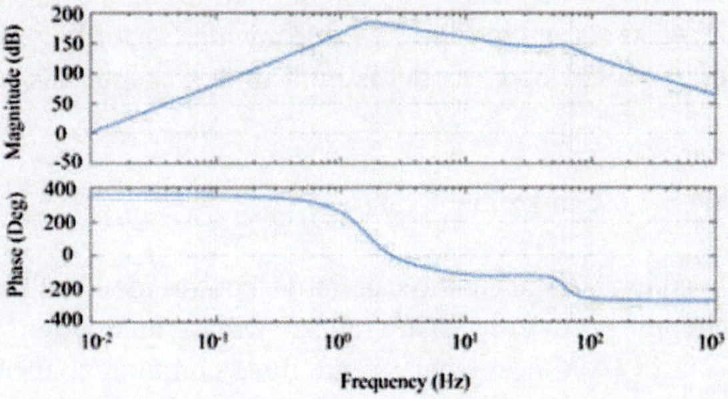

Fig. 8.12 The response curve of EVs based on the unbalanced radial force

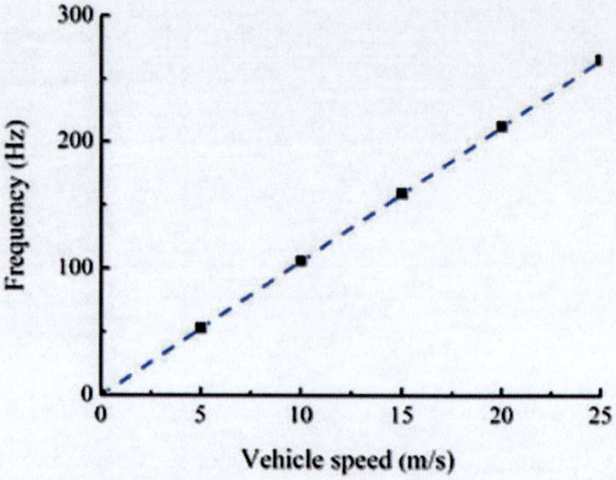

Fig. 8.13 The relationship curve of the excitation frequency and the vehicle speed

increases, the excitation frequency of IWSRM increases linearly. Meanwhile, it can also be seen from the figure that the excitation frequency of IWSRM can be extended to the vehicle resonance frequency domain in the low speed. Thereby, IWSRM may cause uncomfortable vibration to the vehicle at low speed or starting conditions. In addition, the motor excitation frequency is close to the resonance frequency of the tire noise in a high range, directly leading to the increase of noise. Therefore, in order to improve the comfort of the vehicle, the control method of IWSRM is necessary to study and implement to suppress these negative effects directly caused by the unbalanced radial force of the vehicle.

8.4.5 The Design of Controller and Results Analysis

The dynamic performance of SRM cannot be reflected through linear or quasi-linear models due to the highly nonlinear characteristics of SRM. Moreover, through the effect analysis of the eccentricity ratio and eccentricity angle on the phase inductance, the torque, and the unbalanced radial force, it can be observed that there is almost no effect of the eccentricity ratio and eccentricity angle on the phase inductance, the torque, and the unbalanced radial force. Thus, in this chapter, the phase inductance, the torque, and the unbalanced radial force are acquired by FEA, as shown in Fig. 8.14. Meanwhile, according to the basic equations of SRM, the IWSRM body model is built in this chapter, as presented in Fig. 8.15.

8.4.5.1 Controller Design

In general, there are three common control methods for SRM, which include current chopping control (CCC), pulse width modulation (PWM), and angle position control (APC). In Chap. 1, the three common control methods are introduced and

Fig. 8.14 The electromagnetic characteristics of IWSRM

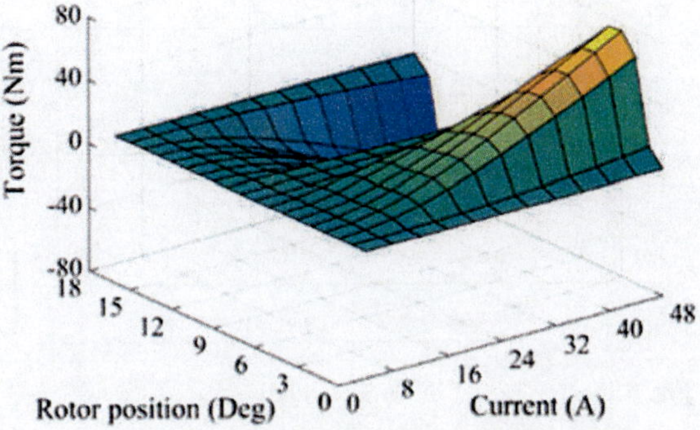

(a) The phase torque curve of IWSRM

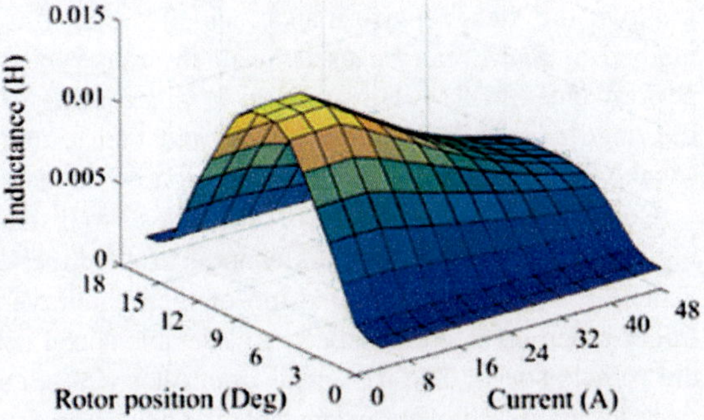

(b) The phase inductance curve of IWSRM

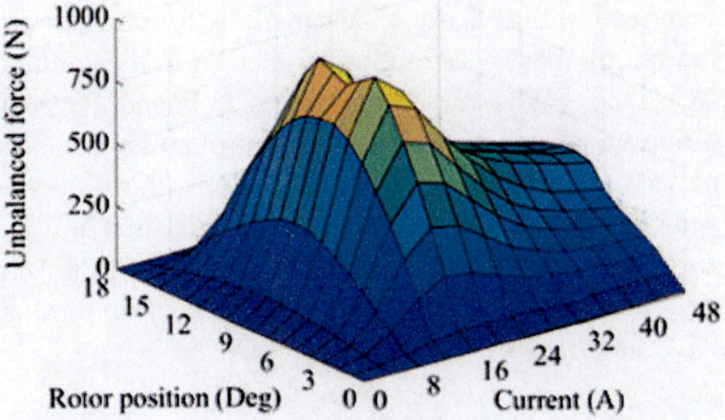

(c) The unbalanced radial force curveof IWSRM

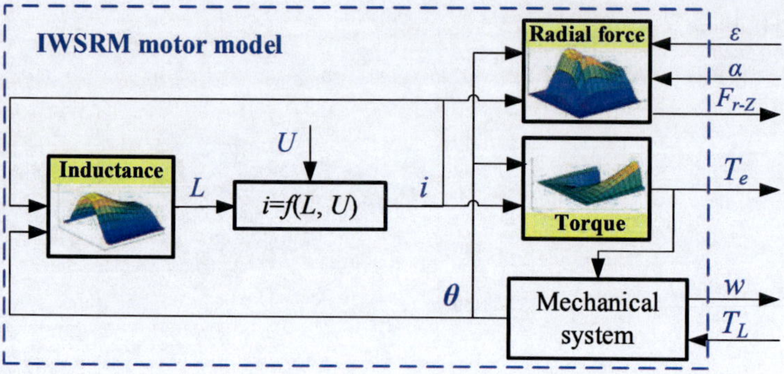

Fig. 8.15 The model of IWSRM body

compared with each other. Therefore, there is a conclusion that when the motor is started, the CCC method can be used to decrease the starting time. Then, when the SRM actual speed exceeds the basic speed of SRM, the PWM method is used to improve the steady performance. Finally, the APC, as a supplementary control method of SRM, can be used in all the range of speed to improve the dynamic performance of SRM. However, in fact, according to different operation condition, the three common control methods and their combinations should be used reasonably to suppress the vertical vibration of the EVs.

Considering all the control strategies, four control methods, as shown in Fig. 8.16, are proposed in this chapter to effectively achieve of the control of the vehicle speed and the suppression of the vehicle vertical vibration. The PID controller based on reference speed and vehicle speed is used to achieve the control of the vehicle speed. The hysteresis controller with a current chopping limit value of Δi is used to achieve the control of CCC. The PWM controller with inputting sampling rate and objective voltage is employed to achieve the control of PWM.

In this chapter, CCC method and PWM method, named as CCC and PWM control, are proposed and employed to achieve the control of the vehicle speed, as described in Fig. 8.16a, c. Meanwhile, for controlling the vehicle vertical vibration, the sprung mass acceleration is selected as the control objective. Then, based on the CCC and PWM methods, the CCC-F and PWM-F methods are designed and proposed as two new feedback control strategies, as shown in Fig. 8.16b, d. Then, under CCC and PWM control methods, the influence of the switch angles on the vehicle acceleration are analyzed, as described in Fig. 8.17. Furthermore, the switch angles are optimized based on the suppress of the vehicle vertical acceleration. The optimization objective function by means of the vehicle vertical acceleration for CCC and PWM method is expressed as

$$F(\theta_{\text{on-opt}}, \theta_{\text{off-opt}}) = \min\{\, a_Z(\theta_{\text{on}}, \theta_{\text{off}})\} \tag{8.28}$$

where a_z represents vertical acceleration of EV, θ_{on} and θ_{off} refer to turn-on and turn-off angles, $\theta_{\text{on-opt}}$ and $\theta_{\text{off-opt}}$ denote optimized turn-on and turn-off angles.

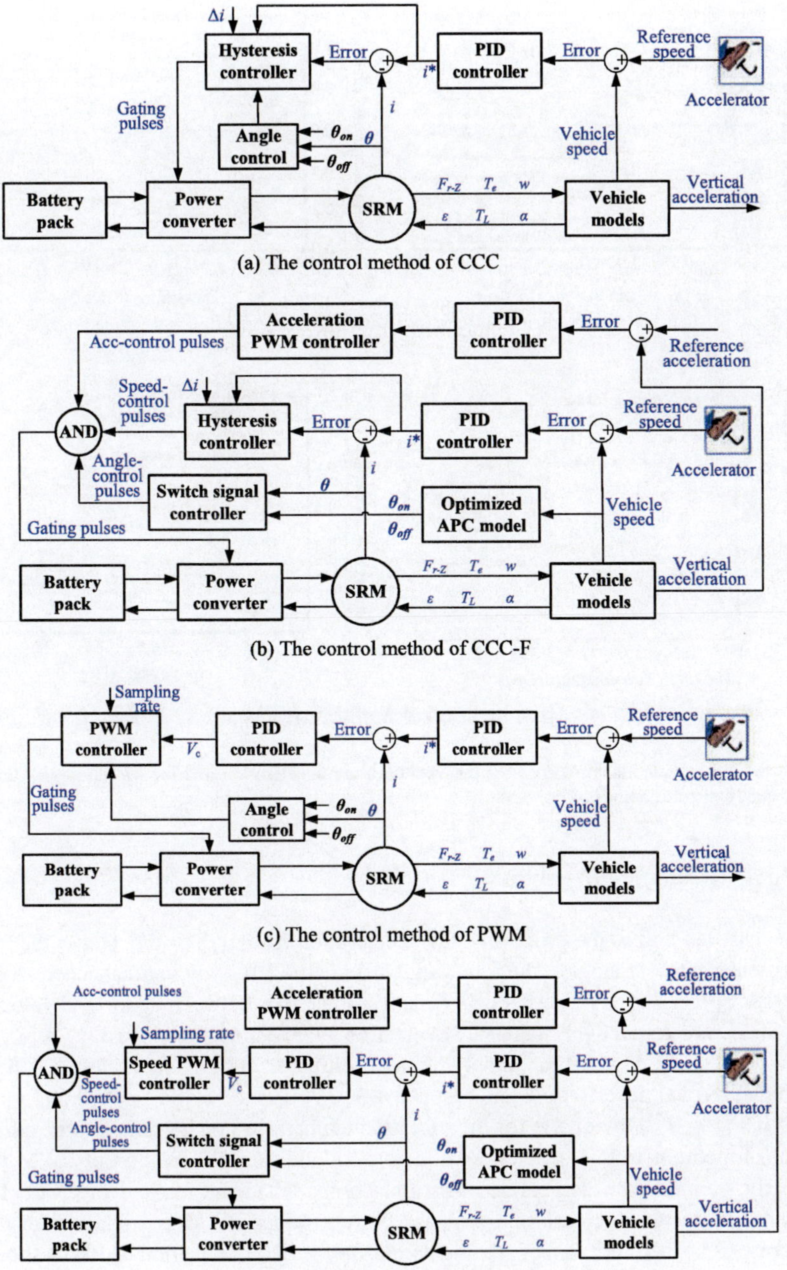

(a) The control method of CCC

(b) The control method of CCC-F

(c) The control method of PWM

(d) The control method of PWM-F

Fig. 8.16 The vehicle system model under the four different methods

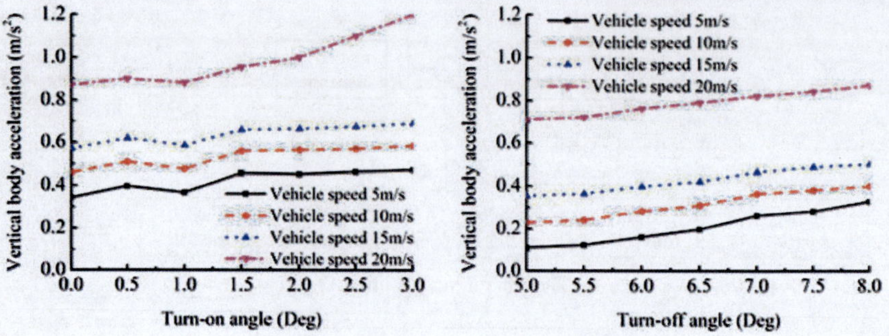

(a) The control method of CCC

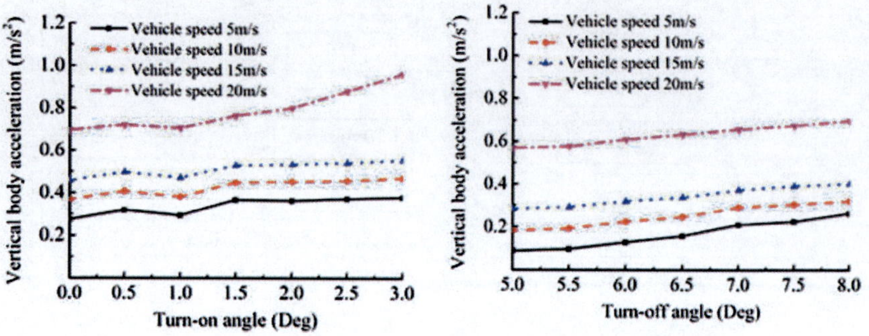

(b) The control method of PWM

Fig. 8.17 The relationship curves of the vehicle body acceleration and the switch angles under different control methods

8.4.5.2 Result and Analysis

This chapter studies the process of EV acceleration from 0 to 5 m/s. Under the four different control strategies, the four variables, i.e. the EV body vertical acceleration, the EV speed, the EV tire deflection, and the air gap between stator and rotor of IWSRM, are respectively taken as evaluation indexes, as shown in Fig. 8.18. As can be seen in Fig. 8.18a, under the four different control strategies, the peak vehicle vertical accelerations are respectively 0.760 m/s^2, 0.484 m/s^2, 0.602 m/s^2, and 0.364 m/s^2. Obviously, for the vibration suppression of EV, the two proposed control methods in this chapter are effective. In addition, the method of PWM can greatly attenuate the vertical acceleration, compared to the method of CCC. The reason is that the increase of the phase current under PWM is smaller than that under CCC. Therefore, under starting condition for EV, the method of PWM should be used to accelerate at low speed for the vibration suppression of EV, while in fact, for low speed condition for SRM, the method of CCC is generally used. Thus, thanks to the smaller peak value of the body vertical acceleration under the method of PWM-F, it can be considered as the most effective strategies for improving the ride comfort of EV. Meanwhile, as can be seen in Fig. 8.18b, c, according to comparisons of peak values under four strategies, the method of PWM-F is also effective to control the tire deflection and air gap offset. Compared to the three control methods of CCC, CCC-F, and PWM, the peak value of the tire deflection under PWM-F control method can reach 0.121 mm, with a decrease of 30.46%, 10.37%, and 22.93%. The peak value of the air gap offset under PWM-F control method can reach 0.029 mm, reducing by 34.09%, 9.37%, and 27.5%, respectively. It can be seen from Fig. 8.18d that the control effect of the vertical vibration is obvious under not only CCC-F but also PWM-F, although the starting speed is almost the same. For the acceleration range from 0 to 5 m/s, under the four different control strategies, the starting time is respectively 3.93 s, 4.29 s, 4.36 s, and 4.86 s. As a result, depending on the drive models of EV, the control strategies could be selected for meeting various performance demands of EV, such as comfort, normal, and sports mode.

In addition, it should be paid attention that when the feedback control methods are used to suppress the vertical vibration of EV, the driving resistance force of the EV could be reduced because of the fall of the EV vertical acceleration. Nevertheless, differing from the normal control methods, thanks to the less conductive time under the feedback control methods, the output torque produced by IWSRM is reduced accordingly, as shown in Fig. 8.19. The results showed that the performance of acceleration under the two feedback control methods, PWM-F and CCC-F, are reduced, which is different from that under the two control methods, PWM and CCC. In addition, due to the action of high frequency unbalanced radial force, the frequency of sprung mass acceleration is slightly higher than that of normal condition. As can be seen in Fig. 8.18c, the reason is that the high frequency of IWSRM airgap offset could directly lead to high frequency of the unbalanced radial force. Furthermore, combined with the consideration of the ride comfort of passengers, in this chapter, the weighted root mean square (RMS) of the EV acceleration under the four different control methods are computed, and the values of them are respectively 0.096 m/s^2, 0.082 m/s^2, 0.098 m/s^2, and 0.093 m/s^2. Thanks to the quite small values of RMS vertical acceleration, the passengers cannot have uncomfortable feeling.

Furthermore, in this chapter, for the two driving conditions with the target speeds of 5 m/s and 15 m/s, the proposed control strategies are studied to quantitatively evaluate the performance of them. In addition, RMS values of the body acceleration, tire deflection, and air gap offset were acquired and listed in Table 8.3. It can be seen that the two control strategies proposed in this chapter (CCC-F and PWM-F) are superior to the other two control strategies (CCC and PWM) in terms of the vehicle vertical vibration suppression, among which PWM-F is the most effective one. Therefore, according to the analysis results, the control method of PWM-F is the best choice for IWSRM to achieve comfortable driving mode. Moreover, for sport driving mode, the CCC method can be chosen because it has the best acceleration performance. Further, CCC-F can be used as a good compromise scheme to achieve the comprehensive balance between vertical vibration and vehicle dynamic performance, and it is suitable for ordinary driving mode.

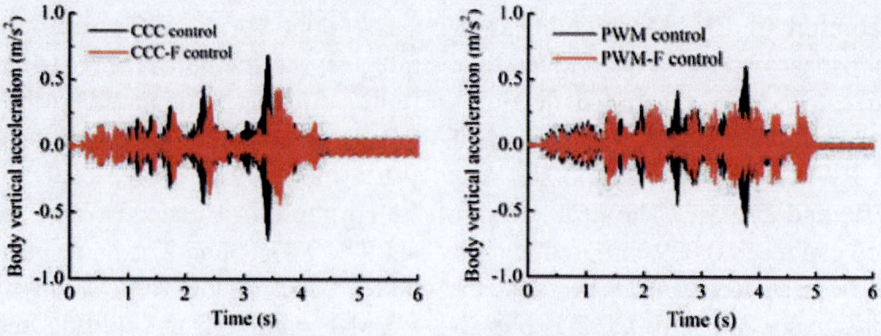

(a) The body vertical acceleration

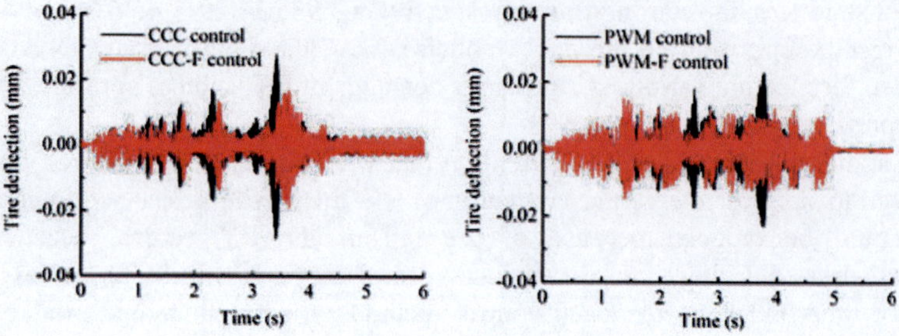

(b) The tire deflection

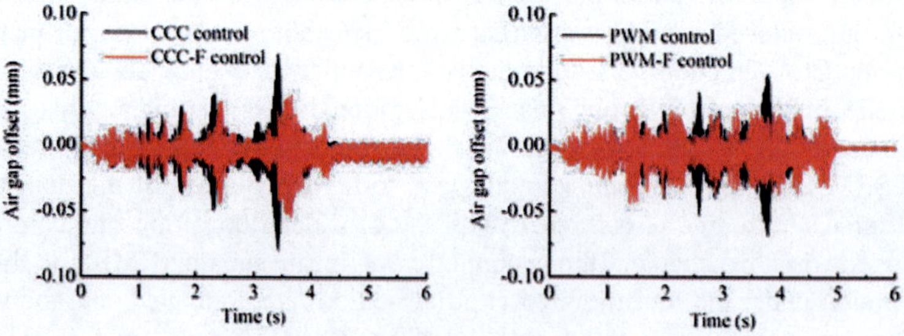

(c) The air gap offset

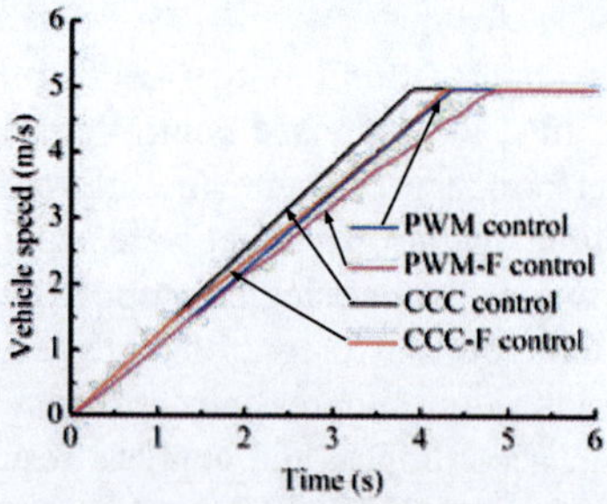

(d) The vehicle speed

Fig. 8.18 The compared results based on four different control methods

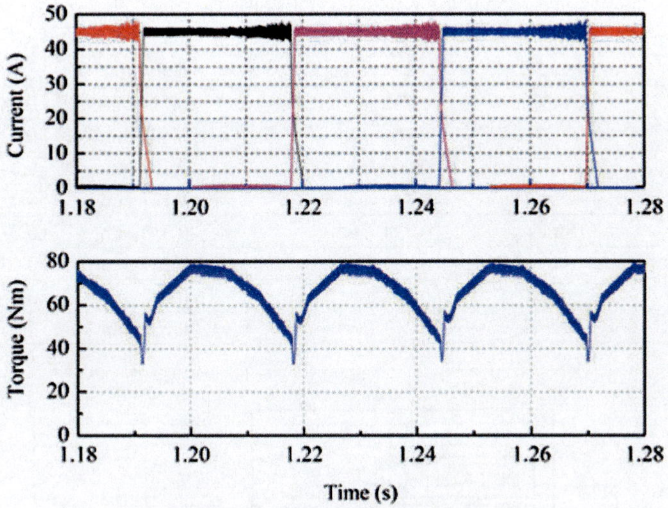

(a)The control method of PWM

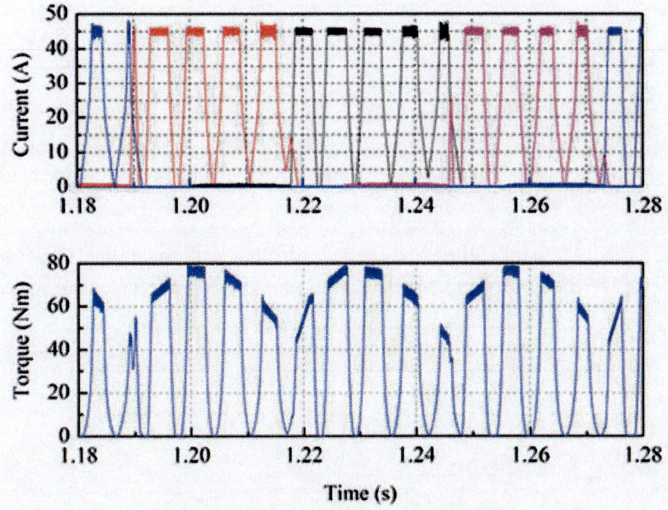

(b) The control method of PWM-F

Fig. 8.19 The output torque and current of IWSRM based on the two control methods

Depending on the performance of various strategies obtained from the above analysis, an IWSRM switching controller, which comprehensively considers various driving modes, is proposed, as shown in Fig. 8.20. It should be noted that, according to different driving modes (sports mode, normal mode, and comfort mode), CCC, CCC-F and PWM-F can be selected respectively to achieve the

Table 8.3 Results comparison under various control strategies

Control strategy	Target velocity	Body vertical acceleration	Tire deflection	Air gap offset
CCC	5 m/s	0.232 m/s^2	0.174 mm	0.044 mm
CCC-F		0.156 m/s^2	0.135 mm	0.032 mm
PWM		0.209 m/s^2	0.157 mm	0.040 mm
PWM-F		0.140 m/s^2	0.121 mm	0.029 mm
CCC	15 m/s	0.394 m/s^2	0.313 mm	0.079 mm
CCC-F		0.270 m/s^2	0.225 mm	0.057 mm
PWM		0.355 m/s^2	0.282 mm	0.071 mm
PWM-F		0.243 m/s^2	0.202 mm	0.051 mm

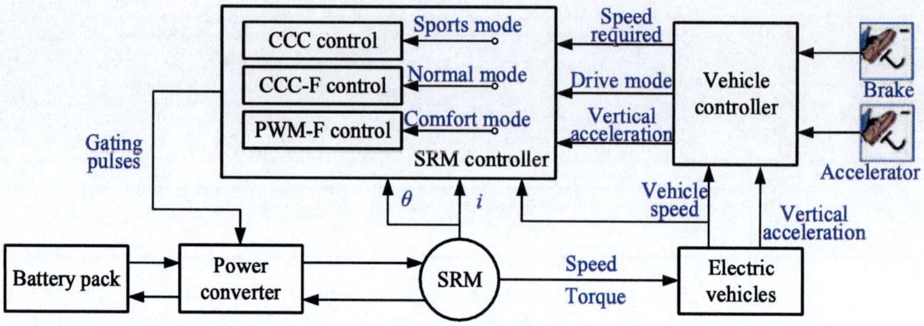

Fig. 8.20 The driving system under different modes based on various control methods

control of IWSRM. Thus, the designed controller can meet the various driving needs and realize the comprehensive balance of vibration performance and dynamic performance of EVs.

8.5 The Vibration Suppression Under Regenerative Braking Condition

8.5.1 The Driving System of SRG

As mentioned above, SRM has the characteristic of the highly nonlinear, thus, its dynamic performance cannot be reflected via linear or quasi-linear models. Meanwhile, through the above analysis, it can be known that the position angle and eccentricity ratios have little effect on inductance, torque, and the unbalanced radial force. Moreover, under the condition of the regenerative braking, for effectively tracking the driver required real-time braking torque and decreasing the braking response time of the SRG driving system, based on four-phase currents and

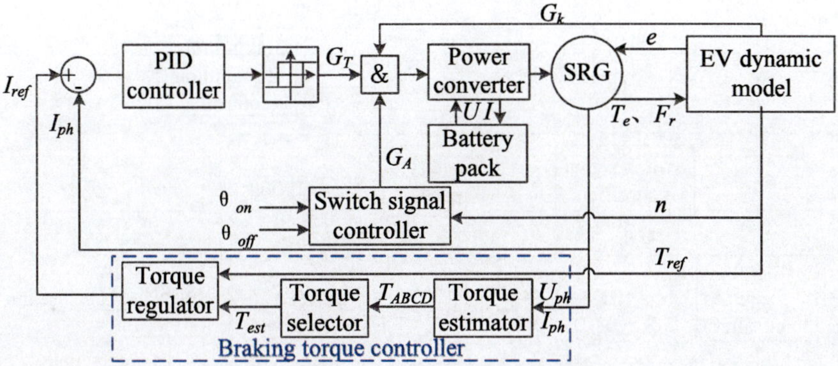

Fig. 8.21 The driving system of SRG

voltages, a braking torque controller of closed-loop is designed in this chapter, which mainly includes torque estimator, torque selector and torque regulator, as shown in Fig. 8.21.

8.5.2 The Braking System of the Vehicle

When the vehicle operates, it needs to overcome a serious of resistances, which include the rolling resistance F_f, the air resistance F_w, the gradient resistance F_i, and acceleration resistance F_j. The driving force F_t can be obtained by the sum of the rolling resistance F_f, the air resistance F_w, the gradient resistance F_i, and acceleration resistance F_j. Therefore, the driving equation of EV can be expressed as

$$F_t = F_f + F_w + F_i + F_j \tag{8.29}$$

In addition, when the vehicle brakes, except the air resistance, the rolling resistance, and the gradient resistance, it is also subjected to a braking force from IWSRM. At this time, the driving equation of EV can be expressed as

$$m\frac{du}{dt} = mgf\cos\alpha + \frac{1}{21.15}C_D A u^2 + mg\sin\alpha + F_b \tag{8.30}$$

where F_b refers to the braking force, which is the sum of the mechanical braking force F_{ff} and the regenerative braking force F_{reg} of SRG. Hence, the braking force can be computed as

$$F_b = F_{ff} + F_{reg} \tag{8.31}$$

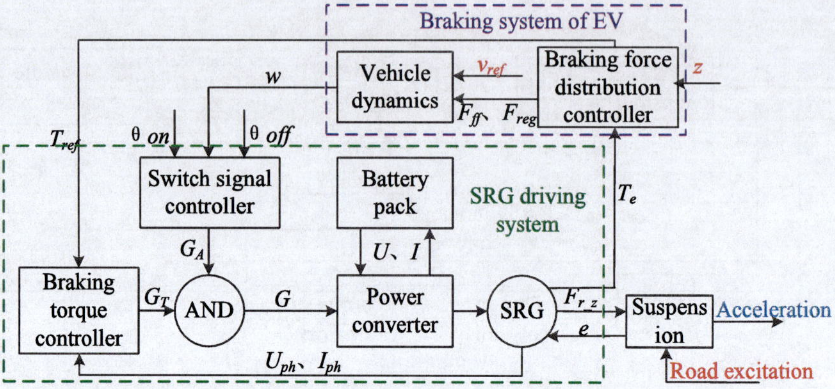

Fig. 8.22 The vehicle dynamic system model under the regenerative braking condition

Thus, under the condition of the regenerative braking, depending on the quarter IWSRM-suspension system, the driving system of SRG, and the braking system of EV, the dynamic system model of EV can be built in this chapter, as described in Fig. 8.22.

8.5.3 The Influence of Eccentricity and Electromagnetic Excitation

8.5.3.1 The Influence of Eccentricity on the Vertical Unbalanced Radial Force

Depending on the quarter IWSRM-suspension system, as shown in Fig. 8.4, the relative eccentricity of SRG can be acquired via the vertical displacements different of the stator and rotor. The relative eccentricity of SRG can be calculated as

$$e = x_{sa} - x_{wa} \tag{8.32}$$

To analyze the influence of eccentricity on the unbalanced vertical force, under the different eccentricities, the unbalanced vertical forces can be obtained by simulation, as shown in Fig. 8.23. It can be seen from the figure that the value of the unbalanced vertical force increases as the value of the relative eccentricity. It should be note that when the value of eccentricity is zero, the total vertical unbalanced vertical force is also zero. Thanks to the value of the relative eccentricity is quite small, the conclusion could be obtained that even a small eccentricity can produce a large vertical unbalanced radial force.

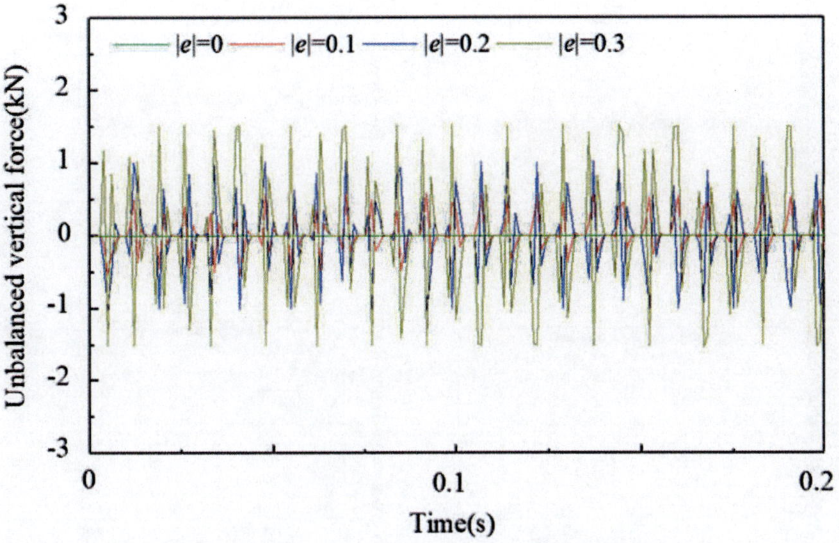

Fig. 8.23 The unbalanced vertical force based on different eccentricities

8.5.3.2 The Influence of Electromagnetic Excitation on Vehicle Dynamics

As seen in Fig. 8.24, which expressed that the effect of the electromagnetic excitation on the vertical acceleration of EV based on different roads under the regenerative braking system. Meanwhile, as seen in Fig. 8.25, which expressed that the effect of the electromagnetic excitation on the wheel dynamic load of EV based on different roads under the regenerative braking system. These dates are obtained by the vehicle dynamic system simulates, as shown in Fig. 8.22. The initial simulation parameters are that the braking strength is set as 0.1 and the initial braking speed is set as 40 km/h.

It can be seen from Fig. 8.24 that the value of the vehicle vertical acceleration with the unbalanced electromagnetic excitation is obviously larger than that without the unbalanced electromagnetic excitation under different road conditions, which could sharply worsen the ride comfort of the EV. However, it can be seen from Fig. 8.25 that the value of the vehicle wheel dynamic load with the unbalanced electromagnetic excitation is almost equal with that without the unbalanced electromagnetic excitation under different road conditions. Moreover, with the road surface roughness increases, the vibration acceleration and the wheel dynamic load amplitudes of EV both augments, which could mean that the adhesion effect of the wheel will be worse. Thus, there is a conclusion that the vertical unbalanced radial force is the key factor that aggravates vehicle vibration.

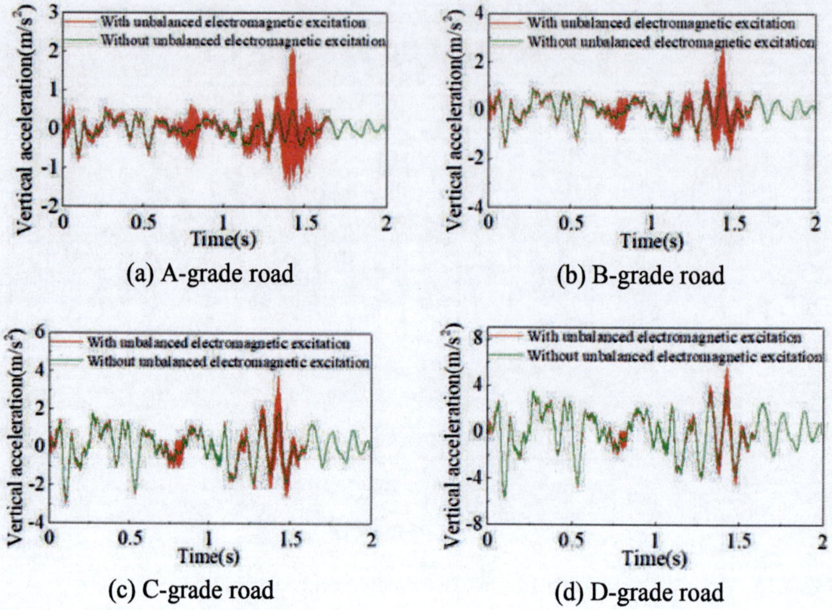

Fig. 8.24 The vertical acceleration with or without electromagnetic excitation

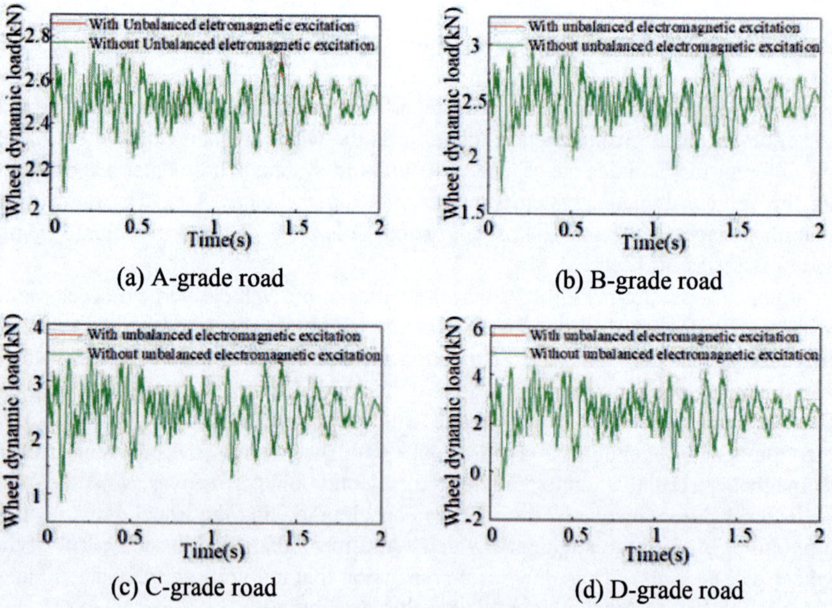

Fig. 8.25 The wheel dynamic load with or without electromagnetic excitation

8.5.4 The Design of Controller and the Results Analysis

8.5.4.1 The Design of Controller

Based on the previous effect analysis, there is a conclusion that the influence of the unbalanced radial force by electromagnetic excitation on the vertical vibration of EV is quite obvious. Therefore, to achieve the vibration control of EV, the unbalanced radial force should be considered as a control subject, further a vertical unbalanced radial force feedback control (VFC) method is presented in this chapter. For the feedback control method, whose feedback is the vertical unbalanced radial force of IWSRM, which is controlled by difference of a PID controller depended on the reference and the actual vertical unbalanced radial force. Then, based on the difference of the PID controller, as shown in Fig. 8.26, a vertical unbalanced radial force PWM controller is designed and used. Meanwhile, under the regenerative braking consideration, not only the vibration of EV, but also the regenerative braking energy should be considered. Thus, for the condition of the regenerative braking, for further suppressing the vertical vibration of EV without obviously reducing regenerative braking energy, depending on the vertical acceleration and regenerative braking energy, a switch angles optimization objective function is presented. The optimization objective function can be expressed as

$$F(\theta_{on}, \theta_{off}) = \min\left\{ \omega_1 \frac{1/E_{reg}}{1/E_{reg_{\max}}} + \omega_2 \frac{a_z}{a_{z\max}} \right\} \tag{8.33}$$

where ω_1 and ω_2 respectively refer to the weight of the index, E_{reg} means the regenerative braking energy of EV under the regenerative braking condition, a_z is vertical acceleration of EV. When the vehicle brakes, it is no doubt that the vertical vibration suppression is quite important due to the obvious effect on the ride comfort of passengers. Meanwhile, the regenerative braking energy recovery is also vital because of the endurance mileage of EV. Thus, considering the vertical acceleration of EV and the regenerative braking energy comprehensively, the weight factors of the regenerative braking energy and the vertical acceleration of EV can be respectively defined as 0.4 and 0.6.

For the SRG driving system, due to the characteristic of the nonlinearity, the traditional optimization methods are hard to optimize the switch angles. Differing from the traditional optimization algorithms, the genetic algorithm (GA), an algorithm of random search based on natural selection and natural genetic mechanism in the biological world, starts from a randomly generated initial solution and iterates progressively through certain selection, crossover, and mutation until a predefined evolutionary generation or the threshold is reached. The GA is suitable for dealing with the problems of the complex nonlinear because of it's the feature that it does not depend on the gradient information. In this chapter, the GA is used to optimize switch angles based on the regenerative braking energy and the vertical acceleration

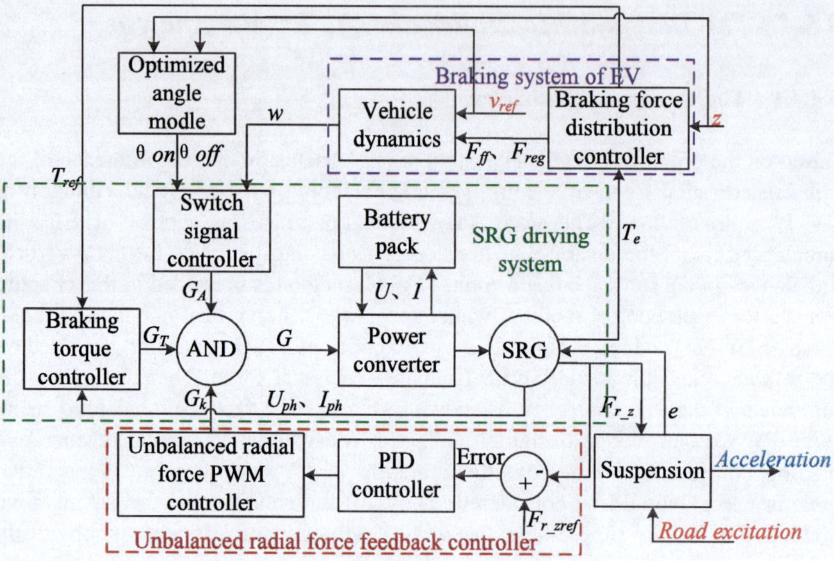

Fig. 8.26 The VFC system under the regenerative braking condition

of EV. Taking two conditions as an example, one is that the braking strength is set as 0.2 and the initial braking speed is set as 50 km/h, and the other is that the braking strength is set as 0.5 and the initial braking speed is set as 90 km/h. The optimized progress is shown in Fig. 8.27 under the two conditions. Thus, based on different combinations of the vehicle speed u and the braking strength z, the models of the optimized turn-on and turn-off angle are acquired, which are be described in Fig. 8.28.

8.5.4.2 The Results Analysis

In this chapter, taking several typical regenerative braking conditions as an example to verify the effectiveness of the proposed control method by the comparison and analysis. The different combinations of the medium-low speed and medium-high speed, and the low and medium braking strength of vehicle are selected as four typical regenerative braking conditions. It should be noted that the high braking strength is not considered because of the reason that when the EV brakes with the high braking strength (the emergency braking condition), the barking force is only supplied by mechanical braking system.

When the EV brakes at the initial speed of 30 km/h with the braking strength of the low and medium ($z = 0.1$ and $z = 0.5$), the simulation results are acquired and shown in Figs. 8.29 and 8.30, where it should be noted that (a), (b), (c), and (d) represent the EV vertical acceleration, the regenerative braking energy, the

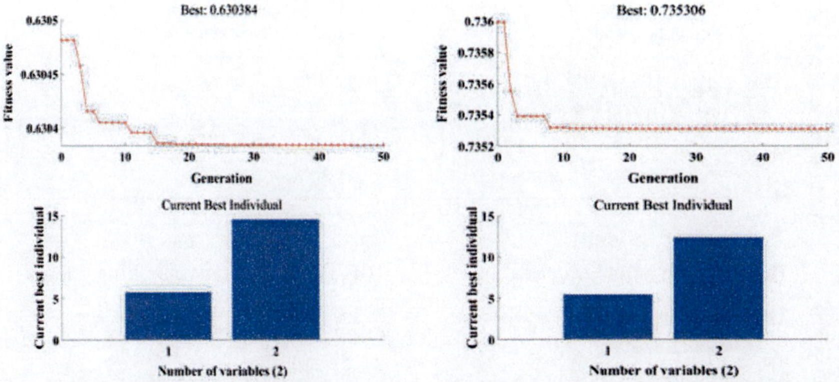

(a) $z = 0.2$ and $u = 50$ km/h (b) $z = 0.5$ and $u = 90$ km/h

Fig. 8.27 The optimized process for switch angles based on GA under the two conditions

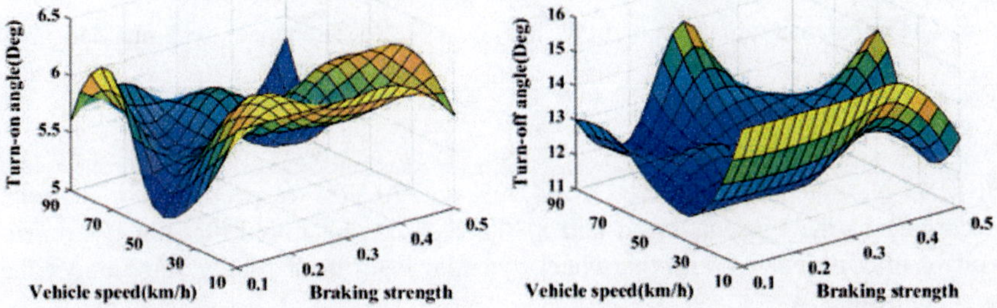

(a) Optimal turn-on angles (b) Optimal turn-off angles

Fig. 8.28 The optimal switch angles based on GA

vertical unbalanced radial force of IWSRM, and the wheel dynamic load, respectively. Based on the Fig. 8.29a, c, it can be seen that, under the controller without VFC, the peak values of the EV vertical acceleration and the vertical unbalanced radial force of IWSRM is respectively 2.63 m/s^2 and 3133.67 N when the EV brakes with low braking strength. However, under the controller with VFC, the peak values of the EV vertical acceleration and the vertical unbalanced radial force of IWSRM is respectively 1.69 m/s^2 and 2279.57 N, with a fall of 35.7% and 27.3%, respectively. In the same way, based on the Fig. 8.30a, c, it can be seen that when the EV brakes with medium braking strength, the peak values of the EV vertical acceleration and the vertical unbalanced radial force of IWSRM under the controller without VFC is respectively 1.87 m/s^2 and 2571.24 N. Nevertheless, under the controller with VFC, the peak values of the EV vertical acceleration and the vertical unbalanced radial force of IWSRM is respectively 1.04 m/s^2 and

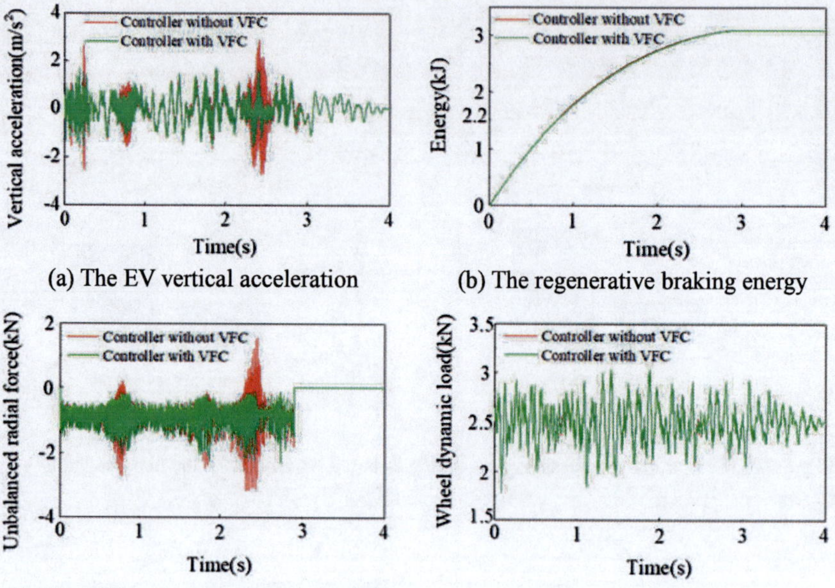

(a) The EV vertical acceleration (b) The regenerative braking energy

(c) The unbalance radial force of IWSRM (d)The wheel dynamic load

Fig. 8.29 Simulation results (u = 30 km/h and z = 0.1)

2042.45 N, which are decreased by 44.39% and 20.57%, respectively. In addition, according to the Figs. 8.29b, d and 8.30b, d, it can be found that the EV regenerative braking energy and the wheel dynamic load are basically same under the two controllers.

In addition, to further verify the designed controller performance of the vehicle vibration suppression, RMS values of the vehicle vertical acceleration and the vertical unbalanced radial force of IWSRM are calculated and listed in Table 8.4. As can be seen in the Table, when the EV brakes with the low braking strength under controller without VFC, the RMS values of the vehicle vertical acceleration and the vertical unbalanced radial force of IWSRM are respectively 0.57 m/s^2 and 974.79 N. Nevertheless, under controller with VFC, the RMS values of the vehicle vertical acceleration and the vertical unbalanced radial force of IWSRM are respectively 0.5 m/s^2 and 945.29 N, with a decrease of 12.28% and 3.03%, respectively. Similarly, when the EV brakes with the medium braking strength under controller without VFC, the RMS values of the vehicle vertical acceleration and the vertical unbalanced radial force of IWSRM are respectively 0.45 m/s^2 and 956.16 N. However, under controller with VFC, the RMS values of the vehicle vertical acceleration and the vertical unbalanced radial force of IWSRM are respectively 0.42 m/s^2 and 940.78 N, which are reduced by 6.67% and 1.61%, respectively.

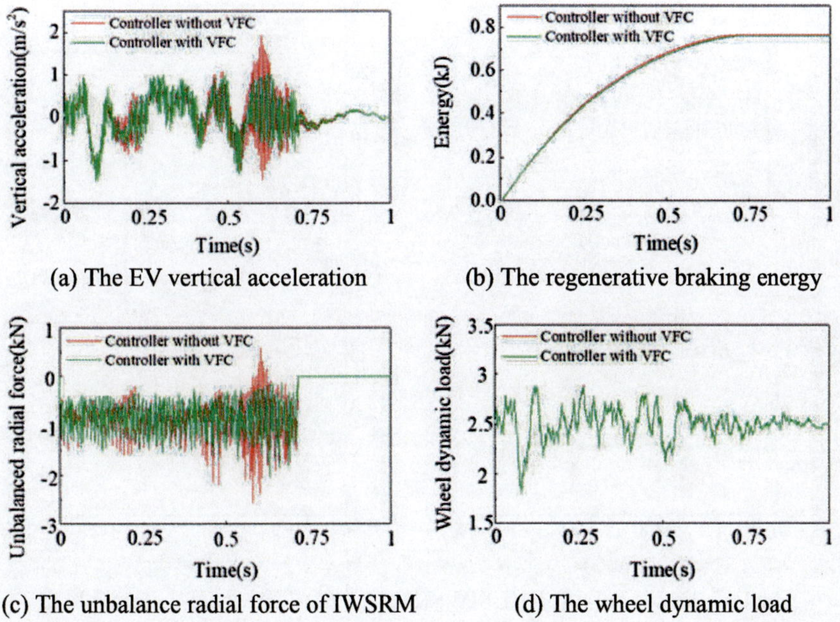

Fig. 8.30 Simulation results (u = 30 km/h and z = 0.5)

Table 8.4 Root mean square (RMS) comparisons under two controllers

Braking strength	Control strategy	Vertical acceleration	Unbalanced radial force
z = 0.1	The controller without VFC	0.57 m/s^2	974.79 N
	The controller with VFC	0.5 m/s^2 ↓12.28%	945.29 N ↓3.03%
z = 0.5	The controller without VFC	0.45 m/s^2	956.16 N
	The controller with VFC	0.42 m/s^2 ↓6.67%	940.78 N ↓1.61%

Moreover, When the EV brakes at the initial speed of 70 km/h with the braking strength of the low and medium (z = 0.1 and z = 0.5), the simulation results are acquired and shown in Figs. 8.31 and 8.32. Meanwhile, it should be noted that (a), (b), (c), and (d) represent the EV vertical acceleration, the regenerative braking energy, the vertical unbalanced radial force of IWSRM, and the wheel dynamic load, respectively. Based on the Fig. 8.31a, c, it can be seen that, under the controller without VFC, the peak values of the EV vertical acceleration and the vertical unbalanced radial force of IWSRM is respectively 3.76 m/s^2 and 3285.51 N when the EV brakes with low braking strength. However, under the controller with VFC, the peak values of the EV vertical acceleration and the vertical unbalanced radial

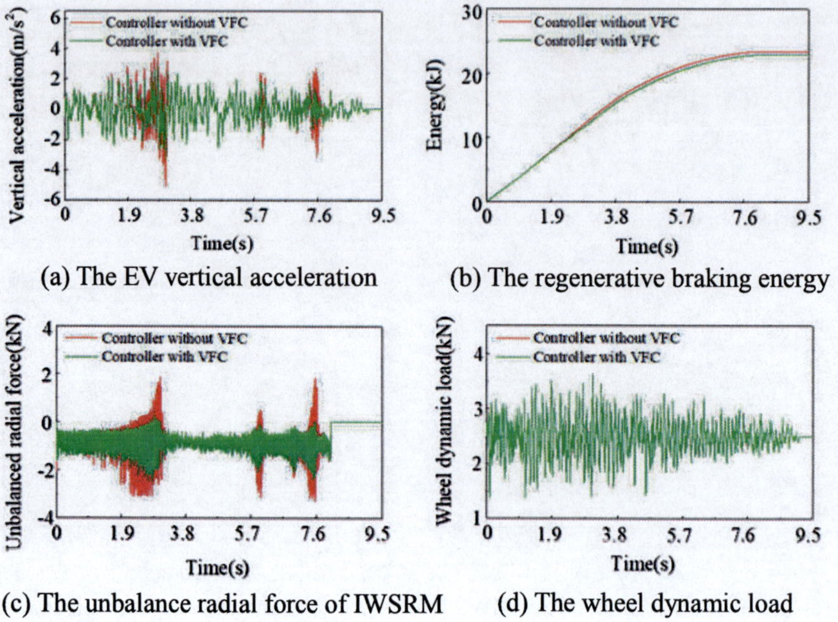

(a) The EV vertical acceleration (b) The regenerative braking energy

(c) The unbalance radial force of IWSRM (d) The wheel dynamic load

Fig. 8.31 Simulation results (u = 70 km/h and z = 0.1)

force of IWSRM is respectively 2.9 m/s^2 and 2494.04 N, which are decreased by 22.9% and 24.1%, respectively. Similarly, based on the Fig. 8.32a, c, it can be seen that when the EV brakes with medium braking strength, the peak values of the EV vertical acceleration and the vertical unbalanced radial force of IWSRM under the controller without VFC is respectively 2.25 m/s^2 and 3056.12 N. Nevertheless, under the controller with VFC, the peak values of the EV vertical acceleration and the vertical unbalanced radial force of IWSRM is respectively 1.69 m/s^2 and 2232.36 N, with a decrease of 24.89% and 26.95%, respectively. In addition, according to the Figs. 8.31b, d and 8.32b, d, it can be found that the EV regenerative braking energy and the wheel dynamic load are basically unchanged under the two controllers.

Meanwhile, it can be found from Table 8.5 that the RMS values of the vehicle vertical acceleration and the vertical unbalanced radial force of IWSRM are significantly reduced under the designed controller with VFC. When the EV brakes with the low braking strength under controller without VFC, the RMS values of the vehicle vertical acceleration and the vertical unbalanced radial force of IWSRM are respectively 0.89 m/s^2 and 1014.29 N. Nevertheless, under controller with VFC, the RMS values of the vehicle vertical acceleration and the vertical unbalanced radial force of IWSRM are respectively 0.79 m/s^2 and 933.35 N, which are decreased by 11.24% and 7.98%, respectively. Similarly, when the EV brakes with the medium braking strength under controller without VFC, the RMS values of the

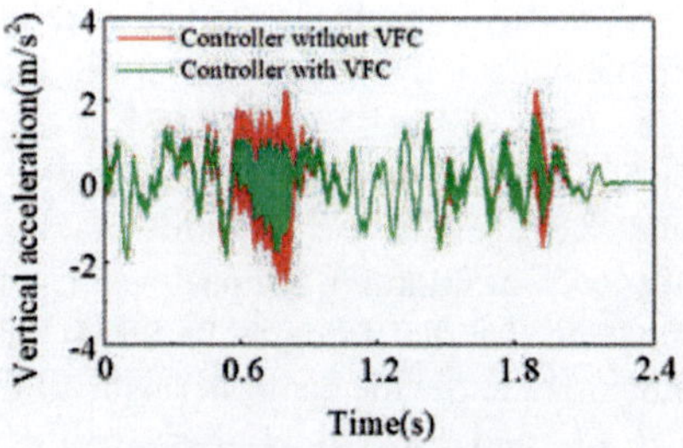

(a) The EV vertical acceleration

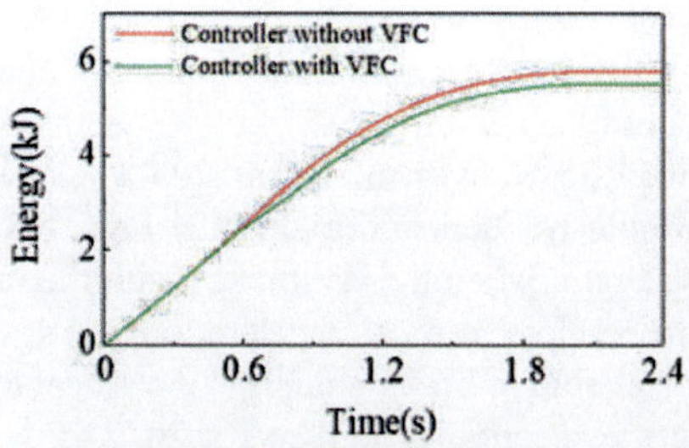

(b) The regenerative braking energy

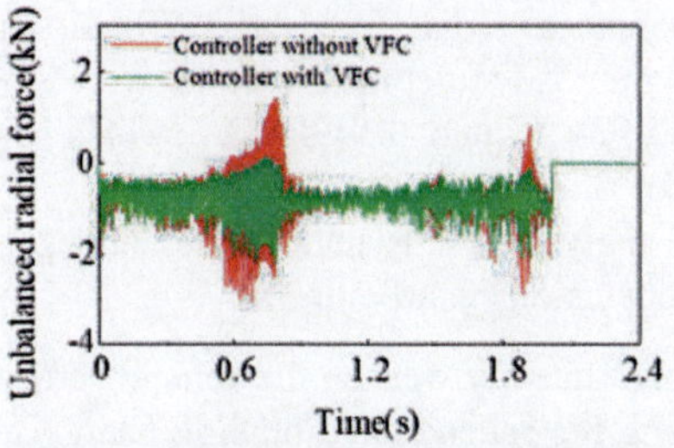

(c) The unbalance radial force of IWSRM

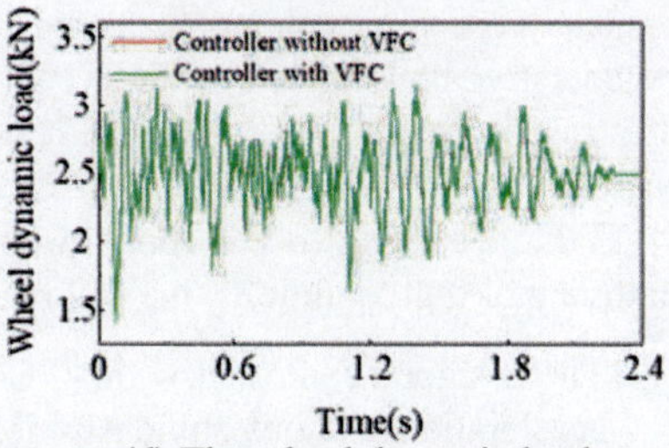

(d) The wheel dynamic load

Fig. 8.32 Simulation results (u = 70 km/h and z = 0.5)

Table 8.5 Root mean square (RMS) comparisons under two controllers

Braking strength	Control strategy	Vertical acceleration	Unbalanced radial force
z = 0.1	The controller without VFC	0.89 m/s^2	1014.29 N
	The controller with VFC	0.79 m/s^2 ↓11.24%	933.35 N ↓7.98%
z = 0.5	The controller without VFC	0.75 m/s^2	1002.48 N
	The controller with VFC	0.66 m/s^2 ↓12%	936.87 N ↓6.54%

vehicle vertical acceleration and the vertical unbalanced radial force of IWSRM are respectively 0.75 m/s^2 and 1002.48 N. However, under controller with VFC, the RMS values of the vehicle vertical acceleration and the vertical unbalanced radial force of IWSRM are respectively 0.66 m/s^2 and 936.87 N, with a fall of 12% and 6.54%, respectively.

In addition, according to the Figs. 8.29 and 8.31, under the condition of the low braking strength, when the EV brakes with the initial speeds of 30 km/h, the peak values of the vehicle vertical acceleration, vertical unbalanced radial force of IWSRM, the regenerative braking energy, the wheel dynamic load under the controller with VFC are respectively 1.68 m/s^2, 2279.57 N, 3.09 kJ, and 3036.93 N. However, when the EV brakes with the initial speeds of 70 km/h, the peak values of the vehicle vertical acceleration, vertical unbalanced radial force of IWSRM, the regenerative braking energy, the wheel dynamic load under the controller with VFC are respectively 2.89 m/s^2, 2494.04 N, 22.85 kJ, and 3611.56 N, with a rise of 72.02%, 9.41%, 506.67%, and 18.92%, respectively. Similarly, depending on the Figs. 8.30 and 8.32, under the medium braking strength, when the EV brakes with the initial speeds of 30 km/h, the peak values of the vehicle vertical acceleration, vertical unbalanced radial force of IWSRM, the regenerative braking energy, the wheel dynamic load under the controller with VFC are respectively 1.04 m/s^2, 2042.45 N, 0.76 kJ, and 2873.49 N. Nevertheless, when the EV brakes with the initial speeds of 70 km/h, the peak values of the vehicle vertical acceleration, vertical unbalanced radial force of IWSRM, the regenerative braking energy, the wheel dynamic load under the controller with VFC are respectively 1.69 m/s^2, 2232.36 N, 5.52 kJ, and 3149.41 N, which are increased by 62.5%, 9.3%, 626.32%, and 9.6%, respectively.

Thus, according to the comparison and analysis results under different regenerative braking conditions, the following conclusions can be drawn.

(1) The designed controller with VFC can significantly weaken the vehicle vertical acceleration without obviously decreasing the regenerative braking energy by suppressing the vertical unbalanced radial force of IWSRM, which will greatly improve the vehicle ride comfort.
(2) When the initial braking speed is the same, the vibration performance during medium braking strength is better than that the low braking strength.
(3) When the braking strength is same, the vibration performance at medium-low speed is better than that at medium-high speed because of the reduction of the vertical unbalanced radial force.

References

1. Öksüztepe E. In-Wheel switched reluctance motor design for electric vehicles by using a pareto-based multiobjective differential evolution algorithm. IEEE Transactions on Vehicular Technology, 2017, 66(6): 4706–4715.
2. Lin J, Schofield N, Emadi A. External-rotor 6-10 switched reluctance motor for an electric bicycle. IEEE Transactions on Transportation Electrification, 2015, 1(4): 348.356.
3. Li Z, Zheng L, Ren Y, et al. Multi-objective optimization of active suspension system in electric vehicle with in-wheel-motor against the negative electromechanical coupling effects. Mech Syst Signal Process 2019, 116: 545–565.
4. Mikail R, Husain I, Sozer Y, Islam MS, Sebastian T. Torque-ripple minimization of switched reluctance machines through current profiling. IEEE Transactions on Industry Applications. 2013, 49(3): 1258.1267.
5. Xu A, Shang C, Chen J, Zhu J, Han L. A new control method based on DTC and MPC to reduce torque ripple in SRM. IEEE Access. 2019, 7: 68584–68593.
6. Xing Z, Duan X, Liu L. MPPT for wind power system with switched reluctance generator. Proc 13th IEEE Conf Ind Electron Appl ICIEA 2018. 2018: 1420–1424.

7. Dong L, Liu Y, Ling L, et al. Design of converter topology for switched reluctance wind power generator. AIP Conf Proc; 1839. Epub ahead of print 2017. DOI: https://doi.org/10.1063/1.4982402.
8. Zhu Y, Yang C, Yue Y, Wei W, Zhao C. Design and optimization of an In-wheel switched reluctance motor for electric vehicles. IET Intelligent Transport Systems. 2019, 13(1): 175–182.
9. Xue XD, Cheng KW, Ng TW, Cheung NC. Multi-objective optimization design of In-Wheel switched reluctance motors in electric vehicles. IEEE Transactions on Industrial Electronics. 2010, 57(9): 2980–2987.
10. Liang J, Lee DH, Ahn JW. Direct instantaneous torque control of switched reluctance machines using 4-level converters. 2009, 3(4): 313–323.
11. Cai W, Yi F. An integrated multiport power converter with small capacitance requirement for switched reluctance motor drive. IEEE Transactions on Power Electronics, 2015, 31(4): 3016–3026.
12. Kuai S, Zhang H, Xia X, Li K. Unipolar sinusoidal excited switched reluctance motor control based on voltage space vector. IET Electric Power applications, 2019, 13(5): 670–675.
13. Chen H, Chen H, Yang Z. Average torque control of switched reluctance machine drives for electric vehicles. 2015, 9(7): 459–468.
14. Hu C, Wang R, Yan F, et al. Differential steering based yaw stabilization using ISMC for independently actuated electric vehicles. IEEE Trans Intell Transp Syst 2018, 19: 627–638.
15. Mao Y, Zuo S, Wu S, et al. High frequency vibration characteristics of electric wheel system under in-wheel motor torque ripple. Journal of Sound and Vibration. 2017, 400: 442–456.
16. Wang Y, Li Y, Sun W, et al. Effect of the unbalanced vertical force of a switched reluctance motor on the stability and the comfort of an in-wheel motor electric vehicle. Proceedings of the Institution of Mechanical Engineers, Part D: Journal of Automobile Engineering. 2015, 229(12): 1569–1584.
17. Wang R, Jing H, Yan F, et al. Optimization and finite-frequency H∞ control of active suspensions in in-wheel motor driven electric ground vehicles. Journal of the Franklin Institute, 2014, 352(2): 468.484.
18. Shao X, Naghdy F, Du H. Reliable fuzzy H∞ control for active suspension of in-wheel motor driven electric vehicles with dynamic damping. Mechanical Systems & Signal Processing. 2016, 87(6): 365–383.
19. Hredzak B, Gair S, and Eastham JF. Control of an EV drive with reduced unsprung mass. IEE Proc. Electr. Power Appl., 1998, 145(6): 600–606.
20. Nagaya G, Wakao Y, and Abe A. Development of an in-wheel drive with advanced dynamic-damper mechanism. JSAE Rev., 2003, 24(4): 477–481.
21. Sun W, Li Y, Huang J. Vibration effect and control of in-wheel switched reluctance motor for electric vehicle. Journal of Sound and Vibration. 2015, 338: 105–120.
22. Li Z, Zheng L, Gao W, and Zhan Z. Electromechanical coupling mechanism and control strategy for in-wheel-motor-driven electric vehicles. IEEE Transactions on Industrial Electronics, 2019, 66(6): 4524–4533.

Chapter 9
Temperature Filed Analysis and Optimization for the SRM

9.1 Introduction

For switched reluctance motors (SRM), the relative research makes good progress in many aspects, but the topics on temperature filed still need further research. Iron loss and copper loss generated during SRM operation will eventually be converted into heat energy to increase the internal temperature of the motor, which will lead to the decrease of motor efficiency, accelerate insulation aging, and cause motor failure or other accidents, thus greatly shortening the service life of the motor [1]. Therefore, when designing the structure of SRM, it is necessary to comprehensively consider the electromagnetic performance and internal temperature distribution of the motor. In conclusion, it is very important to analyze and study the temperature field of SRM for electric vehicle (EV) to improve its performance and operation reliability.

Scholars all over the world have carried out in-depth research on SRM and achieved considerable research results in many aspects, but the research on core loss of SRM is still limited. Due to its unique structure, the magnetic flux density in the motor core is affected by structural parameters, control strategy, and speed, and the nonlinear magnetic flux density waveform also makes it difficult to accurately calculate the iron loss of the motor [2, 3]. The research of iron loss mainly includes the experimental method, the magnetic circuit analysis method, and the finite element method (FEM). Among them, due to its comprehensive factors and high calculation accuracy, the FEM can obtain the distribution of iron loss in the whole motor through modeling and simulation. Therefore, the FEM is the main method to calculate the electromagnetic parameters and core loss of the motor, currently [4–6].

In addition, since the 1970s, researchers have gradually paid more attention to the distribution of motor temperature rise, and proposed a variety of methods for calculating motor temperature rise, such as simplified equation method, equivalent heat circuit method and FEM [7, 8]. However, with the improvement of work requirements, the simplified equation method has been unable to meet design needs,

Y. Zhu, *The Key Technologies for Powertrain System of Intelligent Vehicles Based on Switched Reluctance Motors*, Recent Advancements in Connected Autonomous Vehicle Technologies 1, https://doi.org/10.1007/978-981-16-4851-9_9

such that the current research methods of temperature field are mainly equivalent thermal circuit method and FEM [9, 10].

In this chapter, the temperature field of SRM is studied and the motor nonlinear mathematical model is established. Based on ANSYS Maxwell, combined with finite element simulation of the transient field of SRM, the magnetic field distribution, the flux density wave form, and the iron loss density of various parts of the motor iron core are analyzed. On the basis, the loss of motor is calculated, and the mathematical model of thermal analysis of a prototype is built according to heat transfer theory. Then, in the ANSYS Workbench, the thermal analysis of prototype is done to more fully understand the loss and the temperature distribution inside the motor which provides a reference for optimization design of the motor. Finally, the structure of the motor is optimized to reduce motor temperature rise.

9.2 Finite Element Analysis of SRM

The special structure of SRM will cause complex magnetic field distribution, and the magnetic field of SRM is affected by magnetic pole distribution and control method. The SRM studied in this chapter is unidirectional excitation with the winding mode of NNNN–SSSS as shown in Fig. 9.1. The rated speed n of the SRM prototype is 1500 rpm, and the main angle position parameters when simulating are fixed as follows, the turn-on angle θ_{on} is 0 degree, and the turn-off angle θ_{off} = is 15 degrees.

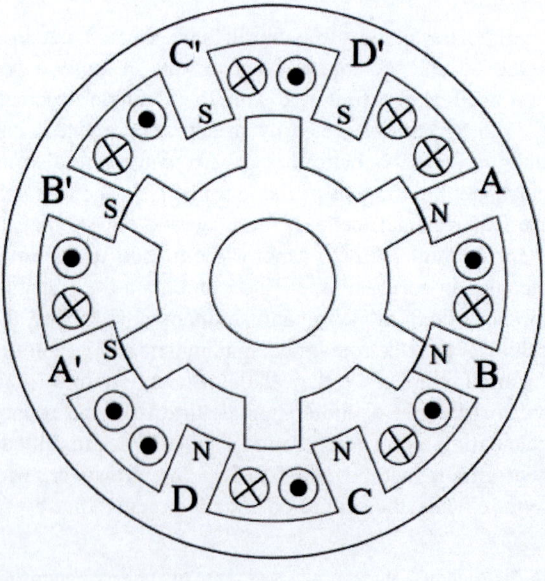

Fig. 9.1 The typical winding distribution

The typical phase current waveform of SRM is shown in Fig. 9.2, and the flux linkage waveform is shown in Fig. 9.3. It can be seen from the figure that the four phases, A, B, C, and D, are turned on in sequence to realize the unidirectional excitation of SRM. The four-phase winding current waveform is basically the same as the flux linkage waveform, with a period of 60 degrees and a phase difference of 15 degrees. When the phase-A winding is turned on, a current is generated in the phase-A winding, and then a drag torque is generated, which drags the rotor to rotate by an angle. When the phase A is turned off, the phase-A winding current begins to decrease, and the phase current of this phase will feed part of the residual energy back to the power supply. In addition, the phase B starts to conduct as long as the phase A is turned off. In this way, when the clockwise conduction is repeated one by one, and the rotor of SRM rotates counter clockwise around the shaft because of the principle of the minimum magnetic reluctance.

Figure 9.4 shows the typical output torque waveform of SRM. It is obvious to see from the Fig. 9.4 that there is a large torque ripple in the output torque of the SRM prototype, which is the main disadvantage for SRM. This is due to the special structure of the SRM and the non-linear characteristics of the inductance curve of

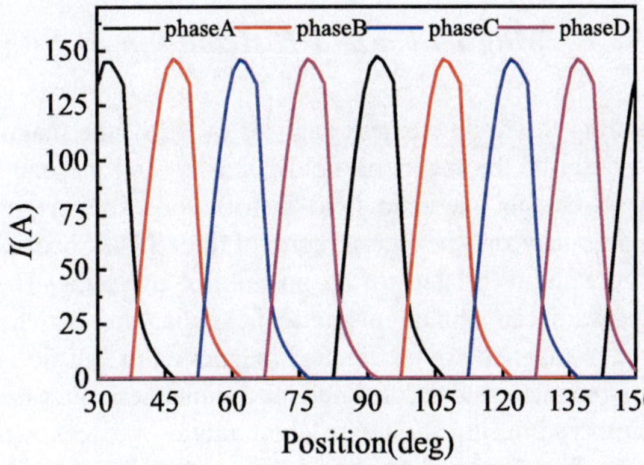

Fig. 9.2 Phase current waveform

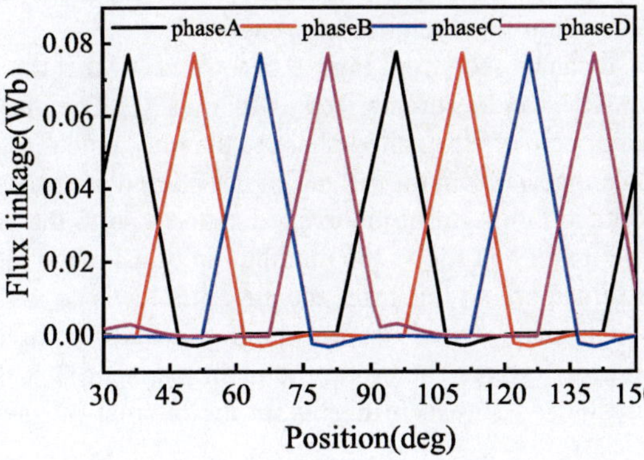

Fig. 9.3 Flux link waveform

Fig. 9.4 Output torque waveform

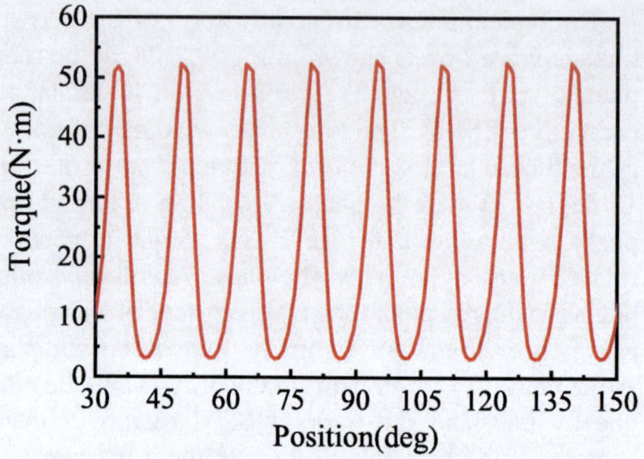

the SRM, which makes its magnetic circuit have a high non-linear saturation. Therefore, the torque ripple of the motor should be improved by optimizing the structure of the motor.

9.2.1 Magnetic Field Distribution of SRM

During the finite element analysis of SRM, the magnetic flux density is often used to describe the magnetic field intensity, and the magnetic induction line is used to describe the magnetic field distribution. The magnetic density is not uniformly distributed on the stator core of the SRM, because the magnetic density and direction are related to the position of the rotor. The post-processing of the FEA software can obtain magnetic field diagrams, such as magnetic induction lines, vector magnetic potential, and magnetic flux density. Figures 9.5, 9.6 and 9.7 show the magnetic induction line and the magnetic flux density distribution of the SRM prototype during the phase commutation process, where the direction of the magnetic flux lines represented by the red-yellow lines refer to the counterclockwise direction, while the direction of the magnetic flux lines represented by the blue-green lines refer to clockwise.

It can be seen from Figs. 9.5, 9.6 and 9.7 that the magnetic flux distribution rules of SRM can be summarized as follows. (1) The main magnetic flux exists from the stator pole of the excited phase to the rotor pole through the air gap; (2) The leakage flux appears from the air gap of the interpolar to the stator yoke; (3) The interphase leakage flux is from the excited stator pole to the adjacent stator pole.

Figure 9.5a shows the distribution of magnetic induction lines when the phase D is turned on. At this time, the air gap reluctance is small, and most of the magnetic flux forms a closed curve through the stator yoke, stator pole, air gap, rotor pole, and rotor yoke. Therefore, the main magnetic flux is larger and the leakage flux is smaller. As shown in Fig. 9.5b, the internal magnetic flux density of the SRM is

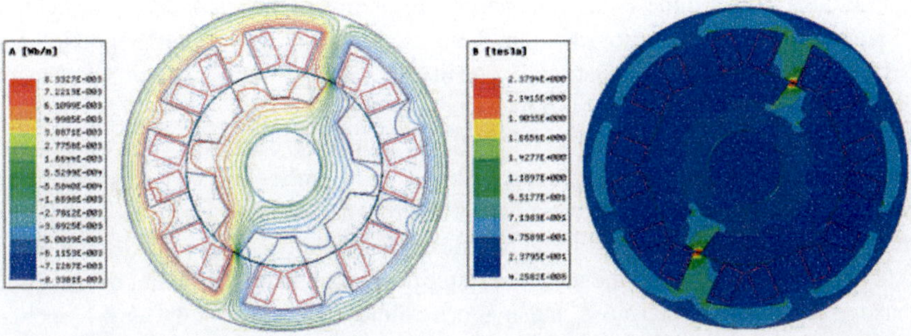

(a) Distribution of magnetic induction lines (b) Distribution of magnetic flux density

Fig. 9.5 Distributions of magnetic induction lines and magnetic flux density when phase D is excited

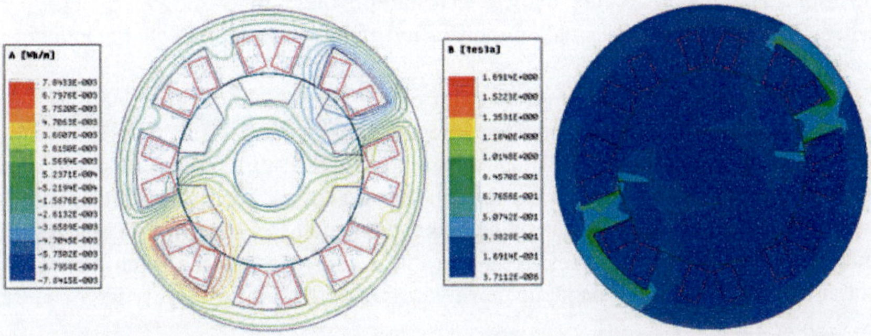

(a) Distribution of magnetic induction lines (b) Distribution of magnetic flux density

Fig. 9.6 Distributions of magnetic induction lines and magnetic flux density when phase D is excited and phase A is freewheeling

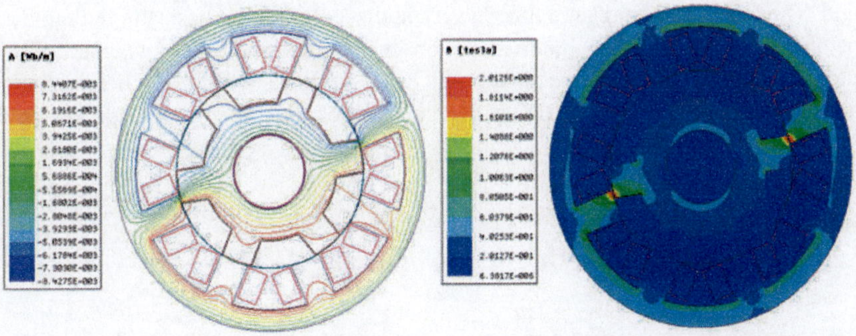

(a) Distribution of magnetic induction lines (b) Distribution of magnetic flux density

Fig. 9.7 Distributions of magnetic induction lines and magnetic flux density when phase A is excited mainly distributed in the stator yoke and the tooth tips of the stator and rotor. At this time, due to the special rotor position, the available air gap for going magnetic lines is limited such that the magnetic flux density at the tooth tips of the rotor and stator poles is larger, and it is in a slightly saturated state.

Figure 9.6a is the distribution of magnetic induction lines when the phase D is turned on and phase A is under condition of freewheeling. At this time, the reluctance in air gap is large, and a part of the magnetic flux reaches the stator yoke through the air gap between the poles. Meanwhile, there is also a small amount of flux lines passing through the adjacent stator pole, such that the amount of magnetic leakage is large. Furthermore, the magnetic field changes as long as the excited phase changes from phase D to phase A, such that the magnetic pole changes from S pole to N pole, because the motor winding distribution is NNNN–SSSS. At this time, the direction of the magnetic lines in the yoke between the two adjacent poles changes, resulting in the magnetic domain direction of the yoke of the motor also changes correspondingly, which can cause a larger magnetic density value at this position. The magnetic flux density distribution diagram is shown in Fig. 9.6b, where the air gap has a small permeability and the overall magnetic induction intensity is small.

Figure 9.7a is the distribution diagram of the magnetic induction line when the phase A winding is turned on and the freewheeling current of the phase D winding ends. At this time, due to the small air gap reluctance, most of the magnetic flux lines are evenly closed through the stator pole, air gap, and rotor pole of the phase A winding, and the amount of magnetic leakage is small. As shown in Fig. 9.7b, the magnetic density is mainly distributed in the stator yoke and the tooth tips of the stator and rotor. And the magnetic density of the tooth tips is relatively large, which is slightly saturated.

During the commutation process from phase D to phase A shown in Figs. 9.5, 9.6 and 9.7, it can be seen that since the movement of SRM follows the principle of minimum reluctance, when the rotor of the motor rotates from the unaligned position to the aligned position, the magnetic reluctance of the air gap gradually decreases with the less magnetic leakage and the greater saturation of the magnetic field. The changes in magnetic flux density in these figures are basically the same as the distribution trend of magnetic induction lines, which shows that when the SRM rotor rotates, the air gap reluctance will exist nonlinear changes. Therefore, the internal magnetic field of SRM will change, such that then the magnetic field intensity will also change.

9.2.2 *Magnetic Flux Density Analysis of SRM*

The size and direction of the magnetic flux density of the core inside the SRM change with the rotor position, which can be expressed by Eq. (9.1), where, $\vec{B}_{(t)}$ represents the magnetic flux density vector at time t; $B_{r(t)}$ denotes the radial component of magnetic density at time t, $B_{t(t)}$ refers to the magnetic close component at time t, $\vec{e}_r$ represents the radial unit vector, $\vec{e}_t$ denotes the tangential unit vector.

$$\vec{B}_{(t)} = B_{r(t)} \cdot \vec{e}_r + B_{t(t)} \cdot \vec{e}_t \tag{9.1}$$

Since the size and direction of the magnetic flux density of the iron core inside the SRM change with the variations of the rotor position, it is particularly important to analyze the changing law of the magnetic flux density at each position in the iron core. This chapter selects the key finite element nodes in the stator tooth tip, stator tooth center, stator tooth root, stator yoke, rotor tooth tip, rotor tooth center, rotor tooth root, and rotor yoke of the SRM prototype. It should be noted that the magnetic flux density changes in these points are used to replace the changes of magnetic flux density in corresponding areas. Then, the changing law of the magnetic flux density in these areas are analyzed. The selection points of the corresponding parts are shown in Fig. 9.8.

In Fig. 9.8, points A1, A2, and A3 respectively correspond to the magnetic density of the stator tooth tip, stator tooth center, and stator tooth root of the phase A. The same area of phase B, C, and D are selected to take finite element nodes. In addition, Points DA, AB, BC, and CD correspond to the magnetic density of the stator yoke section DA, AB, BC, and CD, respectively. Points 1, 2, 3, and 4 correspond to the magnetic density of the rotor tooth tip, rotor tooth center, rotor tooth root, and rotor yoke, respectively.

The electromagnetic coupling simulation analysis of the SRM prototype is carried out in the transient state analysis of Maxwell 2D when the motor is rotating

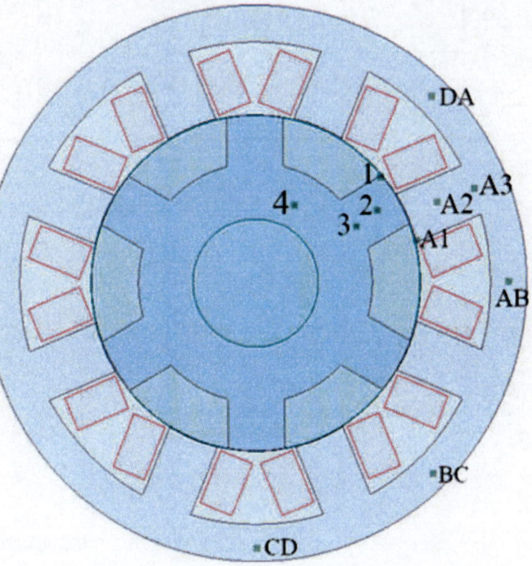

Fig. 9.8 Picking points of corresponding parts

counterclockwise. The composite magnetic density waveform of each finite element node of SRM and its tangential component B_t and radial component B_r are shown from Figs. 9.9, 9.10, 9.11, 9.12 and 9.13.

Figures 9.9, 9.10 and 9.11 show the magnetic flux density waveforms of the tooth tip, tooth center, and tooth root of each phase of the stator, and the waveform diagrams of the tangential component B_t and the radial component B_r. From the simulation results of Figs. 9.9, 9.10 and 9.11, it can be seen that no matter it is the composite magnetic density, the radial component, or the tangential component, each node changes periodically and the frequency is the same with the change period of 60 degrees. In addition, the shapes of the magnetic flux density waveforms in the same area of the stator teeth corresponding to each of A, B, C, and D are basically the same, but the phase difference is 15 degrees. The magnetic flux density of each area takes the radial component B_r as the main component, which gradually decreases from the tip of the tooth to the root of the tooth. The peak of the composite magnetic flux density in the tooth tip is the largest, which is the most saturated area in the stator core. Among them, since the tips of the stator teeth have pole shoes, the magnetic field lines enter and exit the air gap along the two sides of the pole shoes, and the tooth root is connected with the stator yoke, causing the magnetic field lines to bend. Therefore, the tooth tip and tooth root also contain the radial component B_r and the tangential component B_t. Since the SRM operation follows the principle of minimum reluctance, the magnetic flux in the middle of the tooth is almost all outward or inward along the direction of the stator tooth, which makes the middle of the tooth hardly contain the tangential component B_t. The direction of the magnetic density component of each phase is the same. This is

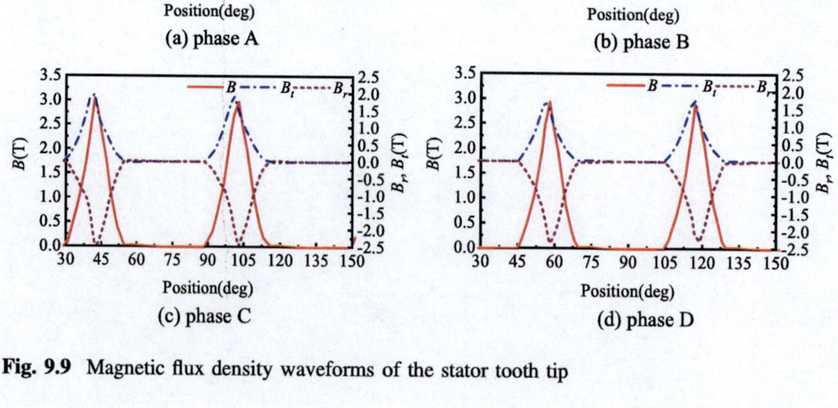

Fig. 9.9 Magnetic flux density waveforms of the stator tooth tip

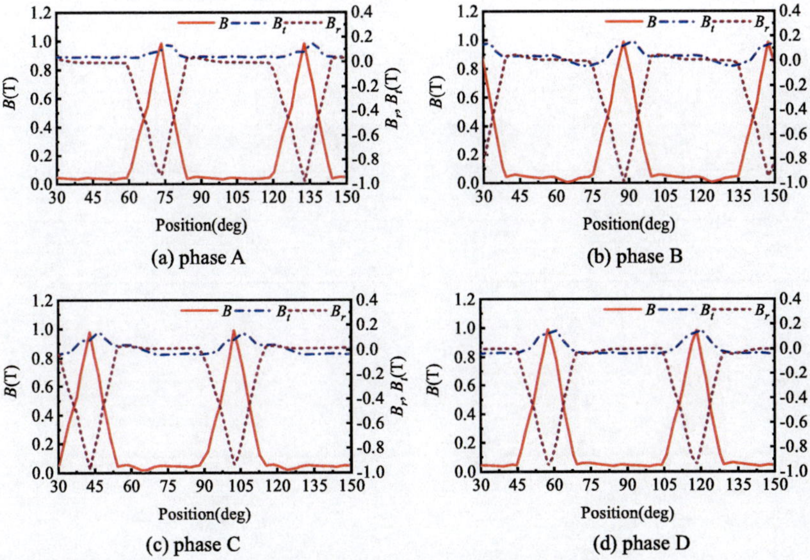

Fig. 9.10 Magnetic flux density waveforms in the middle of stator teeth

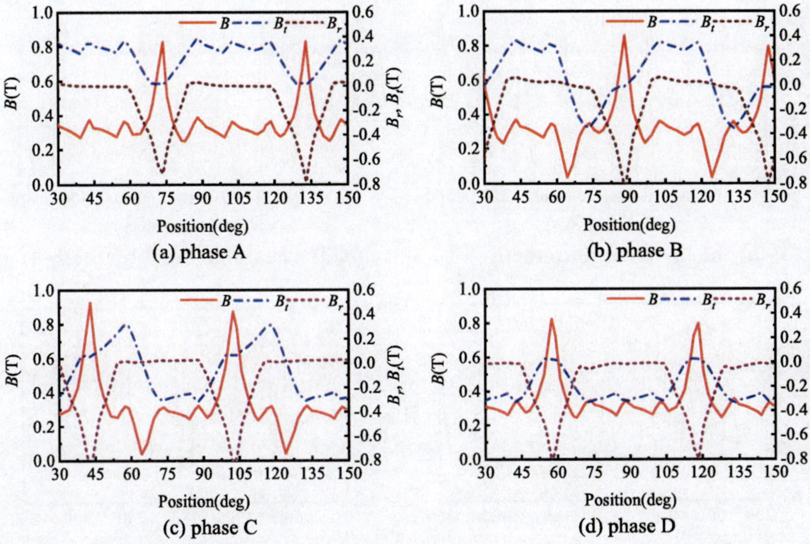

Fig. 9.11 Magnetic flux density waveforms of stator tooth root

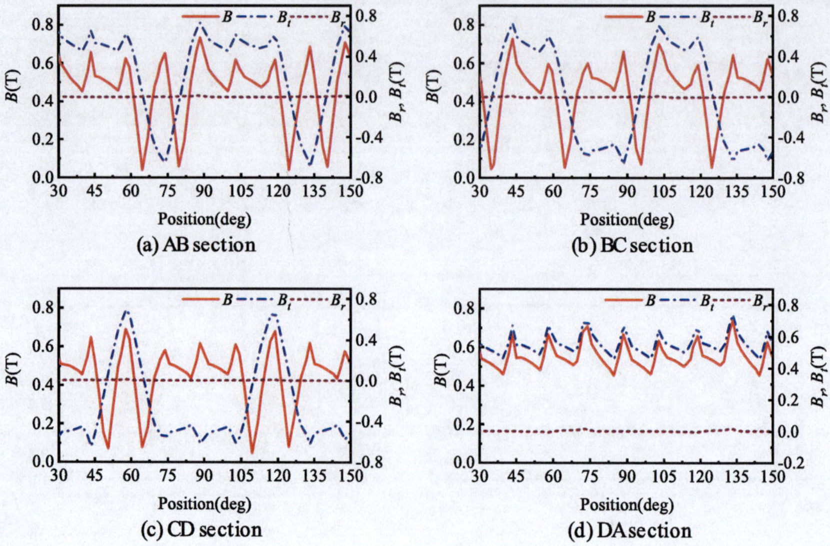

Fig. 9.12 Magnetic flux density waveforms in stator yoke

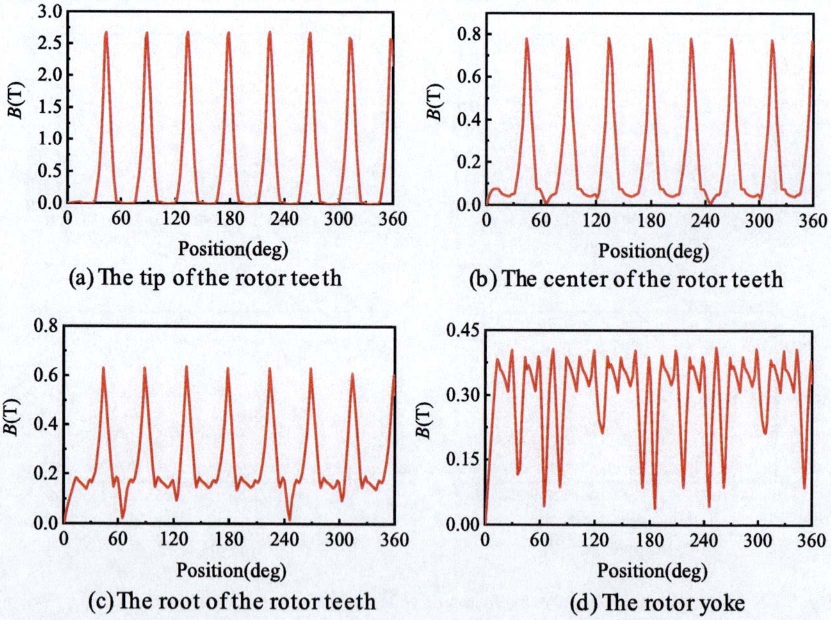

Fig. 9.13 Magnetic flux density waveforms in different areas of the rotor

because the windings of the motor are wound in the manner of NNNN–SSSS, such that they have the same magnetic field polarity.

The magnetic flux density waveforms of each segment of the stator yoke are shown in Fig. 9.12. Analyzing the simulation results of Fig. 9.12, it can be seen that there are great differences among the waveforms of the magnetic flux density of each segment of the stator yoke as a whole. In the AB section, BC section, CD section, and DA section of the stator yoke, the change period of the magnetic density component is consistent with the stator teeth and corresponds to a rotor pole pitch of 60 degrees. From the component point of view, each segment is dominated by the tangential component B_t and almost no radial component B_r. From the waveform point of view, the composite magnetic flux density waveforms of the AB segment and the CD are the same. This is because the tangential components of the AB segment and the CD have the same values, although the directions are opposite. The magnetic flux density components of the AB section, BC section, and CD section in the stator yoke have both positive and negative half cycles, which indicates that the magnetic flux density direction will change. The magnetic flux density component of the CD segment is only a positive half cycle, indicating that the magnetic density of the DA segment is only along one direction.

The magnetic flux density waveforms in different areas of the rotor are shown in Fig. 9.13. The change period of the magnetic flux density in the tooth tip, tooth center, tooth root, and rotor yoke of the rotor is 180 degrees, which is due to the symmetrical distribution of each phase winding of the prototype. The magnetic flux density gradually increases from tooth root to tooth tip, which is consistent with the previous analysis of the larger magnetic density at the stator tooth tip.

9.3 Calculation and Analysis of the Losses for SRM

9.3.1 Overview of Iron Loss

When SRM runs in a constant magnetic circuit, its stator and rotor cores do not produce power loss. However, when an SRM runs in an alternating magnetic circuit, its stator and rotor cores will cause power loss, such that it is called iron loss. The losses due to the action of hysteresis, eddy current, and additional parts are called hysteresis loss P_h, eddy current loss P_e, and additional loss P_c, respectively, and these losses jointly constitute iron loss.

The specific model calculation equation is expressed as follows.

$$P_{Fe} = P_e + P_h + P_c \tag{9.2}$$

where P_{Fe} denotes the total iron loss per unit volume (W/m^3); P_e means the eddy current loss per unit volume (W/m^3); P_h represents the hysteresis loss per unit volume (W/m^3); P_c is the additional loss per unit volume (W/m^3). In the low-frequency magnetic field, the core loss is mainly composed of eddy current loss P_e and hysteresis loss P_h. Therefore, in the process of calculating the core loss, the influence of the additional loss P_c on the calculation of the total core loss P_{Fe} is generally ignored.

9.3.1.1 Eddy Current Loss P_e

According to the basic principle of electromagnetic induction, the loop-shaped closed current line formed by the induced current flowing as a vortex in a plane perpendicular to the magnetic flux is called the eddy current. The eddy current will produce a magnetic effect in the SRM core, resulting in uneven magnetic flux distribution. In addition, when the eddy current flows in the core of the SRM, the resistance will hinder the eddy current flow. As a result, the Joule heating effect is generated, and this energy loss in the cores of the stator and rotor is called eddy current loss P_e.

Because the resistance of the monolithic iron core is very small, a strong magnetic effect and Joule heating effect will be generated under alternating magnetic flux, resulting in a large amount of eddy current loss P_e. Therefore, in electromagnetic devices, such as motor, a monolithic iron core is generally not employed. In general, the thin laminated steel sheet coated with an insulating material is used to composite the entire motor body. The purpose of adding silicon material to ferromagnetic material is to increase its resistivity. Moreover, a layer of insulating material is evenly coated between each thin silicon steel sheet. The laminated stator and rotor cores not only have very good magnetic permeability, but also greatly reduce the eddy current loss P_e of the core.

In a sinusoidal magnetic field, without considering the skin effect, Eq. (9.3) can be used to calculate the eddy current loss in the core, where, K_e denotes the eddy current loss coefficient; f means the magnetic flux frequency (Hz); B_m is the magnetic density amplitude (T). The Eq. (9.3) has been used to calculate the eddy current loss in practical engineering for a long time, which proves that the equation has high accuracy in the calculation of low frequency magnetic flux of iron core.

$$P_e = K_e(fB_m)^2 \tag{9.3}$$

Generally, K_e can be considered as a constant under the low frequency magnetic flux of the iron core. It can be calculated by Eq. (9.4), where, σ and d respectively represent the conductivity and thickness of the silicon steel sheet.

$$K_e = \frac{1}{6}d^2\pi^2\sigma \tag{9.4}$$

9.3.1.2 Magnetic Hysteresis Loss P_h

The cause of hysteresis loss can be explained by magnetic domain theory. The solid metal is composed of metal atoms, and the freely movable electrons inside the metal will flow toward the end of the high potential to form an electric current, thereby generating a magnetic field. A large number of atoms neatly arranged in the same direction of the magnetic field can form a group structure, which is called a magnetic domain. Generally, the magnetic domains in ferromagnetic materials are separated by magnetic domain walls one by one, and the magnetic field directions of two adjacent magnetic domains are not consistent. This makes them different in the direction of their magnetic moments, because the magnetic moments counteract each other such that the material does not show magnetism, as shown in Fig. 9.14a. When an external magnetic field acts on the ferromagnetic material, the internal magnetic domains will change direction or flip. And the direction can gradually become the same as the external magnetic field. At this time, the magnetic field in the magnetic material is strengthened, and magnetism begins to appear externally. This process of change described above is magnetization. Furthermore, the magnetic material reaches magnetic saturation when the direction of the external magnetic field is the same as that of the magnetic domains, as shown in Fig. 9.14b. When the SRM is in operation, the motor core is repeatedly magnetized in the constantly changing magnetic field. The direction of the magnetic domains in the core material will constantly change, and the energy loss caused by a large amount of friction among the magnetic domains is called hysteresis loss P_h.

During the magnetization process of the core materials of the stator and rotor of SRM, the relationship curve between the magnetic field intensity H and the magnetic induction intensity B presents a hysteresis loop shape. The hysteresis loss P_h

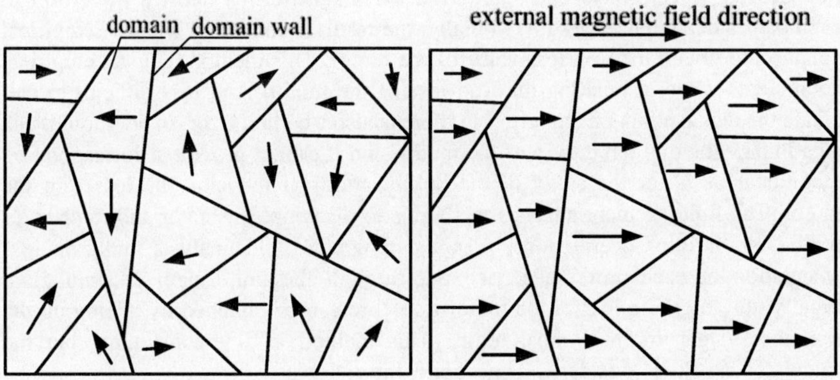

(a) Before magnetization (b) Magnetic saturation reached after magnetization

Fig. 9.14 Magnetization process of magnetic materials

can be expressed by the area enclosed by the hysteresis loop. Then, the hysteresis loss is finally converted into heat energy, which promotes the increase of the temperature of the iron core of the motor. Generally, silicon steel sheets are used as the core material in the manufacture of electromagnetic equipment. This is because silicon steel sheets are soft magnetic materials, and the hysteresis loop of soft magnetic materials is relatively narrow. Moreover, the use of silicon steel sheets can increase the maximum permeability, reduce the coercive force and thereby reduce the hysteresis loss P_h of the core in the SRM.

The hysteresis loss P_h of the motor core has a great relationship with its material and the magnetic field flux, which can be expressed by Eq. (9.5), where, K_h is the core hysteresis loss coefficient, which is related to the manufacturing process, the properties of the core material, and the structure of the electromagnetic equipment. Generally, the value range of α is [1.6, 2.2], and the value can be selected as 1.6 when $B_m < 1$ T and 2 when $B_m > 1$ T.

$$P_h = K_h f B_m^{\alpha} \tag{9.5}$$

9.3.2 Calculation Method of Iron Loss

The methods of calculating SRM iron loss mainly include experimental method, magnetic circuit analysis method, and finite element analysis method. The experimental method is a simple method to calculate the iron loss of the motor, which is also the earliest measurement method. However, this method has a large error, such that it cannot accurately analyze the iron loss of the motor. The magnetic circuit analysis method is to establish the magnetic circuit balance equation for each part of the motor, then to calculate the average magnetic flux density or average magnetic flux density waveform of each part. And the magnetic flux density waveform of each part is decomposed by FFT. Finally, the result is combined with the empirical equation to obtain the iron loss value of the motor. This method requires empirical coefficients when calculating the iron loss of the motor. And it is difficult to calculate the leakage flux effectively, and there is also a certain error. In addition, it can also impact the optimal design of the motor, and it cannot provide a corresponding theoretical basis for the study of the optimization for reducing the losses of the motor. The finite element analysis method is to discretize the stator and rotor cores of the motor into several units, then the magnetic field analysis and iron loss calculation of each unit are carried out through the finite element simulation. Finally, the total iron loss of the motor iron core can be obtained by adding up the iron loss values of these small units. This method is time-consuming, but the calculated result is considered as accurate and reliable.

After comprehensive comparative analysis, this chapter uses finite element analysis to analyze the iron loss of the SRM. The magnetic field distribution of the SRM is analyzed by using ANSYS Maxwell 2D transient analysis. On this basis, the iron loss of the motor can be calculated by combining with typical iron loss calculation equations. The typical iron loss calculation methods normally include traditional calculation methods, dual frequency method, and ellipse method, which are introduced in detail below.

9.3.2.1 Experimental Method

The magnetic circuit analysis method or finite element method should be used firstly to obtain the magnetic flux density waveforms of the stator and rotor core. And then FFT decomposition is performed to obtain the amplitude and frequency of the magnetic flux density harmonics in the stator and rotor cores of the motor. Furthermore, the experimental methods are used in a sinusoidal magnetic field with different amplitudes and frequencies, and the eddy current loss per unit volume P_{ei} and the hysteresis loss per unit volume P_{hi} of the core of the SRM are measured. Finally, the total iron loss of the motor core is obtained by summing all the units. The experimental method can be carried out by means of combination with Eq. (9.6).

$$\left.\begin{aligned} P_e &= \sum_{i=1}^{N}\left\{\sum_{k=1}^{n} Pei(B_k,f_k) \times V_i\right\} \\ P_h &= \sum_{i=1}^{N}\left\{\sum_{k=1}^{n} P_{hi}(B_k,f_k) \times V_i\right\} \end{aligned}\right\} \tag{9.6}$$

where P_e and P_h respectively represent the total eddy current loss and total hysteresis loss of the iron core (W), P_{ei} and P_{hi} respectively mean the eddy current loss and hysteresis loss per volume of the i-th element in the core elements (W/m^3), B_k refers to the magnetic flux density amplitude of the k-th harmonic (T), f_k represents the k-th harmonic frequency (Hz), V_i represents the volume of the i-th element (m^3), N denotes the total number of iron core elements of the SRM model after discretization, n is the highest harmonic order.

If the influence of harmonics is not considered and only the fundamental wave of the magnetic flux density waveform of the motor core is analyzed, the eddy current loss per volume P_{ei} and hysteresis loss per volume P_{hi} of the motor core at the fundamental frequency and the peak magnetic flux density can be measured through experiments. Then, the total eddy current loss P_e and total hysteresis loss P_h of the motor at this time can be calculated by the following Eq. (9.7).

$$\left.\begin{aligned} P_e &= \sum_{i=1}^{N}\{P_{ei}(B_{\max},f) \times V_i\} \\ P_h &= \sum_{i=1}^{N}\{P_{hi}(B_{\max},f) \times V_i\} \end{aligned}\right\} \tag{9.7}$$

where B_{max} represents the peak value of the magnetic flux density (T), and f means the frequency of the fundamental magnetic flux density (Hz). This experimental method has a small workload because it only needs to calculate the core loss per volume of a sinusoidal magnetic field with different amplitudes at the fundamental frequency. But its accuracy and precision are lower than the experimental method based on FFT decomposition and transformation, because the influence of the magnetic flux density harmonic component is ignored. However, no matter which experimental method is used, the eddy current loss per volume P_{ei} and hysteresis loss per volume P_{hi} of the motor core in different sinusoidal magnetic fields must be measured through experiments. On the other hand, it is very difficult to accurately measure P_{ei} and P_{hi}, such that the experimental methods are rarely used during the calculation of iron loss for actual motors.

9.3.2.2 Traditional Calculation Method

In the area of low-frequency magnetic field, the core loss in a sinusoidal alternating magnetic field can be obtained by means of combining Eqs. (9.3) and (9.5), which is expressed as Eq. (9.8).

$$P_{Fe} = P_e + P_h = K_e(fB_m)^2 + K_h f B_m^{\alpha} \tag{9.8}$$

However, considering the influence of harmonics on core loss, the Eq. (9.8) cannot be used directly to calculate motor core loss, because the magnetic flux density waveform in the core of the SRM is non-sinusoidal. In addition, ferromagnetic materials have nonlinear characteristics, the iron loss cannot be directly obtained by the superposition principle. However, the magnetic flux density waveforms of each element of the SRM core can be obtained through finite element simulation. The hysteresis loss correction coefficient C_h and the eddy current loss correction coefficient C_e of the stator and rotor cores can be used to correct P_h and P_e respectively. Therefore, C_h and C_e can be expressed by Eqs. (9.9) and (9.10), respectively.

$$C_h = 1 + \frac{b}{B_p}\sum_{k=1}^{N} \Delta B_k \tag{9.9}$$

$$Ce = \left(\frac{B1}{Bp}\right)^2 \sum_{k=1}^{n} \left(\frac{kBk}{Bp}\right)^2 \tag{9.10}$$

where B_k is the amplitude of the fundamental magnetic flux density (T); B_p denotes the peak magnetic flux density (T); k refers to the number of harmonics; N means the number of pulsations generated by the magnetic flux density waveform in one cycle; n represents the highest order of all harmonics; b means a constant with a value range of (0.6, 0.7); ΔB_k is the peak value of the magnetic flux density of the kth harmonic.

After introducing the hysteresis loss correction coefficient C_h and the eddy current loss correction coefficient C_e, the iron loss calculation equation under non-sinusoidal magnetic flux density is expressed as Eq. (9.11).

$$P_{Fe} = C_e K_e f^2 \sum_{k=1}^{n} k^2 B_m^2 + C_h K_h f \sum_{k=1}^{n} k B_m^\alpha \tag{9.11}$$

9.3.2.3 Dual Frequency Method

Through the traditional calculation method mentioned above, it can be seen that the core eddy current loss P_e is proportional to the square of the magnetic field frequency f, and the core hysteresis loss P_h is directly proportional to the magnetic field frequency f. Therefore, when the magnetic flux density of the motor is B, the core loss P_{Fe} per volume of the core can be expressed by Eq. (9.12).

$$P_{Fe} = a(B)f + b(B)f^2 \tag{9.12}$$

where $a(B) = K_h B^\alpha$ and $b(B) = K_e B^2$ are constants independent of the frequency of the magnetic field. Therefore, when the core magnetic flux density B is constant, Eq. (9.12) can be transformed into Eq. (9.13).

$$\frac{P_{Fe}}{f} = a(B) + b(B)f \tag{9.13}$$

Substituting the iron loss P_{Fe} at two different magnetic field frequencies into Eq. (9.13), the constants $a(B)$ and $b(B)$ can be calculated. Then, the Fourier series decomposition is performed on the magnetic flux density waveform. As a result, the eddy current loss P_e and the hysteresis loss P_h are obtained by applying Eq. (9.12).

The iron loss calculation in the traditional methods requires the eddy current loss coefficient K_e and the hysteresis loss coefficient K_h. However, manufacturers of motor materials generally do not provide the values of these two coefficients. It is also difficult to inquire the values of these two coefficients from the design manual, and the assembly process of the motor can also affect these two coefficients. This increases the difficulty of calculating the motor core loss. However, the dual frequency method can obtain the eddy current loss P_e and the hysteresis loss P_h of the motor core through analysis without knowing the eddy current loss coefficient and hysteresis loss coefficient.

9.3.2.4 Ellipse Method

Under the operating conditions of the motor, the motor core contains both an alternating magnetic field and a rotating magnetic field. Between them, the amplitude of the magnetic flux density of the alternating magnetic fields can change, but their directions are the same. On the contrary, the rotating magnetic fields does not change their magnetic flux density amplitude, while their directions are different.

The core loss caused by any magnetic field in the silicon steel sheet of motor core can be decomposed into the core losses generated by two orthogonal alternating magnetic fields. Therefore, for the core loss of the SRM, its magnetic field can be orthogonally decomposed into a radial magnetic field and a tangential magnetic field. The losses caused by them are called radial component magnetic core loss and tangential magnetic component core loss, which are then separated into eddy current loss P_e and hysteresis loss P_h for calculation.

According to the principle of harmonic analysis and the principle of orthogonal equivalence, the vector trace of the k-th harmonic magnetic flux density can be drawn. As shown in Fig. 9.15, $B_{k\max}$ and $B_{k\min}$ in the figure are the k-th harmonic amplitudes of the radial and tangential magnetic density components, respectively.

According to the ellipse method, the core loss density of SRM can be calculated by Eqs. (9.14) and (9.15).

$$P_{ei} = K_e f^2 \sum_{k=1}^{n} k^2 \left(B_{k\max}^2 + B_{k\min}^2\right) \tag{9.14}$$

$$P_{hi} = K_h f \sum_{k=1}^{n} k \left(B_{k\max}^{\alpha} + B_{k\min}^{\alpha}\right) \tag{9.15}$$

where P_{ei} and P_{hi} respectively denote the eddy current loss and hysteresis loss per volume of the i-th split element (W/m^3); f represents the core magnetic flux density

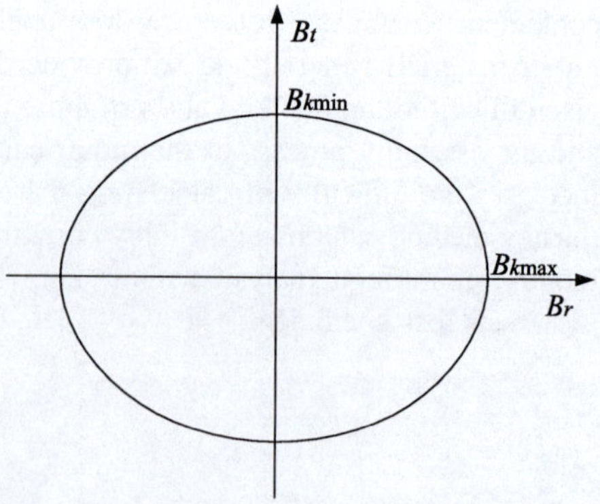

Fig. 9.15 The elliptical trajectory of the k-th harmonic magnetic flux density vector

frequency (Hz), which is related to the speed of the motor and the number of poles of the stator and rotor; n represents the highest harmonic order; $B_{k\max}$ and $B_{k\min}$ respectively mean the k-th harmonic amplitudes of the radial and tangential magnetic density components (T); α generally takes 2 when calculating.

Therefore, the total iron loss of SRM can be expressed as follows.

$$P_{Fe} = \sum_{i=1}^{N} (P_{ei} + P_{hi}) V_i \tag{9.16}$$

where N represents the total number of division elements after the core is discrete.

9.3.3 Calculation and Distribution of Iron Loss

The traditional equation, Fourier transform, and empirical correction coefficient are widely used to calculate iron loss jointly. However, it is not easy to obtain the empirical correction coefficient in the calculation process. Therefore, this chapter uses finite element analysis to analyze the iron loss of the stator and rotor of the SRM. Based on the results of the magnetic field analysis, the ellipse method and the iron loss calculation equation are used to calculate the iron loss.

9.3.3.1 Calculation of Magnetic Flux Density Frequency of Motor Core

The iron loss of the motor is related to the frequency of the magnetic flux density of the motor core, such that the magnetic density frequency of the SRM needs to be calculated. A magnetic field is generated by the excitation of the stator winding, such that the stator flux density frequency f_s is related to the switch frequency. In one electrical cycle, the four-phase windings are cyclically conducted. When the rotor rotates through a rotor pitch angle θ_r, the magnetic flux density of the stator also completes a cycle. Therefore, the changing frequency of the stator flux density can be expressed by Eq. (9.17).

$$f_s = \frac{1}{T_s} = \frac{\omega}{\theta_r} = \frac{nZ_r}{60} \tag{9.17}$$

where f_s represents the stator flux density frequency (Hz), ω refers to the angular velocity (rad/s), θ_r represents the pitch angle (deg.), T_s denotes an electrical cycle (s), Z_r represents the number of rotor poles, n represents the speed of the SRM (r/min).

The magnetic density frequency of the rotor is related to many factors, and the magnetic density frequency is expressed by the Eq. (9.18).

$$f_r = \frac{KnZ_s}{60} \tag{9.18}$$

where f_r means the magnetic density frequency of the rotor, and the unit is Hz, K refers to the number of pole pairs of the motor field, Z_s represents the number of stator poles.

9.3.3.2 Finite Element Analysis of the Iron Loss of SRM

In this chapter, ANSYS Maxwell software is used to simulate the transient field of the SRM to observe the iron loss distribution in the motor core, and the Bertotti's iron loss separation model is used to calculate the iron loss. The core loss material properties are described by selecting the material properties in the material library or inputting the core loss characteristic curve of the used core material. By fitting the iron loss characteristic curve P_{Fe} (B, f) of the core material, the loss coefficients K_e and K_h of eddy current and hysteresis are obtained. Then combined with the ellipse method, the core loss of SRM can be calculated.

The core of the SRM prototype uses 50W470 silicon steel sheet. When the motor speed is 1500 r/min, the simulated iron loss curve varying with the rotor position angle is shown in Fig. 9.16.

From the iron loss curve in Fig. 9.16, it can be seen that the change period of iron loss is 60 degrees, corresponding to a rotor pole pitch. Since the motor generally changes excited phases in every 15 degrees, there is a peak value every 15 degrees in the iron loss curve, corresponding to the phase changing cycle. By comparing to the winding current curves in Fig. 9.2, it can be found that phase A is excited with a obvious peak in each electrical cycle. This is because the motor uses the NNNN–SSSS winding distribution. When the excited phase is changed from phase D to phase A, the magnetic pole changes from S pole to N pole, such that the

Fig. 9.16 Iron loss curve of the SRM

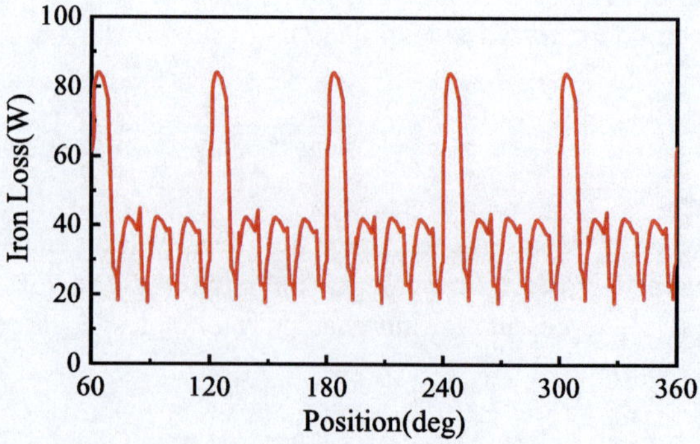

magnetic field can change. The direction of a large amount of magnetic flux changes instantaneously, such that the direction of the magnetic domain of the motor yoke changes accordingly, resulting in an increase in iron loss.

The iron loss density distribution cloud diagram of the motor during the commutation process is shown in Fig. 9.17. Analyzing Fig. 9.17, it can be seen that the iron loss is not uniformly distributed in the body, and its distribution is roughly the same as the magnetic density distribution in Figs. 9.5, 9.6 and 9.7. It is mainly distributed in the stator yoke and the tooth tips of the stator and rotor. And the stator tooth tip and rotor tooth tip are the positions with the highest iron loss density, which is also the same as the magnetic density saturation position analyzed above. This is because the iron loss is an exponential function of the peak magnetic density. It can be seen from the magnetic density distribution cloud diagram in Fig. 9.17 that the iron loss distribution in the stator is relatively uneven and mainly concentrated in the stator yoke and stator tooth tip. However, the iron loss distribution in the rotor is more even and the iron loss density at the rotor tooth tip is relatively large. The iron loss density at the tip of the stator and rotor is relatively large, because the motor always follows the principle of minimum magnetic reluctance. The greater iron loss density at the stator yoke is due to the fact that when the motor is in commutation operation, the magnetic pole changes from S pole to N pole. The magnetic induction line in the yoke between the two phases can change direction with 180 degrees. As a result, the direction of the magnetic domain in the motor yoke is also changed accordingly, which results in a larger iron loss density value.

9.3.3.3 Ellipse Calculation Method for Iron Loss

When using the ellipse method, the magnetic field information of each element of the motor core is required. In the previous section, ANSYS Maxwell was used to perform finite element analysis to obtain the magnetic density information of each typical point of the motor. Figure 9.18 shows the elliptical magnetic field diagram of the stator tooth tip, stator tooth center, and stator tooth root of the SRM, where the radial component B_r of the magnetic flux density in the motor core is selected as the ordinate, and the tangential component B_t is determined as the abscissa.

The vector from point (0, 0) in the figure to a certain point in the figure represents the composite magnetic density vector. It can be seen from the figure that the size and direction of the composite magnetic flux density vector are constantly changing. An alternating magnetic field and a rotating magnetic field exist in the core of the SRM prototype at the same time, and the motor core will produce a certain loss under the action of the magnetic field. The ellipse method can be used to convert the loss caused by the complex synthetic magnetic field into the core loss of the radial and tangential magnetic field components.

Figures 9.19, 9.20 and 9.21 are the FFT decomposition diagrams of the radial component B_r and tangential component B_t of the magnetic flux density in the tooth tip, tooth center, and tooth root elements of the stator tooth. It can be seen in the

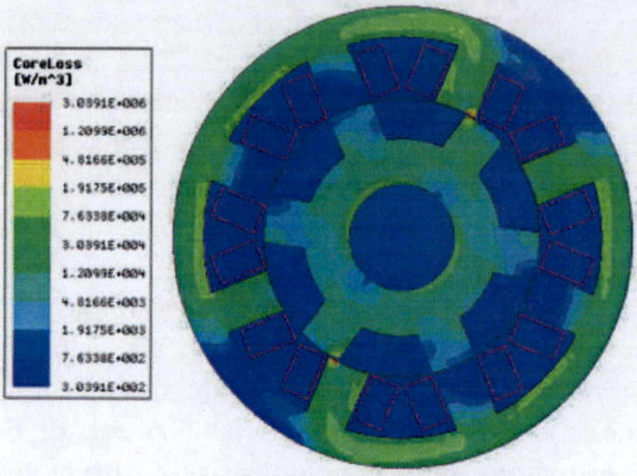

(a) phase D is excited

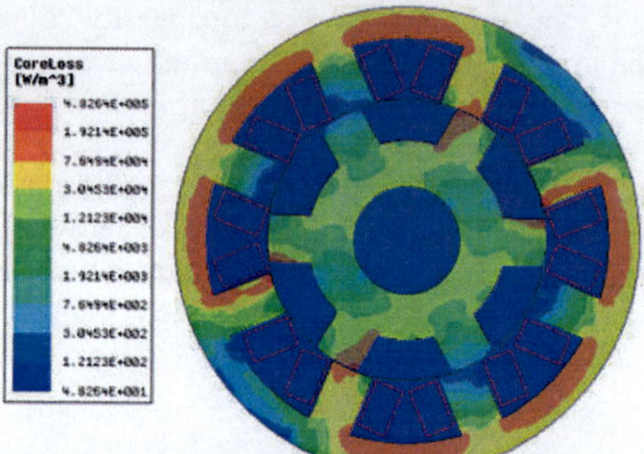

(b) Phase A starts to be excited with freewheeling phase D

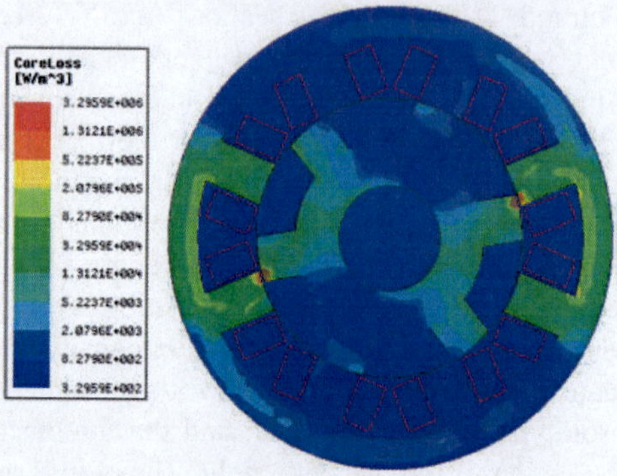

(c) Phase A is excited

Fig. 9.17 The iron loss density distribution diagram of the motor during the commutation process

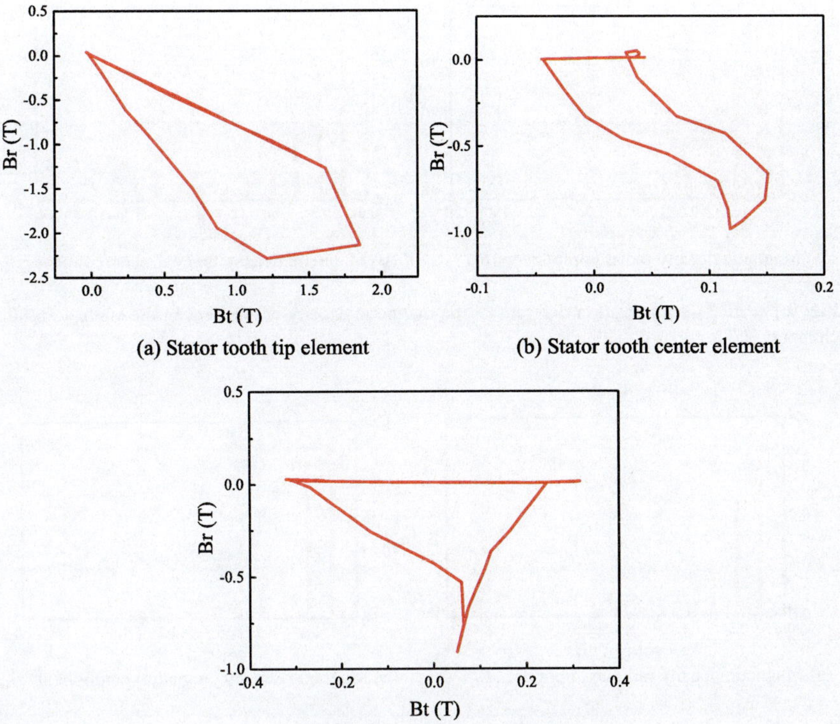

(a) Stator tooth tip element

(b) Stator tooth center element

(c) Stator tooth root element

Fig. 9.18 The elliptical magnetic field diagram of each typical element in the stator tooth

figure that the change frequency of the magnetic flux density of the stator is around 150 Hz, and the magnetic density contains non-negligible harmonic components.

Table 9.1 shows the frequency and amplitude of each harmonic in each typical element of the stator teeth. It can be seen from Figs. 3.19, 3.20 and 3.21 that the harmonic components above the 10th order are very small, such that only the first 10th harmonic components are counted. Among them, $B_{k\max}$ and $B_{k\min}$ respectively denote the k-th harmonic amplitudes of the radial and tangential components B_r and B_t of the magnetic flux density.

According to the amplitude of each harmonic component after FFT decomposition for the magnetic flux density of stator teeth in typical elements shown in Table 9.1, the elliptical magnetic flux density vector diagram of the k-th harmonic for the typical elements in the stator teeth is illustrated, when ignoring the initial phase angle difference of each vector, and taking the k-th harmonic amplitudes $B_{k\max}$ and $B_{k\min}$ of the radial component B_r and the tangential component B_t of the magnetic flux density as the major axis and the minor axis, which is shown in Fig. 9.22.

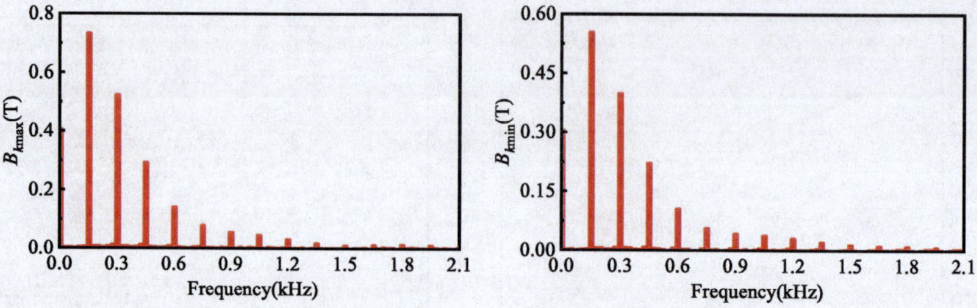

(a) Magnetic density radial component B_r (b) Magnetic density tangential component B_t

Fig. 9.19 FFT decomposition diagram of the magnetic density component in the stator tooth tip element

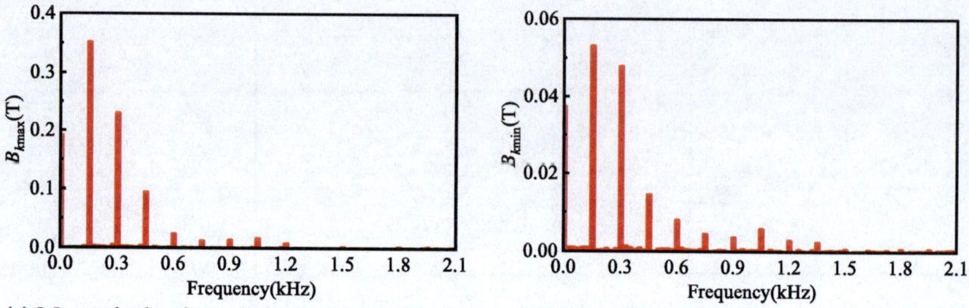

(a) Magnetic density radial component B_r (b) Magnetic density tangential component B_t

Fig. 9.20 The FFT decomposition diagram of the magnetic density component in the stator tooth center

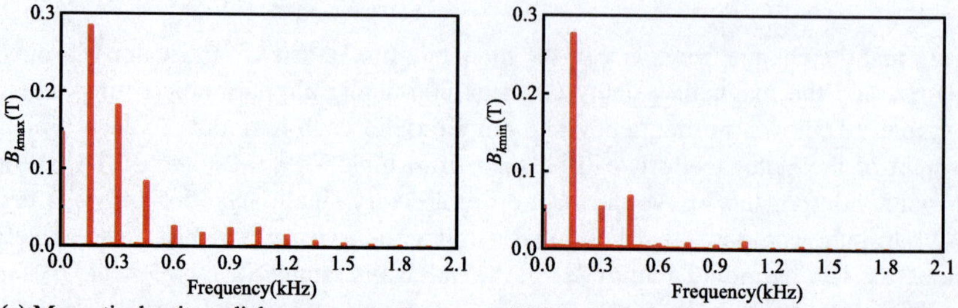

(a) Magnetic density radial component B_r (b) Magnetic density tangential component B_t

Fig. 9.21 FFT decomposition diagram of the magnetic density component in the stator tooth root

With reference to the ellipse method described above, the core loss of each part of the SRM can be obtained as shown in Table 9.2, according to the orthogonal equivalent principle of core loss and Eqs. (9.14), (9.15), and (9.16).

According to the iron loss data of each part in the stator and rotor cores of the motor, shown in Table 9.1, it can be seen that the iron loss of the stator accounts for

Table 9.1 Frequency and amplitude of each harmonic component of the typical elements in stator teeth

Harmonic order k	Frequency f (kHz)	Stator tooth tip		Stator tooth center		Stator tooth root	
		$B_{k\max}$ (T)	$B_{k\min}$ (T)	$B_{k\max}$ (T)	$B_{k\min}$ (T)	$B_{k\max}$ (T)	$B_{k\min}$ (T)
1	0.15	0.7314	0.5533	0.3522	0.0532	0.2852	0.2778
2	0.30	0.5232	0.3984	0.2303	0.0479	0.1828	0.0554
3	0.45	0.292	0.221	0.0954	0.0148	0.0843	0.0697
4	0.60	0.141	0.1058	0.0241	0.0083	0.0265	0.0078
5	0.75	0.0781	0.0568	0.0127	0.0047	0.0181	0.0094
6	0.90	0.0553	0.0429	0.0142	0.0038	0.0244	0.0064
7	1.05	0.0447	0.0384	0.0173	0.006	0.0251	0.0113
8	1.20	0.0309	0.031	0.0089	0.003	0.0161	0.0033
9	1.35	0.0171	0.0217	0.0012	0.0026	0.0088	0.0048
10	1.50	0.01	0.0147	0.0028	0.0007	0.006	0.0027

about 72.2% of the total iron loss. Among them, the iron loss of the stator yoke accounted for 46.9%, and the stator teeth accounted for 25.3%. The iron loss value of the rotor accounts for 27.8% of the total iron loss, in which the rotor yoke accounts for 13.6% and the rotor teeth accounts for 14.2%. Combining the data in the table with the previous finite element analysis results, the loss in each part of the prototype iron core is different. Because, the magnetic fields acting on the stator and rotor cores of the motor are completely different, resulting in different magnetic density distributions and different losses. In the data, the stator has the largest iron loss value. Among them, the iron loss value of the stator yoke is the largest, accounting for the largest proportion, which is also consistent with the previous finite element analysis.

9.3.4 Calculation of Copper Loss

The main heat sources of the SRM are iron loss and copper loss. Copper loss is the loss generated when current flows through the windings. Therefore, it is mainly related to the phase current. If the skin effect on the surface of the winding coil is ignored, the copper loss can be calculated by Eq. (9.19).

$$P_{Cu} = mI_{rms}^2 R_m \tag{9.19}$$

where m denotes the number of phases; I_{rms} means the effective value of the phase current (A); R_m is the resistance of the phase winding (Ω).

Fig. 9.22 The elliptical magnetic flux density vector diagram of the *k*-th harmonic for each typical element in the stator tooth

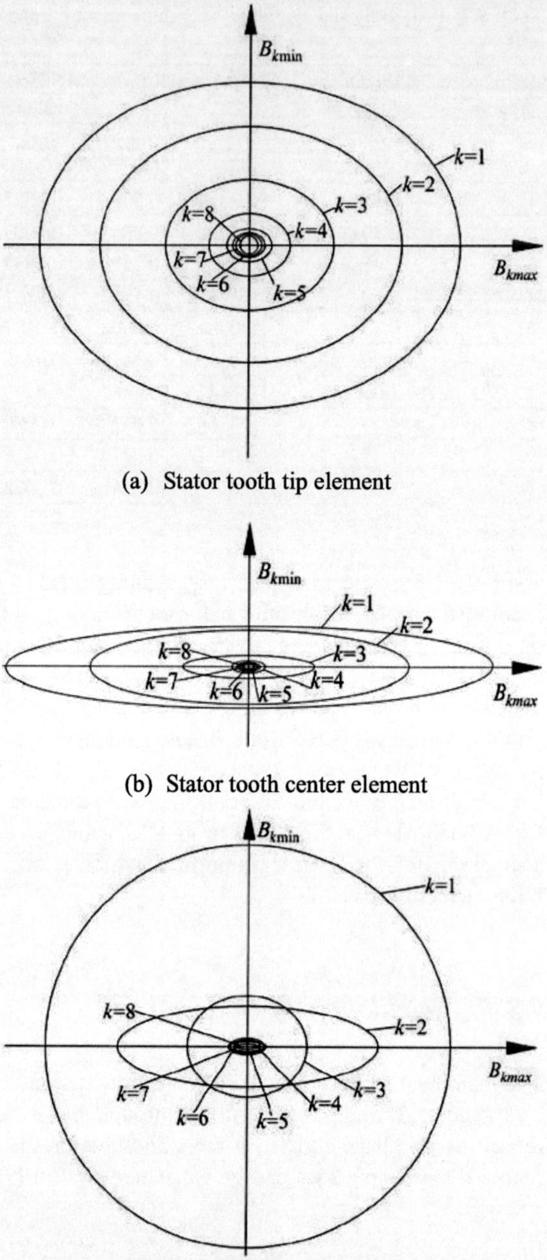

(a) Stator tooth tip element

(b) Stator tooth center element

(c) Stator tooth root element

Table 9.2 Iron loss values of each part of the SRM

Core parts	Stator teeth	Stator yoke	Rotor teeth	Rotor yoke
Iron loss (W)	13.562	25.186	7.644	7.291
Percentage	25.3%	46.9%	14.2%	13.6%

The current waveforms under different control methods will be different with a large extent. If the waveform of the phase current is known, its effective value can be calculated by Eq. (9.20).

$$I_{rms} = \sqrt{\frac{Z_r}{2\pi}\int_{\theta_{on}}^{\theta_p} i^2(\theta)d\theta} \tag{9.20}$$

where Z_r refers to the number of the rotor teeth; θ_p represents the position of the rotor when the phase winding current freewheeling ends; θ_{on} means the opening angle.

In this chapter, the transient simulation is carried out in Maxwell, and the current waveforms of each phase when the prototype speed is 1500 r/min are shown in Fig. 9.9. After processing, the effective value of the phase current is 50.13 A. The internal resistance of the SRM is 0.011 Ω, and the number of phases is 4. Bring these values into Eq. (9.19), then the winding copper loss can be calculated to be 110.572 W.

9.4 Basic Theory of Thermal Analysis of SRM

9.4.1 Heat Transfer Theory

According to the principle of heat transfer, heat transfer can be divided into heat conduction, heat convection, and heat radiation.

9.4.1.1 Heat Conduction

Heat conduction can be considered as the process of internal energy exchange caused by the temperature differences between the two contact objects or the two parts of an object. There is no displacement in this process, which follows Fourier law, as shown in the Eq. (9.21).

$$q = -\lambda \frac{dT}{dx} \tag{9.21}$$

where q represents the heat flux density (W/m^2), T denotes the substance temperature (K), λ is the thermal conductivity coefficient of the material (W/(m K)), and − indicates the direction in which the heat flows to the position with low temperature. In addition, the greater the thermal conductivity coefficient is, the better the thermal conductivity performance of the material can be. Generally, the relationships of the thermal conductivity performance among different substances are $\lambda_{Metal} > \lambda_{\text{Nonmetallic solid}} > \lambda_{\text{liquid}} > \lambda_{\text{gas}}$.

9.4.1.2 Heat Convection

Heat convection is the process of internal energy exchange between the solid surface and the contacting fluid, due to the temperature step between the outer surface of the solid and the fluid. The cooling formula is described and expressed as Eq. (9.22).

$$q = h(T - T_f) \tag{9.22}$$

where q represents the heat flux density of the heat exchange surface (W/m^2); h means the convective and heat transfer coefficient, also known as the heat dissipation coefficient (W/(m^2 K)), which is related to the dynamics of the fluid, the state of flow, the physical properties, the arrangement of the solid, and the shape of the solid surface; and T and T_f are the temperature of the solid surface and the surrounding fluid (K).

9.4.1.3 Heat Radiation

Heat radiation refers to the process in which an object emits electromagnetic energy and is absorbed by other objects into heat energy. The process of heat transfer does not require a medium. The Stefan-Boltzmann equation can be used to calculate the net heat transfer, as shown in Eq. (9.23).

$$q = \varepsilon\sigma A_1 F_{12}(T_1^4 - T_2^4) \tag{9.23}$$

where q represents the heat flow rate (W/m^2); ε represents the emissivity, which is equal to the ratio of the energy radiated by the object and the black body at the same temperature and is usually between 0 and 1; σ represents the radiation coefficient of the black body, which is a constant about 5.67 × 10–8 W/(m^2 K); F_{12} is the shape factor; A_1 is the area of the radiating surface 1; and T_1, T_2 are the thermodynamic temperatures of radiating surface 1 and surface 2, respectively.

As shown in Eq. (9.23), the thermal analysis process is highly nonlinear. During the operation of SRM, the above three heat transfer processes exist at the same time. But the heat radiation transferred at low temperature is very small, and the compute is difficult due to its high nonlinearity. Therefore, the influence of heat radiation is normally not considered when analyzing the internal temperature field of the motor, but the other two heat transfer, that is, heat convection and heat conduction, are mainly considered.

9.4.1.4 Three Types of Basic Boundary Conditions in Heat Transfer

There are three classic types of boundary conditions that are commonly used in engineering, which should be selected according to the specific requirements.

(1) The first type boundary condition

When the temperature on the boundary surface of the object is given, that is, the temperature boundary condition is determined, the temperature of the object can be expressed by Eq. (9.24).

$$T|_S = T_0 \tag{9.24}$$

where T represents the temperature of the object (K), S represents the boundary surface, and T_0 is the temperature on the boundary surface (K). The temperature on the boundary surface is a constant value when conducting heat in a steady state, otherwise it is a function that changes with time.

(2) The second type boundary condition

When the heat flux value on the boundary surface of the object is given, that is, the heat flux boundary condition is determined, the temperature of the object can be expressed by formula (9.25).

$$\frac{\partial T}{\partial n}\bigg|_S = -\frac{q_0}{\lambda} \tag{9.25}$$

where T represents the temperature of the object (K), S is the boundary surface of the object, n refer to the normal vector of the boundary surface of the object, q_0 is the heat flux density through the boundary surface (W/m^2), and λ is the thermal conductivity (W/(m K)). The heat flux density is a constant value when conducting heat in a steady state, otherwise it is a function that changes with time.

(3) The third type boundary condition

When the convective heat transfer coefficient and fluid temperature between the boundary surface of the object and the surrounding fluid are given, that is, the convective heat transfer boundary condition is determined, the temperature of the object can be expressed by Eq. (9.26).

$$-\lambda\frac{\partial T}{\partial n}|_S = h(T - T_f) \tag{9.26}$$

where h is the convective heat transfer coefficient (W/(m^2 K)), and T_f denotes the temperature of the fluid (K).

9.4.1.5 Temperature Rise

Temperature rise refers to the temperature difference between two objects. During the working process of the motor, most of the energy is effectively used, while the other parts of the energy loss that is not effectively used can promote the temperature of each part in the motor to gradually increase. After a period of time, the temperature tends to stabilize. At that time, the temperature of the motor parts is higher than the surrounded air or other environment medium, which is the temperature rise of the motor parts.

The allowable temperature rise due to the different insulation grades of the motor materials is different, because the insulation of the stator windings can withstand the limited temperature. When the motor works over the allowable range of temperature rise, it will reduce the durability of the insulation layer and may cause accidents. In general, the motor failures caused by insulation damage to motor windings are the most common. One of the most important factors affecting the insulation damage of the motor winding is the temperature rise of the winding. Therefore, overheating fault detection, inspection, and protection are very important for reducing accidents and losses, and improving the economic benefits of enterprises. As a result, the motor temperature rise is an important reference index in motor design.

9.4.2 *The Mathematical Model and Boundary Conditions in Temperature Field*

According to the theory of heat transfer, the thermal analysis problem of SRM can be summarized to a boundary value problem, such that the boundary value of its thermal conductivity differential equation can be solved. The thermal analysis mathematical model of the three-dimensional temperature field can be expressed by formula (9.27).

$$\left.\begin{array}{r}\frac{\partial}{\partial x}\left(\lambda_x\frac{\partial T}{\partial x}\right)+\frac{\partial}{\partial y}\left(\lambda_y\frac{\partial T}{\partial y}\right)+\frac{\partial}{\partial z}\left(\lambda_z\frac{\partial T}{\partial z}\right)=\dot{Q}\\ -\lambda_3\frac{\partial T}{\partial n}S_3=h(T-T_f)\end{array}\right\} \tag{9.27}$$

where $\dot{Q}$ is the heat generation rate of the internal heat source (W/m^3), T represents the temperature of the object, λ_x, λ_y, and λz respectively represent the thermal conductivity in the x, y, and z directions (W/(m K)), λ_3 means the thermal conductivity perpendicular to the boundary surface (W/(m K)), h is the heat dissipation coefficient on the boundary surface (W/(m^2 K)), and T_f denotes the temperature of the fluid medium on the boundary surface (K).

9.5 Thermal Analysis Model of SRM

There are two common methods to build the thermal analysis model of SRM. One is to separately establish the stator and rotor models of SRM, and also separately perform simulation analysis on the stator and rotor of SRM. The other is to build the entire mode considering the stator and rotor of the SRM as a whole, and perform the simulation and analysis of the temperature field. After simple analysis for the two methods, they are not different in nature. This chapter adopts the second method, taking into account the influence of the temperature rises of the air inside the motor, which is more realistic.

In order to simplify the problem and facilitate the calculation, the following three assumptions are generally made for the motor when performing thermal analysis on SRM. One is that the thermal conductivity coefficient of the motor core is considered to be isotropic. The second one is that the basic material properties of all parts in the SRM are constant. Finally, the internal parts of the motor are considered to be in good contact.

9.5.1 The Finite Element Model of SRM for Thermal Analysis

This chapter adopts the finite element method in the ANSYS Workbench platform to realize the magnetic-thermal coupling analysis of SRM. A three-dimensional model of the SRM prototype with a four-phase 8/6 pole structure is established as shown in Fig. 9.23a in CATIA, which can be import into ANSYS Workbench. The model includes the stator, the rotor, the stator winding of the SRM, the air, and rotating shaft, as shown in Fig. 9.23b.

After setting up the materials for the established 3D model, the finite element mesh is performed. Generally, triangular meshes divided into regions are commonly used for planar two-dimension geometric model, which can reduce the calculation time without affecting the calculation accuracy. As for the three-dimension model, the tetrahedral mesh has a high density and a complex shape, which has long calculation time, although the accuracy is high. After comprehensive consideration, this chapter uses hexahedral element, which has the ability to describe complex shapes and can accurately describe the geometric shape of the boundary surface of the three-dimension solid. In addition, as the increase of the structure complexity, the number and density of hexahedral grids can also easily change accordingly.

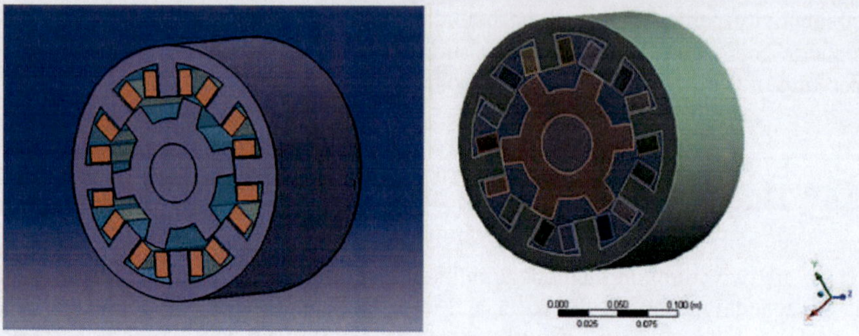

(a) SRM model in CATIA (b) SRM model in ANSYS Workbench

Fig. 9.23 Establishment of a 3D model of a four-phase 8/6 SRM

Fig. 9.24 The finite element model of the SRM for the thermal analysis

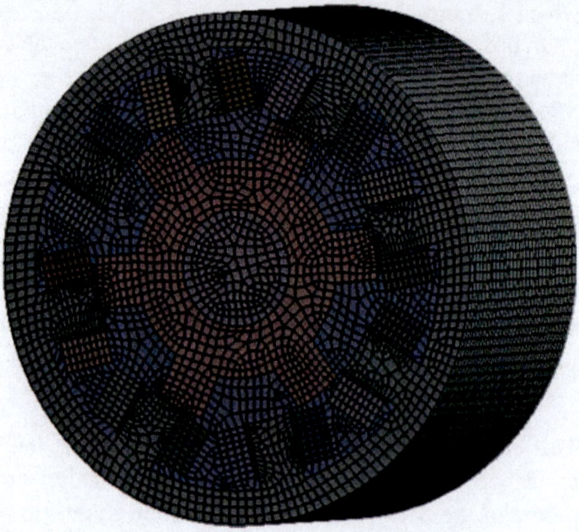

Therefore, compared to tetrahedrons, it can well adapt to structural changes, which greatly saves calculation time and has high accuracy. And The finite element model for thermal analysis of the prototype is shown in Fig. 9.24. The number of elements is 117472 and the number of nodes is 484592.

9.5.2 Determination of Thermal Conductivity

The stator and rotor of the four-phase 8/6 SRM prototype studied in this chapter are manufactured by 50W470 silicon steel sheets. The shaft is made of 45 steel, the material of the air gap is air, and the material of the stator winding includes the

Table 9.3 Thermal analysis parameters of the SRM

Components	Density (Kg/m^3)	Thermal Conductivity (W/(m·K))	Specific heat capacity (J/(kg K))
Stator and rotor	7700	22.6	442
Winding copper wire	8933	401	385
Insulation materials	900	0.26	1200
Shaft	7800	43	473
Air	1.205	0.026	1013

copper wire and the insulating material on the surface. The related parameters of these materials used for thermal analysis are shown in Table 9.3.

9.5.2.1 Equivalent Thermal Conductivity of the Air in the Motor

When SRM is running, the temperature field and fluid field in the motor are coupled with each other, which increases the difficulty of thermal analysis of SRM. In order to simplify calculation and analysis, the rotor in SRM operation is regarded as static processing. The equivalent thermal conductivity of the air gap λ_δ is introduced, and the thermal conductivity of the static fluid is used to describe the heat transfer capacity of the flowing air. As a result, the convective heat transfer in the air gap is equivalent to heat conduction.

Through calculation, the air gap Reynolds value and critical Reynolds value can be obtained, as shown in Eqs. (9.28) and (9.29).

$$\mathrm{Re} = \frac{v\delta}{\gamma} = \frac{\pi D_2 n\delta}{60\gamma} \tag{9.28}$$

$$\mathrm{Re}_{cr} = 41.2\sqrt{\frac{D_{i1}}{\delta}} \tag{9.29}$$

where v represents the circumferential linear velocity of the rotor (m/s), δ represents the length of the air gap (m), D_2 means the outer diameter of the rotor (m), D_{i1} is the inner diameter of the stator (m), n refers to the speed of the motor (r/min), and γ denotes the kinematic viscosity of the air (m^2/s), which is usually 14.8×10^{-6} m^2/s.

When Re is less than Re_{cr}, it is laminar flow. At this time, λ_δ and λ_{air} are approximately equal. When Re is greater than Re_{cr}, it is turbulent flow, such that λ_δ can be calculated by Eq. (9.30).

$$\left.\begin{array}{l} \lambda_\delta = 0.019\eta^{-2.9084}\mathrm{Re}^{0.4614 1/\ln(3.33361\eta)} \\ \eta = \frac{D_2}{D_{i1}} \end{array}\right\} \tag{9.30}$$

In this chapter, it is calculated that Re is 234.585, Re_{cr} is 784.811. Thus, Re < Re_{cr}, such that λ_δ is 0.026 W/(m K).

9.5.2.2 Equivalent Thermal Conductivity of Stator Winding

It can be seen from Table 9.3 that the stator slot of SRM contains copper wire and insulating material on its surface. Since the thermal conductivity of the insulating material is very small and the thermal conductivity performance is very poor, it is impossible to ignore its impact on the temperature field when performing thermal analysis on SRM. As a result, in order to facilitate the analysis before the temperature field analysis, the stator winding coil is often simplified, that is, the entire winding coil is equivalent to a whole. In addition, the following assumptions are always considered. (1) The copper wires in the stator slots are evenly arranged, ignoring the temperature difference between each other. (2) The insulating paint is evenly distributed on the copper wire. (3) The insulating layers are filled with impregnating varnish without gaps. (4) The temperature in the insulating layer and the filling paint changes linearly.

According to the relevant theories of heat transfer, the formula for calculating the equivalent thermal conductivity of multilayer media is shown in Eq. (9.31).

$$\lambda_{eq} = \frac{\sum_{i=1}^{n} \delta_i}{\sum_{i=1}^{n} \frac{\delta_i}{\lambda_i}} \tag{9.31}$$

where λ_{eq} is the equivalent thermal conductivity (W/(m K)), δ_i represents the thickness of the i-th layer of heat conductor (mm), and λ_i denotes the thermal conductivity of the i-th layer of heat conductor (W/(m K)). The diameter of the winding copper wire in the prototype is 1.25 mm, and the thickness of the insulating paint is 0.04 mm. Combining the relevant parameters in Table 9.3 and substituting the data into Eq. (9.28), the equivalent thermal conductivity λeq can be calculated, which is 8.212 W/m K in this chapter.

9.5.3 Determination of the Heat Source

During the operation of an SRM, the internal heat source is mainly the iron loss of the stator and rotor of the motor and the copper loss of the stator winding. According to the calculation method of copper loss and iron loss, these loss values can be obtained. According to these loss values, the heat generation rate of the stator and rotor of the motor is obtained and expressed as

$$\dot{Q} = \frac{P_{loss}}{V} \tag{9.32}$$

where $\dot{Q}$ represents the heat generation rate (W/m^3), P_{loss} denotes the iron loss value of the corresponding part (W), and V means the volume of the corresponding part (m^3).

For the stator winding coil, the calculation formula of the heat generation rate is

$$\dot{Q} = \rho J^2 \tag{9.33}$$

where ρ represents the resistivity of the stator winding, and J represents the current density in the winding coil.

The main consideration in the thermal analysis in this chapter is the iron loss and copper loss of the motor. According to the analysis of motor iron loss and copper loss described above, the heat source parameters of the SRM prototype at n = 1500 r/min is shown in Table 9.4.

9.5.4 Boundary Conditions for Thermal Analysis

The setting of boundary conditions is an important step in the thermal analysis process. The three types of basic boundary conditions in heat transfer are introduced in the chapter about the heat transfer theories. The temperature field analysis of the SRM prototype studied in this chapter only involves the third type of boundary conditions.

During the operation of the motor, through the coupling of the temperature field and the fluid field, the heat dissipation coefficient of each component surface can be accurately calculated, which is also a great problem. With a long time and a lot of practice, scholars in this related fields have summed up some empirical equations that can be used to calculate the heat dissipation coefficient of each part of the motor.

(1) The convective heat transfer coefficient h_s at the end of the stator core can be expressed by the empirical Eq. (9.34), where, v represents the circumferential linear velocity of the rotor (m/s).

$$h_s = 15 + 6.5v^{0.7} \tag{9.34}$$

Table 9.4 Heat source parameters of the prototype

Heat source	Stator teeth	Stator yoke	Rotor teeth	Rotor yoke	Winding
Heat (W)	13.562	25.186	7.644	7.291	110.572

(2) The convective heat transfer coefficient h_r at the end of the rotor core is expressed by Eq. (9.35),

$$\left.\begin{aligned} h_r &= \frac{2Nu_r\lambda_{air}}{D_2} \\ Nu_r &= 1.67\mathrm{Re}_r^{0.385} \\ \mathrm{Re}_r &= \frac{\pi D_2^2 n}{120\gamma} \end{aligned}\right\} \tag{9.35}$$

where D_2 represents the outer diameter of the motor rotor (m), λ_{air} denotes the thermal conductivity of air (W/(m K)), Nu_r represents the Nussett value of the end of the rotor core, and Re_r means the Reynolds value of the airflow on the end of the rotor core.

(3) The convective heat transfer coefficient h_c at the end of the winding can be expressed by Eq. (9.36).

$$h_c = 6.5 + 5.25v^{0.6} \tag{9.36}$$

where v represents the circumferential linear velocity of the rotor (m/s).

(4) The outer surface of the motor adopts natural convection heat exchange, and the heat dissipation system h_k on the surface of the shell can be obtained by the empirical Eq. (9.37).

$$h_k = 14(1 + 0.5\sqrt{\omega_i})\sqrt[3]{\frac{T}{25}} \tag{9.37}$$

where ω_i is the wind speed flowing through the inner wall of the shell, and T is the temperature of the shell surface (K).

After calculation by means of the parameters of the SRM in this chapter, it can be obtained that h_s = 47.395 W/(m^2 k), h_r = 41.549 W/(m^2 k), hc = 27.3 W/(m^2 k), h_k = 16.01 W/(m^2 k).

9.6 Simulation Results and Analysis of Temperature Field

A four-phase 8/6 SRM prototype is simulated and analyzed for the steady-state temperature field and the transient temperature field, where an ambient temperature is 22 °C degrees and a rotation speed is 1500 r/min.

9.6.1 Simulation Analysis Results of Steady-State Temperature Field

The established 3D model can be import into the steady-state thermal analysis module in ANSYS Workbench. Combined the solution of Maxwell electromagnetic analysis module and the steady state thermal analysis module, the magnetic-thermal coupling analysis can be performed. After setting the material and thermophysical parameters of each part, the mesh is divided, the internal thermal load is loaded, the boundary conditions are applied to the model, and finally the solver is set up to simulate the steady-state temperature field of SRM. Figure 9.25 is the simulation result of the steady-state temperature field of the prototype at an ambient temperature of 22 degrees and a rotation speed of 1500 r/min.

Figure 9.25 shows the overall steady-state temperature field cloud diagram of the motor prototype. It can be seen from the figure that the temperature of each part of the motor is different when the temperature is stable after the motor works for a period of time. The maximum temperature difference for the overall motor is 27.349 degrees. The highest temperature of the motor appears in the middle of the stator winding, and the temperature is 93.906 degrees. Long-term operation will accelerate the aging of the insulation, and lead to the motor failure or other accidents. The lowest temperature of the motor appears in the stator yoke, and the temperature is 66.557 degrees. It is obvious that the temperature gradient between the stator winding and the stator is relatively large.

Figure 9.26 is the steady-state temperature field cloud diagram of the stator winding. It can be seen from the figure that the temperature of the winding as a whole is very high. In the steady state, the highest temperature is 93.906 degrees located in the middle of the winding, while the lowest temperature is 83.11 degrees at the end of the winding. The maximum temperature difference is 10.796 degrees. The reason for the highest temperature in the middle part of the winding is that the iron loss of the stator and the coper loss of the winding during the operation of the

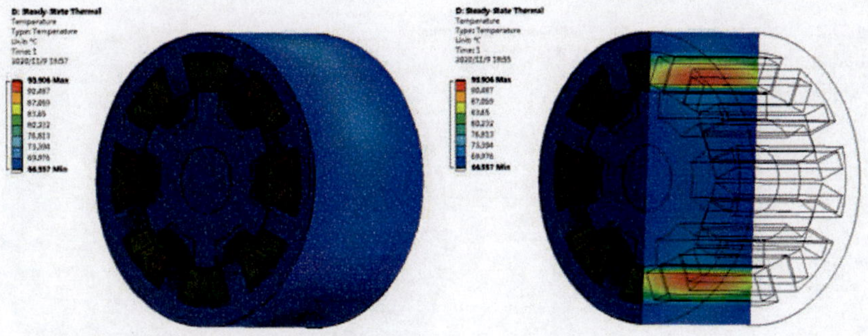

(a) Overall temperature field diagram (b) Axial temperature field diagram

Fig. 9.25 The overall steady-state temperature field of the SRM

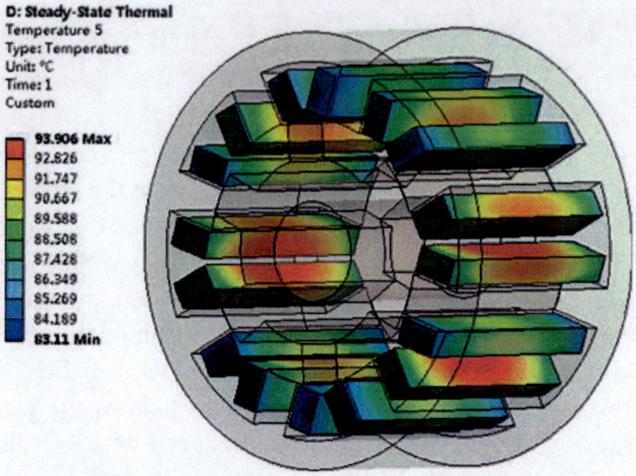

Fig. 9.26 Steady-state temperature field diagram of the SRM stator winding

SRM are combined. In addition, the thermal conductivity of the insulating layer on the winding surface is very small, and the thermal conductivity is poor. The copper wire wrapped in the insulating material and embedded in the stator slot has poor heat dissipation conditions. The air flowing in the air gap conducts convection heat exchange with the winding ends, such that the heat dissipation conditions are better at the winding ends. Therefore, the temperature of the winding ends is lower than that of the middle part of the winding.

Figures 9.27 and 9.28 are the steady-state temperature field cloud diagram of the prototype stator core and the steady-state temperature field cloud diagram of the rotor core, respectively. It can be seen from the figure that the motor generates a temperature gradient along the axial direction, and the temperature gradually decreases from the center of the stator and the rotor to the outside along the axial direction. Because, the axial temperature transfer and the heat dissipation at the ends of the stator and rotor are taken into account when the thermal analysis of the SRM prototype is performed.

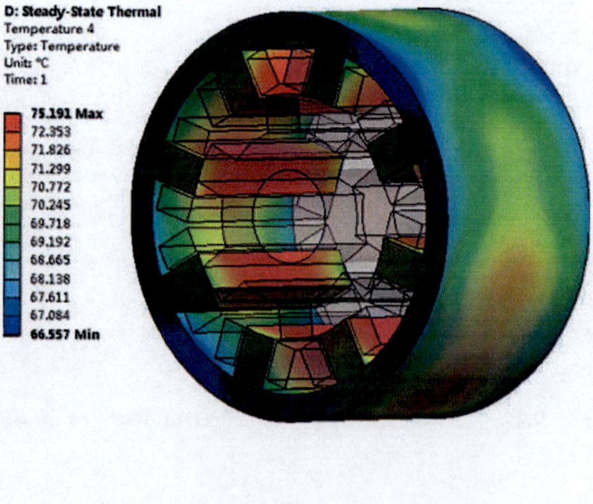

Fig. 9.27 Steady-state temperature field cloud diagram of SRM stator core

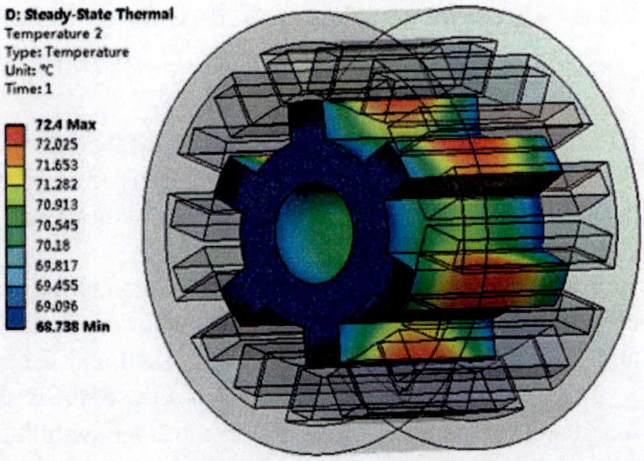

Fig. 9.28 Steady-state temperature field cloud diagram of SRM rotor core

It can be seen from Fig. 9.27 that the overall temperature distribution of the stator core is relatively uniform. In the steady state, the position with the highest temperature is located at the stator teeth, reaching 79.191 degrees, and the position with the lowest temperature is at the stator yoke, with a temperature of 66.557 degrees, and the maximum temperature difference of the stator core is 8.634 degrees. Since the heat is mainly in the radial direction, the heat energy is dissipated to the outside by natural convection heat exchange along the outer surface of the stator yoke, and the temperature of the stator yoke is lower. The winding is connected to the stator slot, and the lost heat of the winding is transferred to the stator teeth through the insulating material under the effect of heat conduction. The cooling speed of the stator teeth is not as fast as heat conduction, and the temperature is higher. Therefore, there is no obvious temperature gradient along the axial direction of the stator core, while the temperature gradually increases from the outside to the inside along the radial direction of the stator core.

As can be seen from Fig. 9.28, the iron loss is less because there is no winding on the rotor. The heat loss of the stator teeth and windings is transferred to the rotor through the air gap in the form of thermal convection, such that the temperature of the rotor teeth is relatively high, reaching 72.4 degrees in the steady state. Since the temperature of the rotating shaft connected to the rotor yoke is relatively low, the heat of the rotor is transferred to the rotating shaft by means of heat conduction, such that the heat dissipation conditions are better. Therefore, the temperature at both ends of the rotor and the position close to the shaft is relatively low. The lowest temperature is 68.738 degrees, and the overall temperature difference is 3.662 degrees.

9.6.2 Simulation Analysis Results of Transient Temperature Field

In order to study the temperature change of the four-phase 8/6 SRM prototype from the initial time to the steady state, the transient temperature field simulation of the prototype is carried out, where the simulation time is set to 14,400 s with the simulation step of 1 s.

Taking the same temperature change range, the transient temperature field distribution cloud diagrams of various parts of the prototype at time of 3600 s, 7200 s, and 14,400 s are shown in Figs. 9.29, 9.30 and 9.31, respectively.

The overall temperature rise curve of the SRM is shown in Fig. 9.32, where, the maximum temperature of the stator, rotor, and windings of the prototype at different times is plotted into a graph. Moreover, the temperature rise curve of each part during the operation of the motor is obtained as shown in Fig. 9.33.

It can be seen from Figs. 9.32 and 9.33 that the temperature of the motor rises quickly before 6000 s. Among them, the temperature of the stator winding rises fastest, followed by the stator, and finally stabilizes around 12,000 s. The temperature change of the rotor core is smaller than that of the winding and the stator.

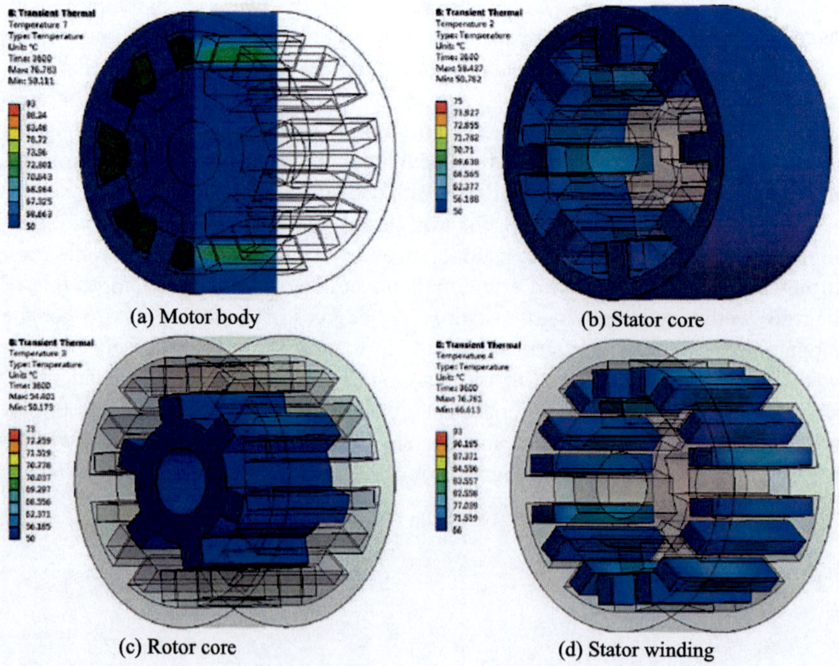

Fig. 9.29 Temperature field distribution diagram of each part of the SRM at 3600 s

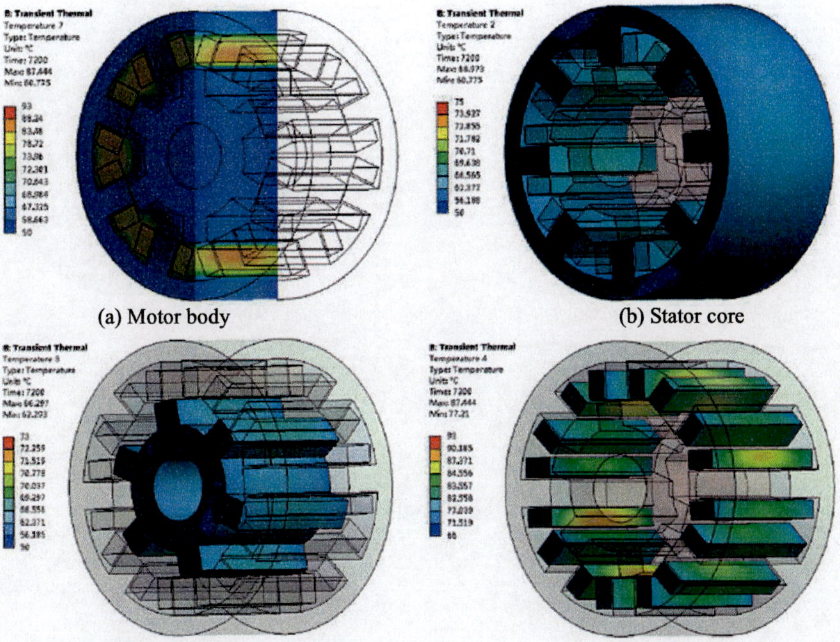

(a) Motor body (b) Stator core

(c) Rotor core (d) Stator winding

Fig. 9.30 Temperature field distribution diagram of each part of the SRM at 7200 s

9.7 Motor Structure Parameter Optimization Considering Temperature Field

The geometric model of the SRM body is a double salient structure, and the topology circuit model adopted is an asymmetric half-bridge structure. Therefore, the geometric structure size of the SRM body has a significant impact on the performance of the motor. The main structural parameters of SRM include the number of stator and rotator poles, inner and outer diameters, yoke height, pole arc coefficient, shaft diameter, air gap size, length of core, and winding turns on the stator poles. For the SRM with the determinate inner and outer diameters of the stator and rotor, and the length of the core, the thickness of the stator yoke, and the stator pole arc coefficient have obvious effects on the torque ripple, noise, and temperature rise of the motor.

Combined with the traditional motor design methods, the suitable range for some structural variables has been put forward, which can be used when designing motor. In order to improve the performance of SRM and expand its scope of application, the optimization focusing on the structural variables has also been done for SRM accordingly, and lots of methods have been verified to be significant. However, with the improvement of motor performance requirements, the temperature rise

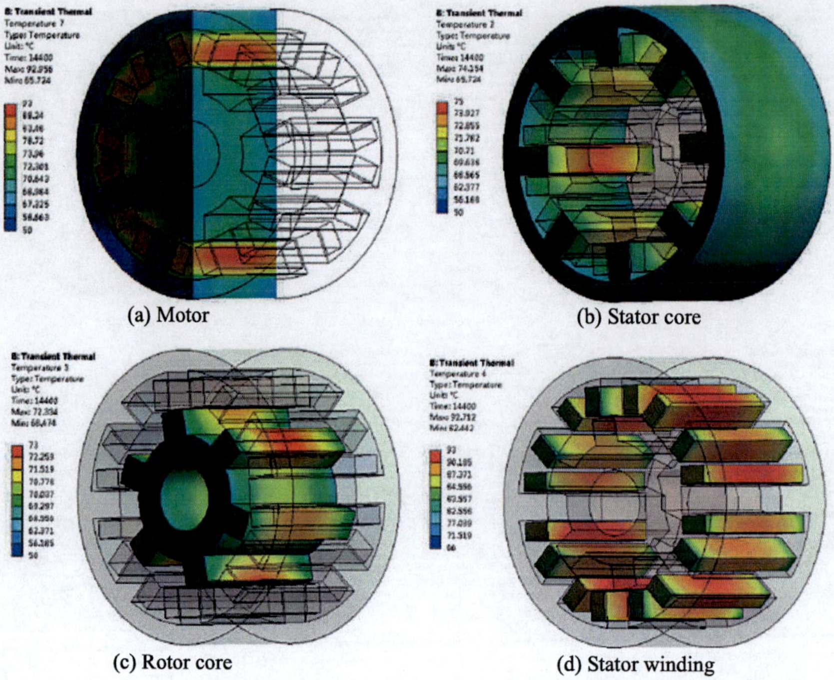

Fig. 9.31 Temperature field distribution diagram of various parts of the SRM at 14,400 s

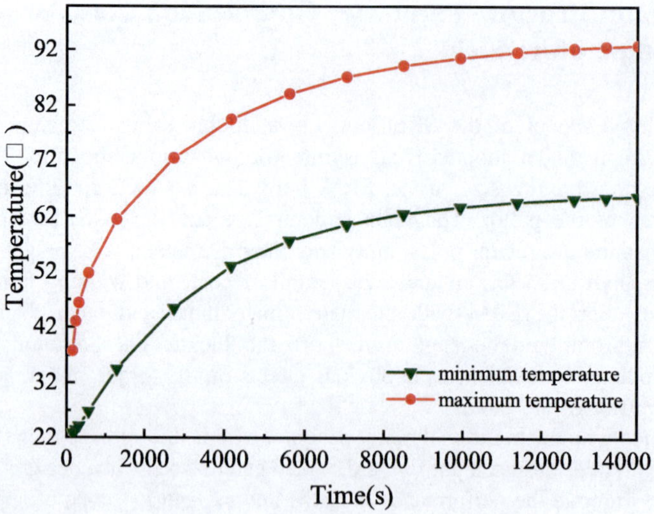

Fig. 9.32 The overall temperature rise curve of the SRM

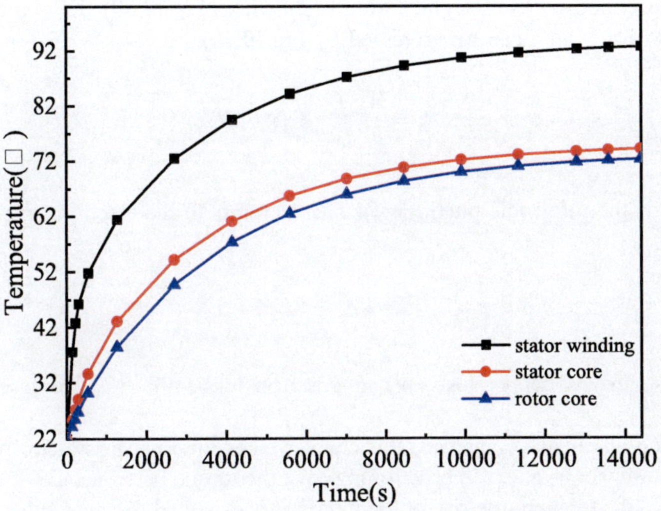

Fig. 9.33 The temperature rise curve of each part during the operation of the SRM

changes of the motor also become extremely important, although the traditional performance should also be greatly significant. As a result, the temperature filed performance should be added into the potential indexes when the SRM is designed and optimized.

In this section, the influences of the motor's stator yoke thickness and stator pole arc coefficient on motor loss and torque are analyzed. Meanwhile, considering the suitable ranges of the thickness of the stator yoke and the stator pole arc coefficient, the geometric parameters of the SRM are optimized to improve the comprehensive performance of the motor through analyzing the simulation results in Maxwell, based on the torque and loss performance indexes.

9.7.1 Optimization Index and Objective Function

Excessive torque pulsation could lead to reduce ride comfort of electric vehicles. The loss during the operation of the motor is the main reason for the temperature rise in the motor. In this chapter, the total loss coefficient per torque is selected as the optimization index, which can be expressed as

$$P_T = \frac{\sum P}{T_a} \tag{9.38}$$

where T_a is the average torque (N·m), which can be defined as the average value of the synthetic torque in an electrical cycle. It should be noted that the average torque can be described by the integral form, as shown in Eq. (9.39). ΣP represents the total loss (W), which can be expressed by Eq. (9.40).

$$T_a = \frac{1}{\theta_r} \int_0^{\theta_r} T_e(i, \theta) d\theta \tag{9.39}$$

where θ_r is the rotor pole pitch (deg.); $T_e(i, \theta)$ refers to the electromagnetic torque (N·m).

$$\sum P = P_{Cu} + P_{Fe} + P_{other} \tag{9.40}$$

where P_{Cu} means copper loss (W); P_{Fe} is iron loss (W); P_{other} represents other losses (W).

After comprehensive analysis, this chapter takes the total loss factor per torque as the optimization index. In order to improve the torque performance of the motor and reduce the temperature rise of the motor, it is expected that this index is as low as possible. Therefore, the objective function is set as

$$F(x) = \min(P_T) \tag{9.41}$$

9.7.2 *Optimization Parameters and Constraints*

One principle for selecting optimization parameters is that the parameters should be greatly relative to the optimization indexes. Based on this principle, the stator yoke thickness and the stator pole arc coefficient are selected as the optimization parameters, which have significant effects on the torque ripple, noise, and temperature rise of the motor. Therefore, the relationship can be expressed as

$$M = [Y_S, \beta_S] \tag{9.42}$$

where Y_S represents the thickness of the stator yoke (mm); β_S refers to the pole arc coefficient of the stator.

The stator yoke thickness Y_S is an important parameter of SRM, which has a great influence on the loss and the torque performance of the motor. If the thickness of the stator yoke is too thin, the magnetic flux density of the motor can easily reach saturation. If the stator yoke is too thick, the weight and economic cost of the motor can be increased. Therefore, the thickness of the stator yoke is generally determined as the following Eq. (9.43), where, b_s represents the pole width of the stator.

$$Y_S = (0.6 \sim 1.2)b_S \tag{9.43}$$

The stator arc coefficient β_S also has a great influence on the current, torque, and loss of the SRM. In order to ensure the self-starting ability of the motor, when designing the pole arc of the SRM, the inductances of two adjacent phases should partially overlap in the inductance rise area. Therefore, the stator arc coefficient should meet the following constraints due to the requirements of the SRM.

$$\left.\begin{aligned} \min(\beta_s, \beta_r) \geq \frac{2\pi}{mN_r} \\ \beta_s + \beta_r \leq \frac{2\pi}{N_r} \end{aligned}\right\} \tag{9.44}$$

where β_r refers to the rotor pole arc coefficient; m represents the number of phases; N_r is the number of rotor teeth.

Using the structural parameters of the SRM in this chapter, the available range of the stator yoke thickness Y_S and the stator pole arc coefficient β_S can be determined as Eq. (9.45), according to the Eqs. (9.43) and (9.44).

$$\left.\begin{aligned} 6 \leq Y_S \leq 20 \\ 0.3 \leq \beta_S \leq 0.5 \end{aligned}\right\} \tag{9.45}$$

9.7.3 Effect Analysis of the Stator Yoke Thickness

Keeping constant for the other geometric structure parameters of the prototype, the thickness of the stator yoke is parametrically modeled in ANSYS Maxwell. It should be noted that the thickness range of the stator yoke is selected to be 6 mm to 20 mm with the value interval of 1 mm. After setting the boundary condition of the motor, the excitation mode, the material properties, the suitable meshes, as well as the available solver, the parameterized motor is finally simulated. According to the simulation results, the influence of the thickness of the stator yoke on the torque performance and losses of the SRM is analyzed one by one. The electromagnetic simulation results of the transient field are shown in Figs. 9.34, 9.35, 9.36, 9.37, 9.38, 9.39, 9.40 and 9.41.

Figures 9.34 and 9.35 show the magnetic flux line distribution and flux density distribution of the SRM with exciting one phase when the thickness of stator yoke is 6 mm, 12 mm, and 20 mm, respectively.

It can be seen from Figs. 9.34 and 9.35 that when one phase of the motor is turned on, the stator tooth tip and rotor tooth tip of the SRM could be saturated. When the thickness of the stator yoke is 6 mm, the magnetic flux lines are arranged very closely to the stator yoke, and the magnetic flux density of the stator yoke is

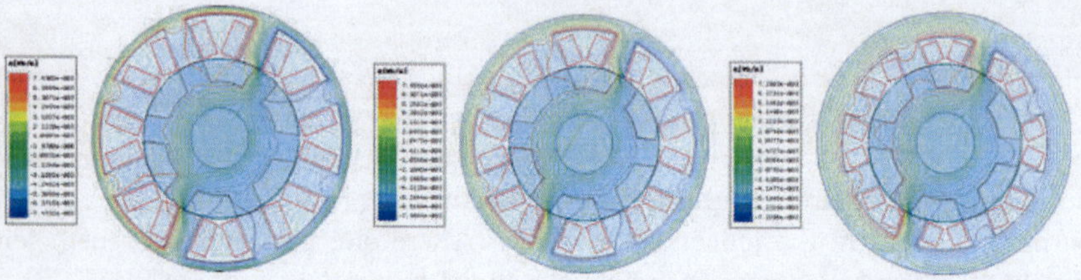

(a) Yoke thickness of 6 mm (b) Yoke thickness of 12 mm (c) Yoke thickness of 20 mm

Fig. 9.34 Distribution of magnetic flux lines with different yoke thicknesses

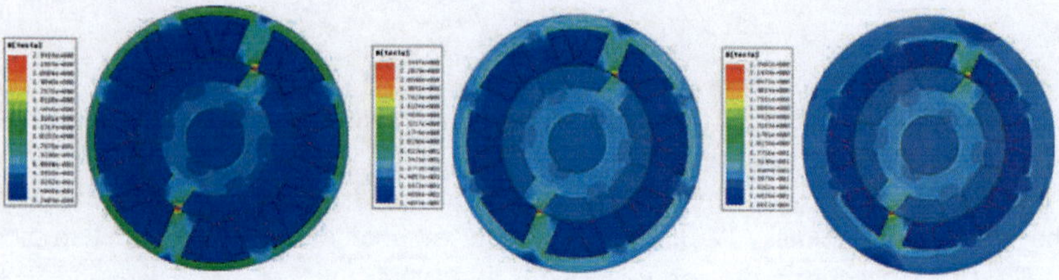

(a) Yoke thickness of 6 mm (b) Yoke thickness of 12 mm (c) Yoke thickness of 20 mm

Fig. 9.35 Magnetic flux density distribution diagram with different yoke thickness

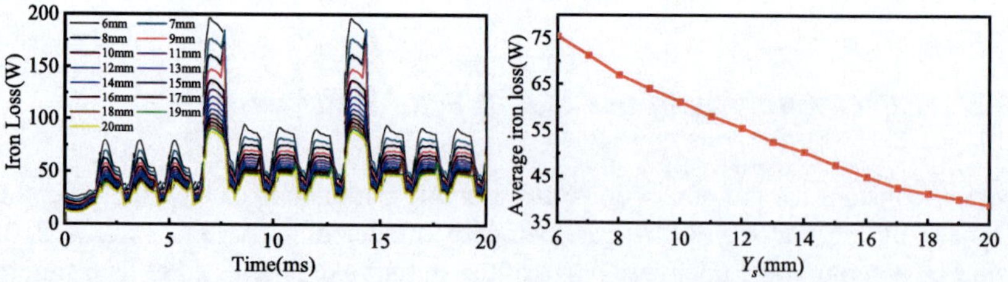

(a) Time domain curves of the iron loss (b) The average iron loss versus yoke thickness

Fig. 9.36 The influence of various stator yoke thickness on the iron loss

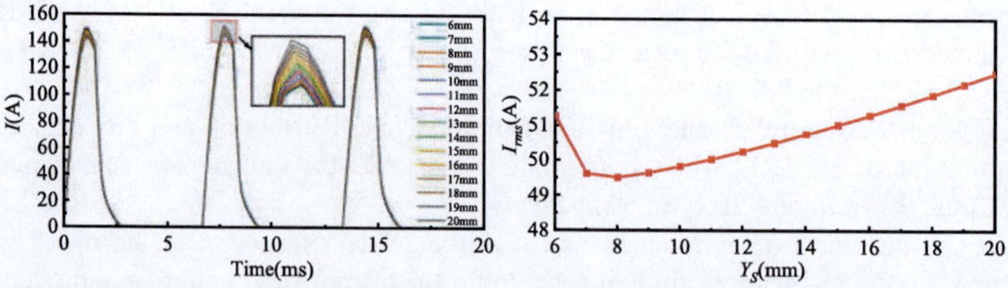

(a) Time domain curves of the phase current (b) The RMS current versus yoke thickness

Fig. 9.37 The influence of various stator yoke thickness on the current

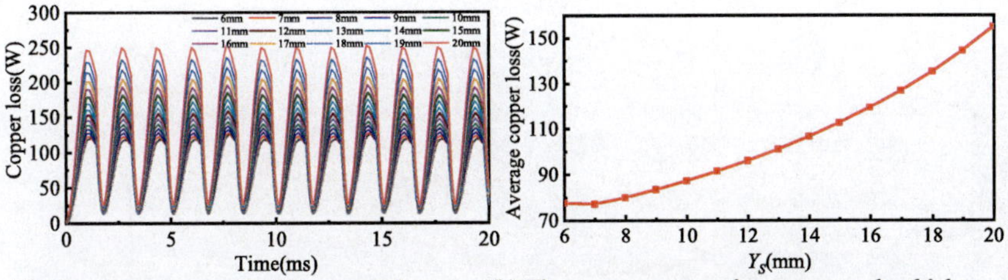

(a) Time domain curves of the copper loss (b) The average copper loss versus yoke thickness

Fig. 9.38 The influence of various stator yoke thickness on the copper loss

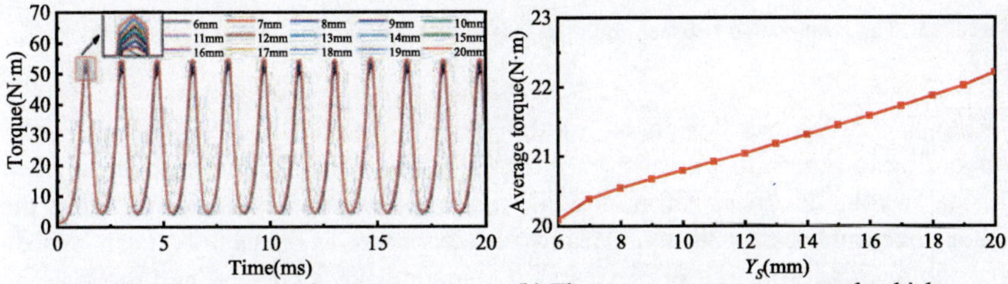

(a) Time domain curves of the torque (b) The average torque versus yoke thickness

Fig. 9.39 The influence of various stator yoke thickness on the torque

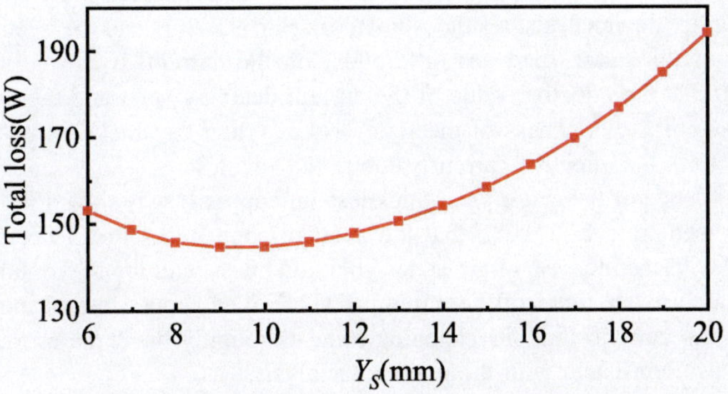

Fig. 9.40 The relationship between the total loss and the yoke thickness

relatively large. It should be noted that as the gradual increase in the thickness of the stator yoke, the magnetic flux lines of the stator yoke gradually become sparse, correspondingly its magnetic flux density also gradually decreases. However, the core loss of the motor is a correlation function of the magnetic density. The smaller the thickness of the stator yoke is, the more easily the magnetic density of the motor stator yoke can be saturated, and the higher the core loss should be.

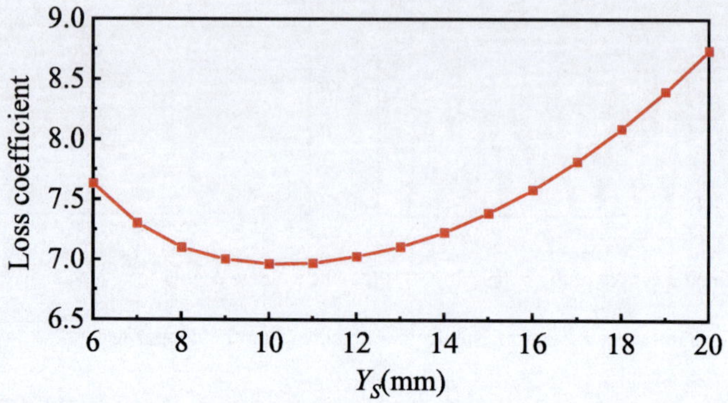

Fig. 9.41 The relationship between the total loss coefficient per torque and the yoke thickness

Figure 9.36 shows the effects of the stator yoke thickness on the iron loss. According to Fig. 9.36a, it can be seen that the iron loss value is the highest when the stator yoke thickness is 6 mm, while the iron loss curve is the lowest when the stator yoke thickness is 20 mm. Meanwhile, according to Fig. 9.36b, it can be seen that the average iron loss decreases as the stator yoke increases, and the average iron loss when the stator yoke thickness is 6 mm is higher by about 36.5 W than that when the stator yoke thickness is 20 mm, which is consistent with the previous analysis.

Figure 9.37 shows the effects of the stator yoke thickness on the current. According to the analysis results shown in Fig. 9.37, it can be seen that the thickness of the stator yoke has little effect on the current. It can be seen from Fig. 9.37b that the effective value of the current decreases first and then rises with the increase of the thickness of the stator yoke. When the thickness of the stator yoke is 8 mm, the effective current value is the smallest.

The influence of the stator yoke thickness on copper loss is shown in Fig. 9.38. It can be seen from Fig. 9.38b that the average copper loss firstly decreases and then rises with the increase of the stator yoke thickness, and the maximum value is approximately twice times of the minimum value. The copper loss of the motor is related to the current, thus, its changing trend is roughly the same as the current, which is also consistent with the previous analysis.

Figure 9.39 shows the effects of the stator yoke thickness on the torque. It can be seen that the torques under various yoke thicknesses shown in Fig. 9.39a are not much different. Figure 9.39b shows that as the thickness of the stator yoke increases, its average torque also increases.

Figures 9.40 and 9.41 are respectively the relationship curve between the total loss and the thickness of stator yoke, and the relationship curve between the total loss coefficient per torque and the thickness of stator yoke. It can be seen from the figure that as the thickness of the stator yoke increases, the total loss and the total loss coefficient per torque both have a slight decline and then rise. Meanwhile, both reach the minimum when the thickness of the stator yoke is 10 mm.

By analyzing the simulation results of the thickness of the stator yoke, there are some conclusions can be drawn as follows. (1) As the thickness of the yoke increases, the total loss decreases. (2) When the yoke thickness is 10 mm, the total loss can reach minimum. (3) The torque gradually increases as the yoke thickness increases. (4) The total loss per torque reaches the minimum when the thickness is 10 mm. Combined with the comprehensive consideration of the limitation from the winding space, the optimized thickness of the yoke is determined as 10 mm in this chapter.

In order to verify whether the torque ripple is optimized, the torque ripple coefficient is used as an index to evaluate the torque ripple, which can be described as

$$T_r = \frac{T_{\max} - T_{\min}}{T_a} \tag{9.46}$$

where $T_{\max}$ refers to the maximum instantaneous torque of SRM, $T_{\min}$ is the minimum instantaneous torque of SRM, and T_a denotes the average torque of SRM.

Figure 9.42 shows the relationship between the torque ripple coefficient and the thickness of the stator yoke. The results show that the torque ripple coefficient increases as the thickness of the stator yoke increases, which indicates that the torque ripple also increases.

9.7.4 Effect Analysis of Stator Pole Arc Coefficient

Similarly, the available range of the pole arc coefficient of the stator is between 0.3 and 0.5, with the value interval of 0.02, according to the analysis above. Then, the

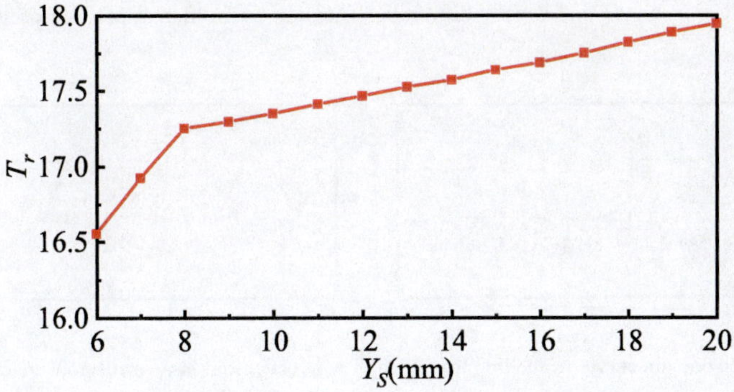

Fig. 9.42 The relationship between the torque ripple coefficient and the yoke thickness

simulation of the parameterized motor is performed. The influence of the stator pole arc coefficient on the torque performance and losses of the motor is analyzed step by step, and the transient electromagnetic simulation results are shown in Figs. 9.43, 9.44, 9.5, 9.46, 9.47, 9.8, 9.49 and 9.50.

Figures 9.43 and 9.44 are the magnetic flux line distribution cloud diagram and magnetic flux density distribution cloud diagram when the stator pole arc coefficient is 0.3, 0.4, and 0.5, respectively. As can be seen from Figs. 9.43 and 9.44, the stator tooth tip and the rotor tooth tip are slightly saturated. As the stator pole arc

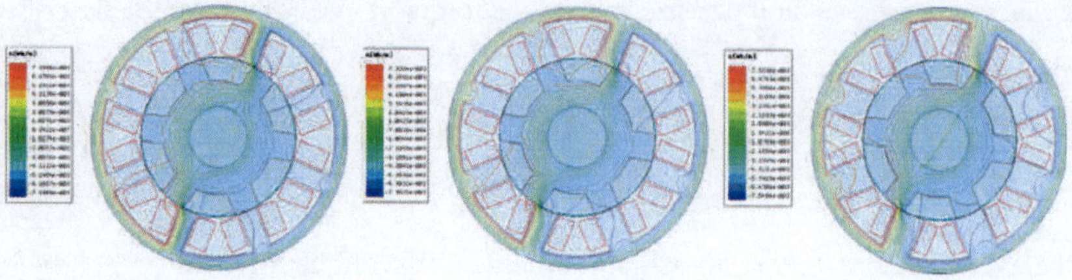

(a) Pole arc coefficient of 0.3 (b) Pole arc coefficient of 0.4 (c) Pole arc coefficient of 0.5

Fig. 9.43 Distribution of magnetic flux lines with different pole arc coefficients

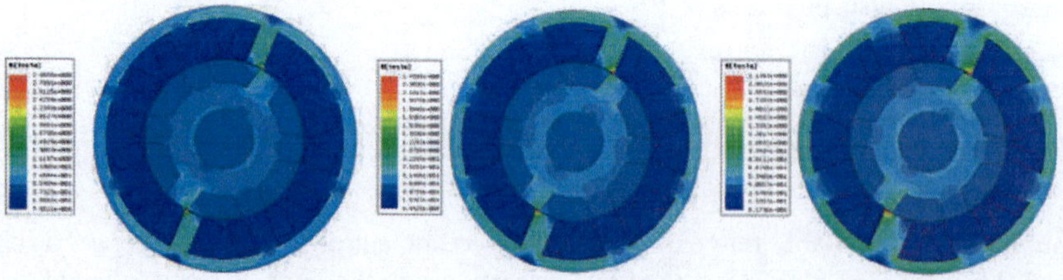

(a) Pole arc coefficient of 0.3 (b) Pole arc coefficient of 0.4 (c) Pole arc coefficient of 0.5

Fig. 9.44 The magnetic flux density distribution diagram with different pole arc coefficients

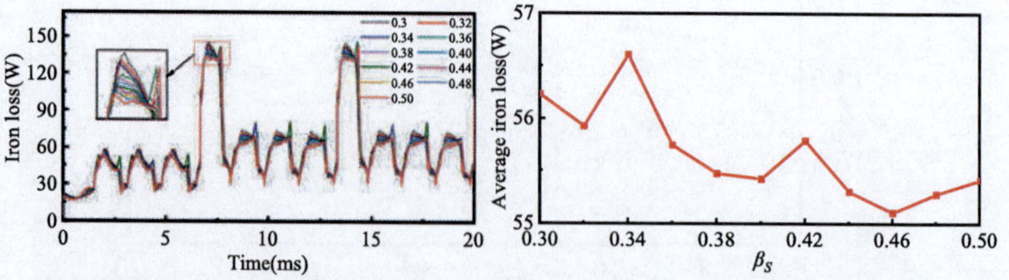

(a) Time domain curves of the iron loss (b) The average iron loss versus pole arc coefficient

Fig. 9.45 The influence of various stator pole arc coefficient on iron loss

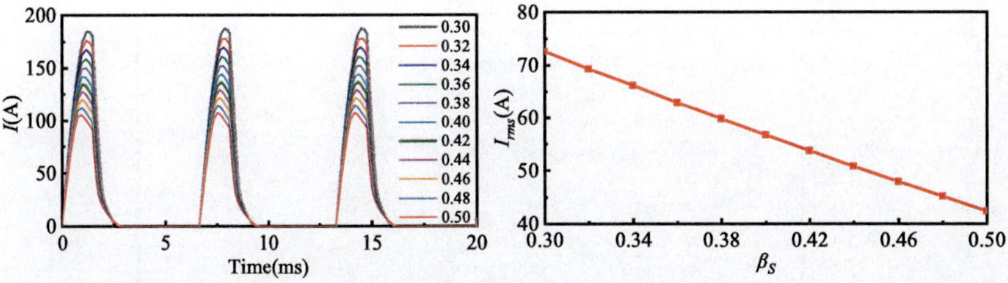

(a) Time domain curves of the phase currents (b) The RMS current versus pole arc coefficient

Fig. 9.46 The influence of various stator pole arc coefficient on current

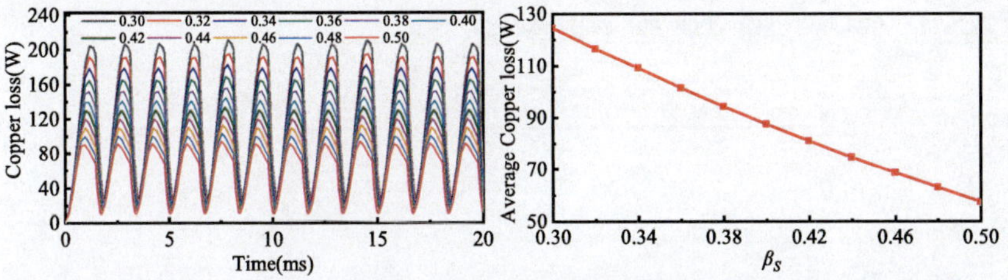

(a) Time domain curves of the copper loss (b) The average copper loss versus pole arc coefficient

Fig. 9.47 The influence of various stator pole arc coefficient on copper loss

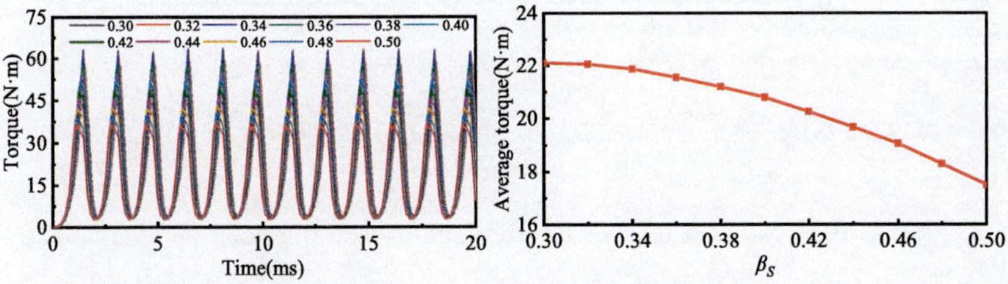

(a) Time domain curves of the torque (b) The average torque versus pole arc coefficient

Fig. 9.48 The influence of various stator pole arc coefficient on the torque

coefficient increases, the stator teeth become wider, and the overlap area of the stator and rotor poles increases, such that the magnetic flux density in this area decreases.

The influence of the stator arc coefficient on the core loss of the SRM prototype is shown in Fig. 9.45. It can be seen from Fig. 9.45a that its influence on the core loss is not as great as the thickness of the stator yoke. According to Fig. 9.45b, it can be seen that the average iron loss value of SRM shows a trend of decline as the pole arc coefficient increases. Its value reaches the maximum when the polar arc

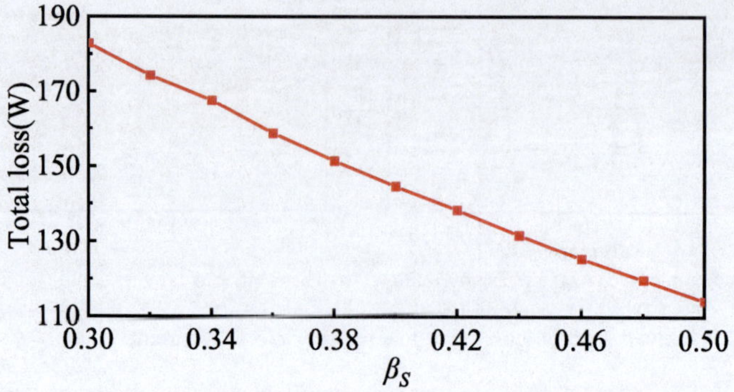

Fig. 9.49 Total loss versus pole arc coefficient

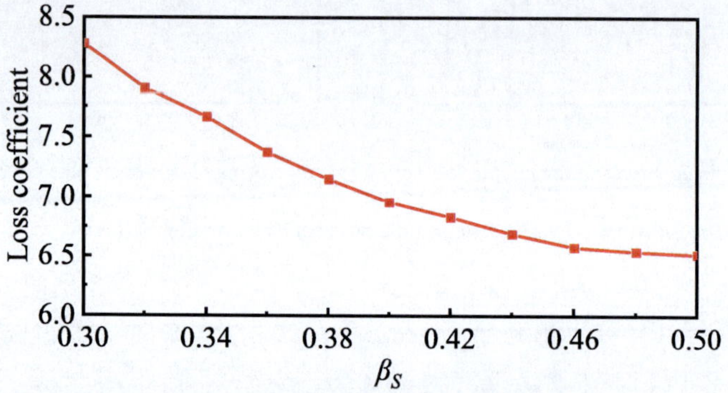

Fig. 9.50 Total loss coefficient per torque versus pole arc coefficient

coefficient is 0.34, while it reaches the minimum when the polar arc coefficient is 0.46.

The influence of the stator pole arc coefficient on the phase current of the SRM prototype and the influence on the copper loss of the SRM are shown in Figs. 9.46 and 9.47, respectively. It can be seen from Fig. 9.46a that, differing from the stator yoke thickness, the stator pole arc coefficient has a greater impact on the phase current of the stator winding. It can be seen from Fig. 9.46b that the effective value of the phase current shows a downward trend as it increases, and the maximum difference is 30 A. From the simulation results shown in Fig. 9.47, it can be seen that the stator pole arc coefficient is inversely proportional to the copper loss of the SRM prototype, which has the same trend as the phase current. Because, the copper loss of the stator winding is related to the phase current. In addition, the copper loss of the stator windings tends to drop sharply as it increases, and the maximum difference is about 67.5 W.

The influence of the stator pole arc coefficient on the output torque of the SRM prototype is shown in Fig. 9.48. From the simulation results in the figure, it can be seen that the stator pole arc coefficient has a considerable influence on the torque, which is opposite to the thickness of the stator yoke. The simulation result of Fig. 9.48b shows that the average torque has a downward trend as the stator pole arc coefficient increases.

Figures 9.49 and 9.50 respectively show the relationship curve between the stator pole arc coefficient and the total loss and the curve between the total loss coefficient per torque and the stator pole arc coefficient. According to the simulation results in the figure, the total loss of the SRM shows a downward trend as the stator pole arc coefficient increases. The total loss coefficient per torque also shows a downward trend, and reaches the minimum value when the rotor pole arc coefficient is 0.5.

The following observations can be drawn by analyzing the simulation results of the stator pole arc coefficient. As the stator pole arc coefficient increases, the current, total loss and torque all show a downward trend. The increase of the stator pole arc coefficient makes the stator tooth pole body wider, such that the flux linkage of the SRM can also increase and the phase current of the stator winding can greatly reduce, which results in the great reduce of the copper loss of the stator winding. Considering the torque performance and losses of the motor, the pole arc coefficient is optimized to be 0.5. The optimized stator pole arc coefficient is substituted into Eq. (9.44) and the results show that it meets the requirements of the motor.

Moreover, Fig. 9.51 shows the relationship curve between the torque ripple coefficient and the pole arc coefficient. The results in the figure show that the torque ripple coefficient firstly rises and then decreases, with the increase of the pole arc coefficient, and it reaches the minimum value when the rotor pole arc coefficient is 0.5.

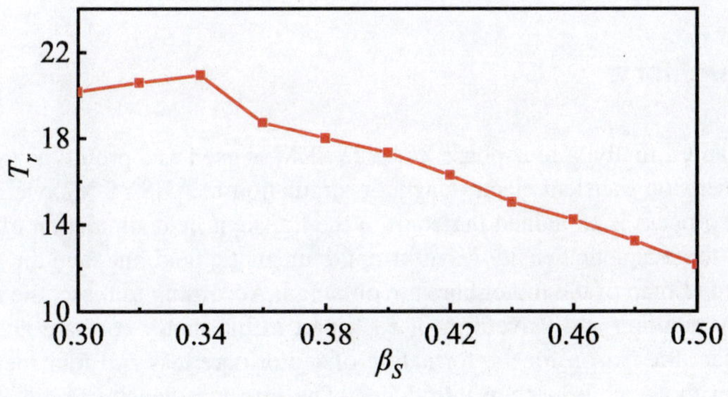

Fig. 9.51 Torque ripple coefficient versus stator pole arc coefficient

Table 9.5 Comparison of variables before and after optimization

Parameter	Initial design	Optimized
Stator yoke thickness Y_S	14 mm	10 mm
Stator pole arc coefficient β_S	0.4	0.5
Unit torque loss factor P_T	7.1	6.44
Total loss ΣP	164.255 W	124.525 W
Torque ripple coefficient T_r	17.57	12.19

9.8 Results Analysis on Three-Dimension Temperature Field

According to the structural variables selected above, combined with the method of calculating loss described previously, the total loss is calculated before and after optimization, and the comparison of motor variables before and after optimization is shown in Table 9.5.

Combined with the method above, the finite element steady-state temperature field simulation of the optimized motor model is performed in ANSYS Workbench, and the simulation results are shown in Fig. 9.52.

It can be seen from the analysis results of the for the optimized motor in Fig. 9.52 that the optimized motor temperature drops significantly. The highest temperature in the motor is 71.112 degrees, which is lower by about 22 degrees than that before optimization.

Figure 9.53 is the overall temperature change curve of the motor before and after optimization. Figure 9.54 is the temperature change curve of each part in the motor before and after optimization. From Figs. 9.53 and 9.54, it can be seen intuitively that the temperature of each part in the optimized motor decreases significantly, which shows that the optimized motor structure can effectively reduce the temperature rise of the motor.

9.9 Summary

In this chapter, firstly, a four-phase 8/6-pole SRM is used as a prototype to conduct a two-dimension transient electromagnetic simulation in ANSYS Maxwell, and the modeling process is explained in detail. In the transient field simulation of ANSYS Maxwell, the magnetic density waveform, the magnetic field line, and the magnetic density cloud map of the motor core are obtained. According to these, the magnetic density distribution and waveform of each part in the motor core are analyzed in detail. Then, the reason for the formation of motor core loss and four methods for calculating motor core loss are introduced. The iron loss density of the stator and rotor during commutation of the SRM is analyzed based on FEA. The reason why the iron loss value becomes large is described, which is that the magnetic flux lines

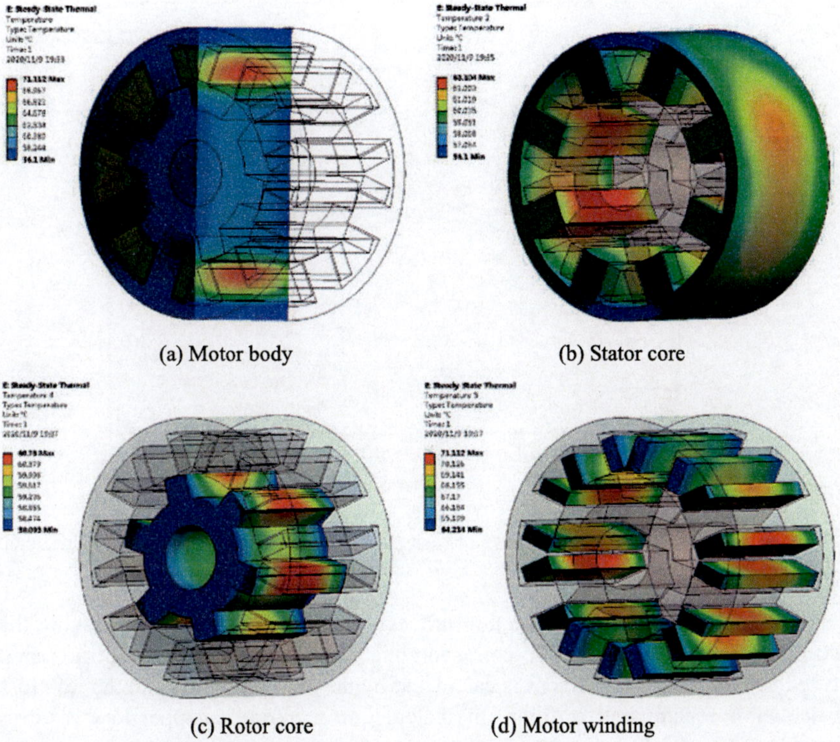

Fig. 9.52 The temperature field distribution of the optimized motor

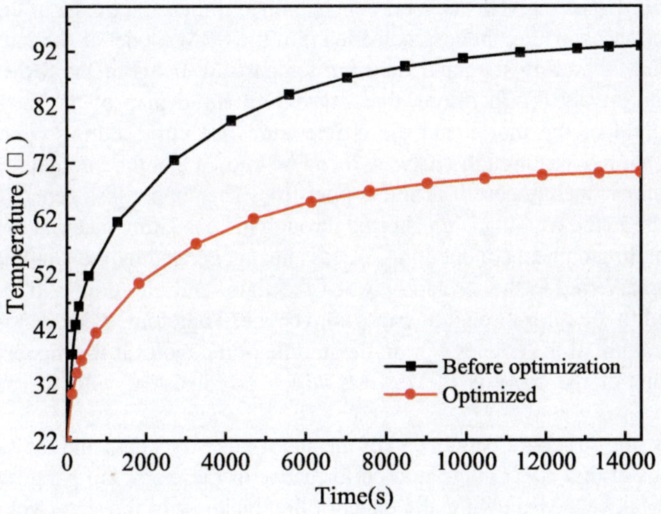

Fig. 9.53 The overall temperature change curve of the motor before and after optimization

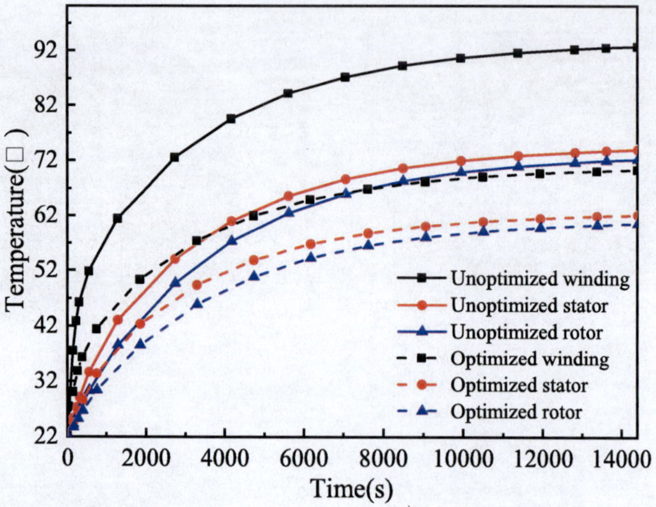

Fig. 9.54 The temperature change curve of each part in the motor before and after optimization

are reversed during the commutation process of the motor. Meanwhile, taking the stator teeth as an example, the corresponding magnetic density waveform is analyzed by FFT, and the iron loss is calculated by the ellipse method and the iron loss calculation equation. In addition, the calculation method of copper loss is introduced and the final copper loss is calculated.

Then, the thermal analysis mathematical model of the SRM is established, and the thermal parameters and boundary conditions of each part in the SRM are determined. Based on ANSYS Workbench, a three-dimension model of the SRM is established. Meanwhile, the loss is loaded into the SRM model as the internal heat source, and the steady-state and transient temperature fields of the SRM are analyzed, respectively. Furthermore, the distribution cloud map of the internal temperature field of the motor and the temperature rise curve during operation are obtained. Moreover, through analysis, it can be known that the internal heat source of the motor is mainly core loss and copper loss. The copper loss generated by the copper wire in the winding is transferred through the insulating material to the stator core depending on heat conduction. The loss heat generated by the winding and the stator is transferred to the air gap between the stator and the rotor, and less heat is transferred to the rotor through thermal convection. Therefore, it can be known that the temperature of the copper wire in the middle of the motor is the highest, and the temperature of the rotor is the lowest, which is consistent with the simulation results.

Finally, the structural parameters of the stator yoke thickness and the stator pole arc coefficient that affect the torque performance of the motor are parameterized in ANSYS Maxwell. Meanwhile, the effect of the thickness of the stator yoke and the stator pole arc coefficient on the torque performance and loss of the SRM is verified through the simulation. Based on comprehensive considerations, these two reasonable parameters are selected as the optimized variables and the optimization is carried out. After that, the temperature field of the optimized motor is simulated in ANSYS Workbench and also compared to the initial design. It can be seen from the comparison results that the total loss of the optimized motor is reduced by 39.73 W and the temperature is reduced by 22 degrees, compared to the initial design. However, it should be noted that the structure optimization considering the temperature field performance is greatly complicated, such that it should comprehensively consider the torque performance, efficiency performance, and even the requirements from the vehicle. Therefore, the structure optimization in this chapter can be regarded as a simple case. The performance indexes and design variables should be selected according to the actual requirements of the design.

References

1. Sun X, Xue Z, Xu X, et al. Thermal Analysis of a Segmented Rotor Switched Reluctance Motor Used as the Belt-Driven Starter/Generator for Hybrid Electric Vehicles. Journal of Low Power Electronics, 2016, 12(3): 277-284.
2. Raulin V, Radun A, Husain I. Modeling of losses in switched reluctance machines. IEEE Transactions on Industry Applications, 2004, 40(6): 1560–1569.
3. Toulabi MS, Torkaman H, Afjei E. Experimental compatison between the electric efficiencies of two different types of switched reluctance generators//2011 2nd Power Electronics, Drive Systems and Technologies Conference, 2011: 157–162.
4. Boivie J, Iron loss model and measurement of the losses in a switched reluctance motor. Electrical Machines and Drives, International Conference on Electrical Machines, 1993.
5. Inamura S, Sakai T, Sawa K. A temperature rise analysis of switched reluctance motor due to the core and copper loss by FEM. IEEE Trans on Magnetics, 2003, 39(3):1554–1557.
6. Yan W, Chen H, Liu Y, et al. Iron loss and temperature analysis of switched reluctance motor for electric vehicles. IET Electric Power Applications, 2020, 14(11): 2119.2127.
7. Ilhan E, Kremers MFJ, Motoasca TE, et al. Transient thermal analysis of flux switching PM machines//Eighth International Conference and Exhibition on Ecological Vehicles and Renewable Energies. Monte Carlo, Monaco, 2013.
8. Marlow R, Schofield N, Emadi A. A continuous toroidal winding SRM with 6- or 12-switch DC converter. IEEE Trans on Industry Applications, 2016, 52(1): 189.198.
9. Namburi NR, Barton TH. Thermal Modelling of an Induction Motor. IEEE Transactions on Power Apparatus & Systems, 1983, PAS-102(8): 2636–2639.
10. Faiz J, Dadgari A. Heat distribution and thermal calculations for switched reluctance motors// International Conference on Electrical Machines & Drives. 1991.

8